JAN BECCALONI

ARACHNIDS

UNIVERSITY OF CALIFORNIA PRESS

Berkeley *Los Angeles*

This book is dedicated to George, my husband – without him I would never have undertaken to do this book! Also to my parents, Audrey and Michael Margerison, for their support in everything I have ever done, and to my pet *Theraphosa blondi* called Tracy, who died during the writing of this book, and whose image is on several of the pages!

University of California Press, one of the most distinguished university presses in the United States, enriches lives around the world by advancing scholarship in the humanities, social sciences, and natural sciences. Its activities are supported by the UC Press Foundation and by philanthropic contributions from individuals and institutions. For more information, visit www.ucpress.edu.

University of California Press
Berkeley and Los Angeles, California

Published simultaneously outside North America by the Natural History Museum, Cromwell Road, London SW7 5BD

Cataloging-in-Publication data for this title is on file with the Library of Congress.

ISBN 978-0-520-26140-2 (cloth : alk. paper).

Copy-edited by Gina Walker
Designed by Mercer Design, London
Reproduction by Saxon Digital Services
Printing by C & C Offset

Manufactured in China

18 17 16 15 14 13 12 11 10 09
10 9 8 7 6 5 4 3 2 1

Contents

1 Spiders aren't the only arachnids

OPPOSITE This Oligocene spider, *Abliguritor niger*, is encased in amber from the Baltic.

'Arachnids – they're spiders, aren't they?' The majority of people who have heard of 'arachnids' know them as spiders, and that's a really great start! The term 'arachnophobia', the fear of spiders, must go a long way to explain this, because it is a word everyone knows. What most people don't realize is that the Arachnida also includes other groups.

So what are arachnids? To answer this we first need to look at the animal kingdom as a whole. It is divided into several phyla or large groups of organisms with shared characteristics. The large phylum Arthropoda, which includes more than three-quarters of all known animal species, contains all animals with an exoskeleton, segmented body and paired, jointed appendages; this phylum includes arachnids, insects, centipedes and millipedes, and crustaceans. Depending on the particular classification, the Arthropoda may be divided into several smaller subphyla, including the subphylum Chelicerata. Other sources have elevated the Chelicerata to phylum; for our purposes it is a subphylum. The Chelicerata is divided into the class Arachnida of eight-legged and mainly terrestrial creatures; the Xiphosura, horseshoe crabs, and the Eurypterida, extinct 'sea-scorpions'. But how can horseshoe crabs and arachnids be related? What specific chelicerate characteristics do they share?

All chelicerates possess a pair of pre-oral jaws called the chelicerae, giving the subphylum its name, and a pair of pedipalps, which may be used in locomotion, or modified as structures to aid in prey capture and feeding, in sperm transfer or as sensory structures. The body is divided into two parts – an anterior prosoma ('front' body), or cephalothorax, and a posterior opisthosoma ('hind' body), instead of a three-part body (with a separate head) as in insects.

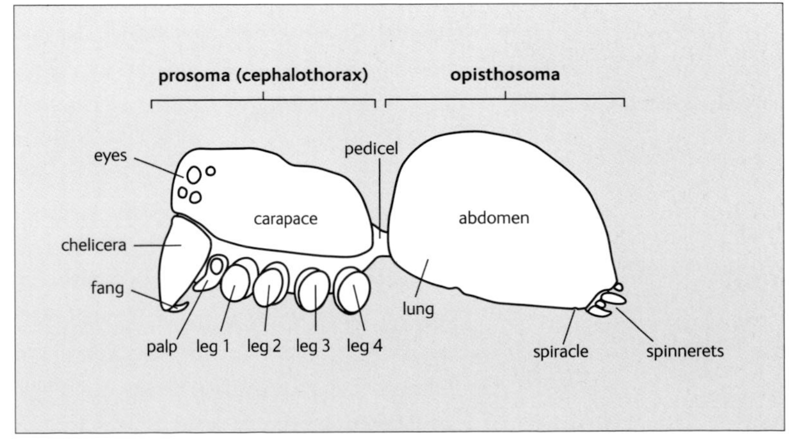

BELOW External body of general arachnid.

The class Arachnida (all members are called arachnids) is divided into 10 orders (depending on the classification system), and one subclass – the Acari. Most classifications then divide the Acari into seven orders (suborders) within three superorders. So, the Arachnida contains not just the spiders (Araneae), but the scorpions (Scorpiones), ticks and mites (Acari), and harvestmen (Opiliones). The lesser-known arachnids are: camel spiders (Solifugae), false scorpions (Pseudoscorpiones), whip spiders (Amblypygi), vinegaroons (Uropygi),

short-tailed whip scorpions (Schizomida), micro whip scorpions (Palpigradi) and hooded tick-spiders (Ricinulei). There are over 90,000 species in total.

As an aside, scientific names are a mixture of both Latin and Greek, and so should be referred to as scientific names, rather than Latin names. Also, the scientific names of the arachnid groups are used throughout rather than the common names, unless they are particularly well known. And one last thing – it is perfectly acceptable to use a shortened form of the scientific names, and this is done throughout. For example, animals in the order Solifugae are referred to as solifugids, animals in the order Amblypygi are referred to as amblypygids, and so on.

Fossils

Arachnids have a long ancestry, with the earliest known arachnid, the scorpion species *Dolichophonus loudonensis*, appearing at the start of the Silurian period, approximately 440 million years ago. Indeed, scorpions have earned the title of 'living fossils'. Amazingly, modern scorpions closely resemble their ancient ancestors; their basic morphology was the same, but they were very different in size. For example, the marine Palaeozoic *Brontoscorpio anglicus* was a whopping 1 m (40 in) in length! But it's not all about scorpions; the first proto-spiders (the new order Uraraneida) are known from the Devonian (approximately 410–360 million years ago), and most of the remaining extant (still living) orders appeared by the Carboniferous (approximately 345–280 million years ago). Arachnids are, therefore, not surprisingly, considered to be one of the most distinctive and diverse groups of living animals.

BELOW This specimen of a fossil *Eurypterus* sp. is from New York, USA.

But why are arachnids that are long since dead so important? Professor Paul Selden, University of Kansas, USA, states: 'In recent years, new interpretations of existing fossils and a few spectacular new finds have filled in the gaps in the records and changed our knowledge and views of the course of arachnid evolution…' Modern studies have incorporated fossil morphological characteristics to help elucidate relationships in modern arachnids, so Giribet et al. (2002) used morphological characters of the extinct Eurypterida, Trigonotarbida and an extinct scorpion genus *Proscorpius*.

The extinct Eurypterida contains the most well-known fossil chelicerate, the giant sea-scorpions that could be over 2 m (6 ft 6 in) long. They lived in the oceans during the Ordovician, Silurian and Devonian periods; after this they were found in semi-terrestrial environments in the Carboniferous and Permian. Eurypterids and scorpions actually co-existed throughout the middle and late Palaeozoic, and some of the giant aquatic scorpions, e.g. *Brontoscorpio* matched many of the bigger eurypterids in size. However, the smaller scorpions were better pre-adapted to a terrestrial environment, so they really left the gigantic eurypterids, with their small limbs, behind in the swamps, rivers and inter-tidal fringe.

The extinct Trigonotarbida are among the first-known land animals. This order is the best-known extinct group because there are some beautifully preserved fossils in the Gilboa mudstones of New York and the Rhynie Chert of Scotland from the Devonian period. The order Haptopoda is known only from a few specimens from the Upper Carboniferous in the UK. It has only one species, *Plesiosiro madeleyi*, and its closest relatives may be the Schizomida, Uropygi and Amblypygi. The Phalangiotarbida order is known exclusively from the Devonian to the Permian periods of Europe and North America. The taxonomic relationships of Phalangiotarbida are obscure. However, most authors think they are most closely related to Opiliones and the Acari (as they resemble opilioacarid mites).

The Arachnids

The two largest orders are the Araneae and Acari. The remaining orders make up just 12% of the arachnids as a whole.

Orders	Number of described species	Size	Key characteristics
Araneae (spiders)	40,024 approx.	0.37–300 mm (1/100–11¾ in)	• produce venom injected into prey through fangs • produce silk from glands in the abdomen • males have copulatory organs on their pedipalps
Amblypygi (whip spiders)	140	5–45 mm (3/16–1¾ in)	• spiny raptorial pedipalps • possess a pair of extremely long and thin whip-like first pair of legs
Uropygi (whip scorpions, vinegaroons)	108	25–80 mm (1–3¼ in)	• posess a pair of large, heavily armoured palps • distinctive whip-like flagellum • spray a noxious-smelling chemical mixture from anal glands, as a form of defence
Schizomida (short-tailed whip scorpions)	258	Majority less than 5 mm (3/16 in), few about 1 cm (½ in)	• unique structure of the dorsal surface of the prosoma, which is divided into platelets • have a short stubby flagellum
Palpigradi (micro whip scorpions)	82	0.65–2.8 mm (3/100–1/10 in)	• colourless and translucent • distinctive multi-segmented flagellum with long setae • use their palps for walking
Ricinulei (hooded tick spiders, tick beetles)	55	4–10.5 mm (1/8–½ in)	• unique in having a hood-like structure or cucullus, that covers the chelicerae and mouth • copulatory organs on the third pair of legs of the adult male • locking device which couples the prosoma and abdomen
Acari (mites and ticks)	45,231	0.1–30 mm (1/250–1¼ in)	• large range of feeding habits • commerical and agricultural importance • medical and veterinary importance
Opiliones (harvestmen, harvest spiders, daddy long legs spiders)	6,411	1–20 mm (1/32–¾ in)	• penis in male and ovipositor in female • possess odiferous glands to repel predators
Scorpiones (scorpions)	1,500 approx.	8.5–229 mm (¼–9 in)	• distinct stinger and claw-like pincers • produce venom • possess sensory pectines
Pseudoscorpiones (false scorpions, book scorpions)	3,380	0.7–12 mm (3/100–¾ in)	• no sting in tail • produce silk from their chelicerae • produce venom from their pedipalps
Solifugae (camel spiders, wind spiders, wind scorpions, sun spiders)	1,100	80–150 mm (3¼–6 in)	• huge jaws • highly aggressive • suctorial organs on pedipalps • possess fan-like racquet organs
Total	**98,289 approx.**		

Phylogeny

Phylogeny is the evolutionary relationship between organisms. Experts still disagree on the phylogenetic relationships of the arachnid orders, even though people have been working on the problem for well over 100 years! However, recent morphological and molecular studies have helped to get closer to discovering the 'holy grail' of how the arachnids are related to one another.

Cladograms are dichotomously (divided in two parts) branching tree diagrams produced using cladistic techniques, which show the hypothesized evolutionary relationships of a group of organisms. Species are situated at the tips of the branches and the more branches separating two species, the more distantly related they are. Common ancestry is represented by the nodes (connection points) from which branches arise, each ancestor giving rise to two branches. In modern 'transformed' cladistics, the nodes have no actual taxa on them and simply represent a hypothesis of common ancestry. The species or groups of species (both are known as taxa) on either side of a node are called 'sister groups'. For example in Giribet et al, 2002, Uropygi and Schizomida are sister groups. A 'clade' is a group of organisms descended from a common ancestor. For example, in Shultz, 1990, the Palpigradi, Araneae, Amblypygi, Uropygi and Schizomida are all grouped into a clade. Cladistics doesn't make any distinction between extant (still living) and extinct species. There are four different cladograms of arachnids shown on the left. Cladograms based on morphology (the form of organisms) may include extinct species, but cladograms based on molecular work don't, because of the great difficulty in doing molecular work on fossils.

BELOW Four different cladograms showing relationships of Opiliones to other chelicerate groups.

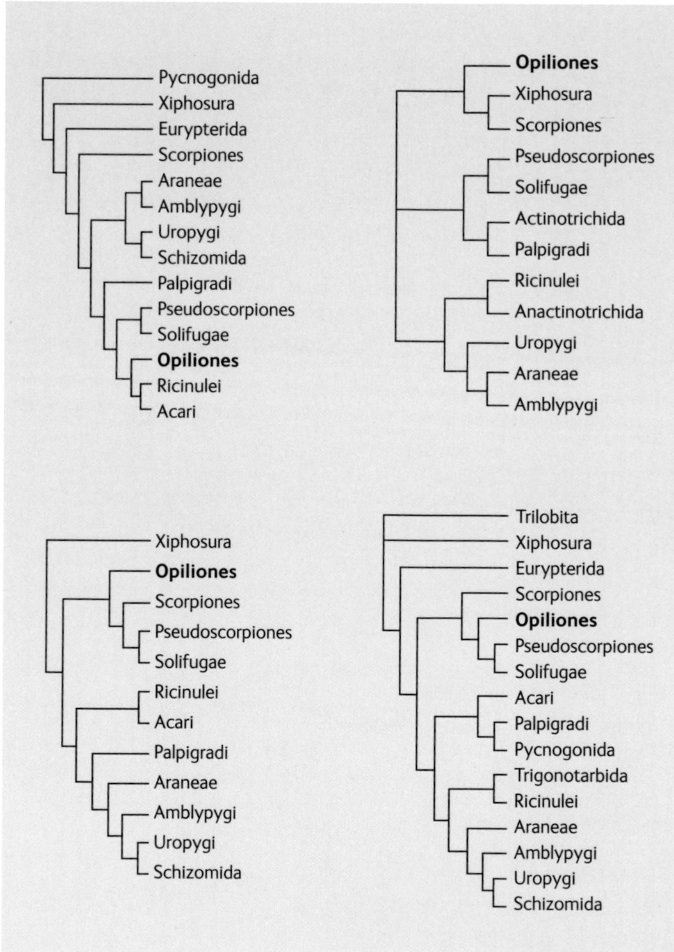

The key texts containing different phylogenies are: Wheeler & Hayashi, 1998; Weygoldt & Paulus, 1979; van der Hammen, 1985; Shultz, 1990 & 2000; Giribet et al., 2002. Some of the phylogenies are based on morphology only, whereas others are based on morphological and molecular data. Wheeler & Hayashi (1998) included analyses with cladograms based on molecules only, morphology only, and a combined approach of morphology and molecules. Giribet et al. (2002) used morphological data from both fossils and recent taxa.

I decided to use the cladogram proposed by Shultz (1990) of the hypothetical relationships of the various arachnid orders to arrange the chapters. However, I did cheat a little, because I wanted to put spiders first (as everyone knows that spiders are arachnids, and because it is, by far, the largest chapter). I therefore moved the Palpigradi after the Uropygi. Although Shultz combines schizomids with uropygids, I decided to keep them separate, as other authorities treat them as their own orders, and I felt that they were sufficiently different to warrant their own chapters. Schizomida is therefore directly after the Uropygida, to reflect their close relationship.

External anatomy

Exoskeleton

As arthropods, arachnids possess a hard external covering called the exoskeleton (integument). The exoskeleton is constructed of various layers, like pasta sheets in a lasagne. The three main layers consist of the outer layer, the epicuticle, which is extremely thin, a much thicker sclerotized exocuticle and the unsclerotized endocuticle. Together these layers are often known as the cuticle or integument, whilst the exocuticle and the endocuticle together are known as the procuticle. These cuticular layers overlay a cellular layer called the hypodermis. The exoskeleton varies in thickness between different arachnids, even within orders, e.g. desert-dwelling arachnids have much thicker exoskeletons than those arachnids in moist areas.

The key structural material of the cuticle is a polysaccharide called chitin. Microfibres of chitin are incorporated into a protein (arthropodin), giving it great elasticity and great strength. Not only does the cuticle cover the body surface, it forms the tendons, joint membranes, and the sensory hairs. It even forms the lining of the reproductive and respiratory organs! Certain parts of the cuticle form in-foldings, called apodernes, which project into the body, and which function as muscle attachment points.

The outermost epicuticle is itself constructed of four distinct layers. The wax layer, which varies enormously in thickness between the arachnid groups, provides an impermeable quality to the exoskeleton and helps to prevent water loss. This operates in a similar way to the layer of wax applied to outdoor clothing. Anyone who owns a Barbour jacket or Australian riding coat knows that unless the garment is waxed from time to time, it will not be effective as a water barrier! The thickness of the wax layer is related to the number of wax secretory cells in the epithelium. The thick exocuticle is made from numerous lamellae (thin plates), which because of the way they are stacked, give a high tensile strength to the whole cuticle. Stiff sensory hairs on the exoskeleton are constructed mainly from exocuticle. The hypodermis is interspersed with gland and sensory cells. The epithelial cells may contain guanine crystals or pigment cells, which help to give an arachnid its colour. The cuticular layers are secreted by the hypodermis. The layers build up on top of the hypodermis.

The cuticle isn't made from unconnected layers however; there is a series of interconnecting narrow channels or pore canals (dermal gland-cell ducts) that link the bottom hypodermis to the various cuticular layers. Once they have passed through the exocuticle, they fan out into finer canals in the epicuticle. If the epicuticle gets abraded, it is replaced by secretions from the pore canals. Wax glands in the epidermis produce a complex mixture of lipids and polysaccharides, which pass through the cuticle via wax canals, get deposited on the surface, and produce a waxy layer.

THE SURFACE The cuticle skeleton certainly isn't smooth; its surface is covered by various types of hairs or setae, which may be sensory. The most common sensory hair is a tactile hair found on both body and legs. These are large, articulated hairs that sit in a socket in the exoskeleton, and are attached to three nerve endings. When the hair moves as a result of being touched, nerve impulses are generated at the base and travel to the central nervous system, where they are converted into a response, such as running away. The trichobothrium is another type of sensory hair that is innervated (attached to nerves). It is only found on certain leg segments and pedipalps, and is much less numerous than the tactile hair. It is

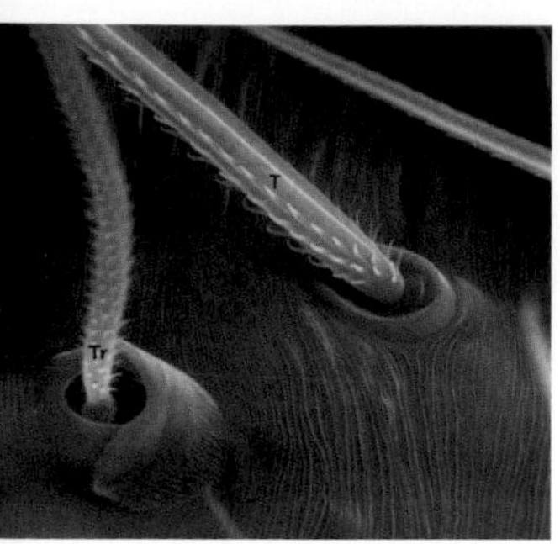

ABOVE The slender sensory trichobothrium hair rises vertically from a prominent socket on the cuticle. The tactile sensory hair emerges from a less-developed socket.

extremely sensitive, because the very fine hair shaft is suspended in a membrane, rather than being firmly attached to the exoskeleton, so it picks up the slightest air movements that then trigger nerve impulses.

Another type of sensory hair is the chemoreceptive hair, which like the others, is innervated, but also has a tip open to the outside. Connections to a nerve cell in the base run up the entire length of the hair shaft to the open tip. These connections detect volatile molecules. These hairs are found mainly on the tarsi of the first legs and on the pedipalps, but some are also found on the mouthparts. Other setae include those used for defence, for provision of colour, for attachment for youngsters and for adhesion.

Arachnids have other sensory structures on their exoskeleton including the slit sensilla. These are slit-shaped structures that are embedded in the exoskeleton, and are found all over the body surface, especially the legs and particularly near the joints. They detect mechanical stress in the exoskeleton, such as those caused by substrate vibrations, and like the sensory hairs, are innervated. Other structures include tubercles, spines, and stridulatory ridges or bristles.

A large proportion of the exoskeleton is hard and usually dark coloured – this is sclerotization. If, however, the whole of the exoskeleton was hard, then there would be no mobility! So there are usually separate segments with soft cuticle in between to allow articulation and expansion after feeding, apart from in certain arachnids such as spiders and mites. These intersegmental membranes are lighter coloured and don't have the tough exocuticle.

The adult arachnid body has between 16 and 18 segments or somites (the prosoma has six, the opisthosoma has 12), depending on the authority. The actual number of segments present can be obscured because of the fusion of segments. The segments are divided into dorsal plates called tergites, and ventral plates called sternites, and there is an intersegmental membrane at the midway line between. The tergites of the prosoma are usually fused to form a tough dorsal shield called the carapace. The sternites are rather variable – in spiders, they have developed into a sternum whilst in Opiliones they are hidden by the coxae, and in Solifugae they are non-existent.

So, the exoskeleton has several functions: it provides structural rigidity for the whole body; it is a protective armour for soft internal parts; it acts as attachment for muscles; it prevents desiccation; it permits very large changes of volume in the body after huge feeding bouts and it has a whole host of sensory structures. Additionally, in a few arachnid species that don't possess respiratory structures, it allows oxygen transfer across the cuticle.

COLOURATION How are colours produced? Every arachnid has an exoskeleton of a certain colour combination. Colour and patterning are produced by one or more of the following: by pigmentation in the actual cuticle or in hairs on the cuticle; by hairs on the cuticle which scatter, diffract (split) or reflect light; by excretory products stored in cells beneath the cuticle, e.g. guanine crystals in spiders located in guanocytes, and by areas of cuticle that reflect light. The majority of arachnids are dull coloured, blending into their backgrounds in a subtle and clever way. Arachnids that live in caves have lost their pigment altogether, rendering them a ghostly white. However, there are some that are very brightly coloured. Colour is used in many ways: for camouflage; mimicry (where a creature has evolved to look like another as a form of defence); as warning colouration (where the animal possesses conspicuous markings which indicate to a predator that the animal is distasteful or venomous) and courtship (where males may have brightly coloured body parts which they move to attract the attention of conspecific females e.g. salticid spiders). Colouration produced by pigmentation, hairs and reflection is permanent. However, colouration produced by excretory products is variable because the guanocytes can be retracted from the surface of the cuticle.

Prosoma

The functions of the prosoma are feeding, locomotion and sensory. The prosoma has six segments, and the tergites are usually fused to form the carapace. The prosoma is fairly similar throughout the Arachnida, apart from in the Schizomida, which is unique in having the dorsal surface divided into platelets. All six segments of the prosoma bear a pair of appendages – the chelicerae on the first, palps on the second, and a pair of legs on each of the other four.

Eyes

Arachnids don't have compound eyes with multiple lenses, like those of insects, instead they have so-called simple eyes, or ocelli, that have only one lens. The light-sensitive parts of the visual cells in the retina called rhabdomeres, provide the image quality. The more rhabdomeres there are, the better the image quality. So, arachnids with good eyesight, such as spiders in the family Lycosidae, have around 4,000 rhabdomeres in their posterior, median and lateral eyes, whereas arachnids with poor eyesight such as the spider genus *Araneus* (Araneidae), have a mere 80. They cannot therefore detect much more than movement.

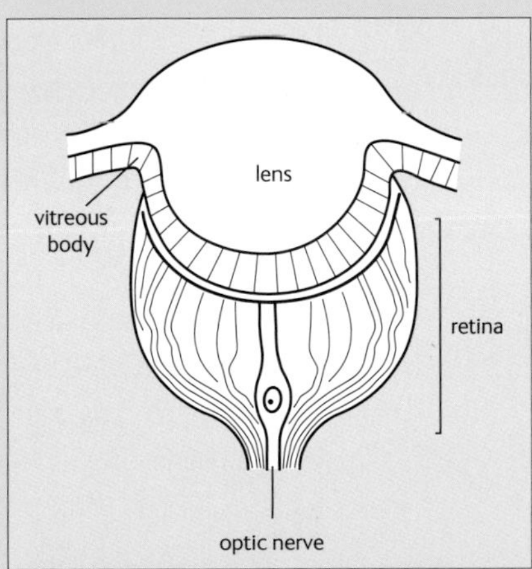

ABOVE The simple or ocellus eye of an arachnid. The lens is formed by a thickening of the cuticle. Beneath the lens is a bed of vitreous cells, called the vitreous body, and the retina, which contains pigment and a few visual cells.

For the most part, vision is a minor sense for arachnids, because many are nocturnal so are more reliant on touch and taste. However, even the most basic arachnid eyes are able to detect subtle changes in light intensity, enabling the arachnid to know whether it is night or day. The eyes can detect movement, so the arachnid knows if danger threatens, and for some arachnids such as hunting spiders, this vision is particularly important for both prey capture and courtship.

Arachnids have a maximum of twelve eyes (in scorpions); eight is a common number (in many spiders), but they may have six, four or two eyes. (Troglobitic arachnids, which live entirely within the dark parts of caves, have none at all.) Why have so many? Having several eyes confers a distinct advantage, because the eyes are arranged so that they have good all-round vision simultaneously. Eyes are on the carapace, usually near the front edge. However, they can be either side of a mound-like structure called an ocularium, which is found dorsally on the midline of the carapace in opilionids.

RIGHT AND FAR RIGHT The chelicerae may have two (right) or three (far right) segments. In spiders, each chelicera consists of two parts – the large, fixed basal part, and the curved, moveable fang that is hinged on the basal part. There is a fine duct within each fang, so that venom can pass through the fang and into the prey. In other arachnids the chelicerae are designed to tear.

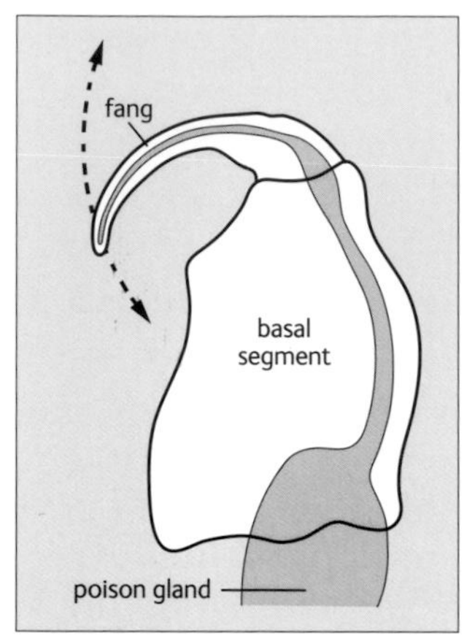

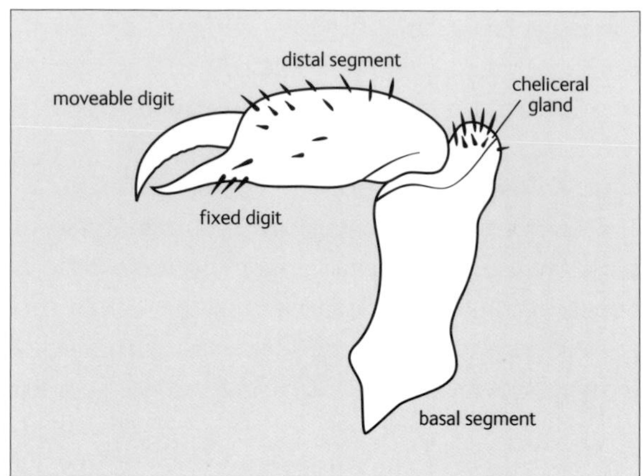

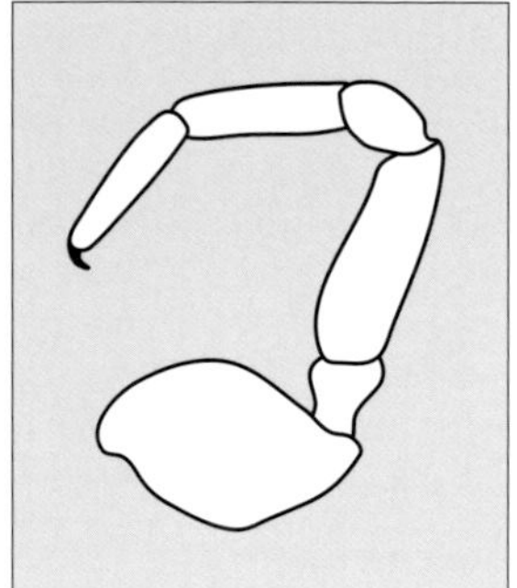

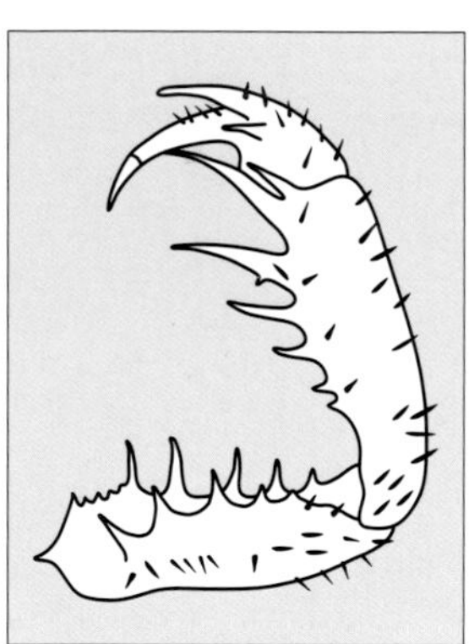

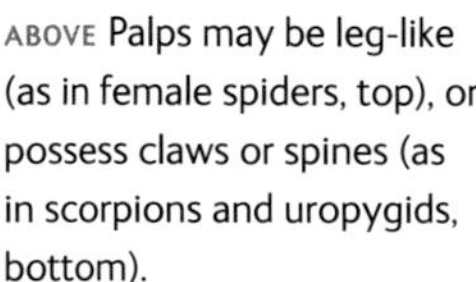

ABOVE Palps may be leg-like (as in female spiders, top), or possess claws or spines (as in scorpions and uropygids, bottom).

Chelicerae

The chelicerae are the first pair of appendages near the mouth (located between the pedipalps), and are modified for grasping and piercing. The word 'chelicera' comes from the Latin 'chelae' meaning 'claws', and they are claw-like (chelate), as in scorpions, or unchelate, as in spider fangs. Chelicerae vary in size comparative to the body, depending on the order, e.g. solifugids have enormous chelicerae.

Pedipalps

Pedipalps are also referred to as palps and usually have the same divisions as the legs, but lack the metatarsus. Depending on the species, they are raptorial and used for grabbing prey – e.g. solifugids have suckers on their palps for this purpose – manipulating prey whilst feeding, in ritualized fighting, in courtship and digging burrows. Adult male spiders have specialized structures on their palps that they use to transfer sperm to the female.

Legs

All arachnids have four pairs of legs that are attached to the underside of the prosoma, at the edge of the sternum. The legs are extended by hydraulic action of the haemolymph. They are not always used for locomotion – the first pair may be carried up and extended forwards and used as tactile organs e.g. in Solifugae. In Amblypygi and Uropygi, the first pair of legs is very long,

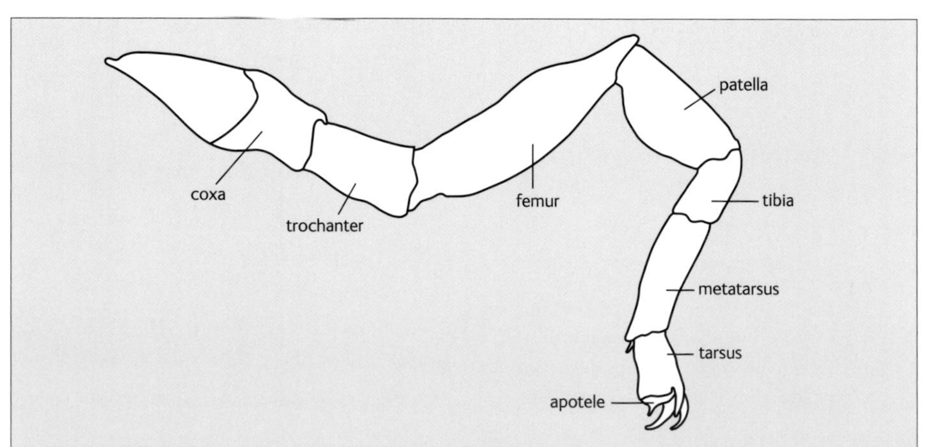

RIGHT A generalized arachnid leg has seven segments, usually referred to as the coxa, trochanter, femur, patella, tibia, metatarsus and tarsus.

thin and whip-like, and cannot support the body's weight, so wouldn't be used for walking. These legs have a sensory function, in the same way as insect antennae, so are known as antenniform legs. They are also very important for communication during fighting and courtship.

Opisthosoma

The key functions of the opisthosoma are digestion, sperm or egg production (as it contains the internal genitalia), and gas exchange. Apart from the spinnerets of spiders, the opisthosoma doesn't have any appendages. It is much more variable than the prosoma, and the only similarity that all opisthosomal designs have, is that the anus is on the last segment! Variations include: unsegmented, globular and sac-like and made from soft cuticle e.g. spiders; unsegmented but not as sac-like as spiders e.g. mites; weakly segmented e.g. opilionids; retention of the ancestral segmented form and elongate, flattened and segmented (consisting of 12 segments) with a combination of hard and soft cuticle e.g. Uropygi, Amblypygi, Pseudoscorpiones, Solifugae; separated into a distinct mesosoma and metasoma, as in scorpions; housing silk glands in spiders; housing odiferous glands in opilionids.

The first segment of the opisthosoma is narrowed in spiders, palpigrades, uropygids, amblypygids, solifugids, schizomids and ricinuleids to form a narrow pedicel (waist), which joins the prosoma to the opisthosoma. The other arachnids (i.e. scorpions, opilionids) don't have a pedicel, and there is a broad join between the prosoma and opisthosoma.

EXTERNAL GENITALIA In general, the genital opening in most arachnids is located on segment eight (of the entire body), i.e. the second opisthosomal segment. It is only in ricinuleids that the external genitalia opens on the pedicel. In some groups, such as tarantulas, you can't see any structures at all because the gonopore (genital opening) is hidden away in a fold in the cuticle on the underside of the opisthosoma near the pedicel. In scorpions, the gonopore is covered by an operculum, which is formed from either a pair of small plates that are partially or totally separate in the male, or a single plate in the female. In other arachnids, the gonopore consists of sclerotized structures, e.g. within the araneomorph spiders there are adult female spiders with complex external genital structures that are sclerotized plates with various in-foldings. These structures are called epigynes and can be easily seen. In some spiders, e.g. Araneidae, the epigyne extends outwards into an elongate structure called a scape, which locks onto the male's palp when the tip is inserted. Because the epigyne is unique to each species, it provides a great identification aid.

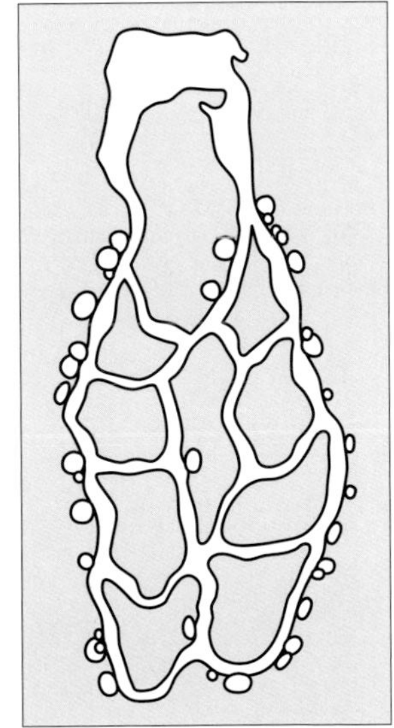

ABOVE The ovary of a female scorpion (here in dorsal view) is very different to the tubular ovary structure of spiders and other arachnids. It is made up of a network of longitudinal and transverse tubes. The male scorpion also has complex reproductive structures.

Sexual dimorphism

Sexual dimorphism means the differences in size or shape between males and females of a particular species. It is much more obvious in some orders than others or may not be present at all. Sexual dimorphism can be demonstrated in several different ways: overall size - females are usually larger than males; robustness - females are often more robust than males; structures - structures such as the palps of amblypygids can vary in size between the sexes; presence of extra features, e.g. palpal bulbs in adult male spiders; some male scorpions have structures or depressions on the palpal chela; surface textures – e.g. male scorpions usually have more granular bodies than females; colour - males are usually much more highly coloured, e.g. the ladybird spider.

Internal characteristics

Central nervous system

The central nervous system (CNS) of arachnids is still relatively unknown compared to that of insects or crustaceans. Some arachnids, such as spiders, have a rather highly developed CNS, whereas it is less developed in other orders. The brain is connected to the ventral nerve cord that swells to a ganglion in each segment. Nerve bundles lead from the CNS to the extremities and organs in the prosoma and opisthosoma, and constitute the peripheral nervous system. In spiders, the sophisticated arrangement of the CNS and peripheral nervous system means that external sensory input can be stored in the brain and recalled later.

BELOW The generalized CNS is a basic brain in the prosoma called the cerebral ganglion (a collection of nerve cells), which surrounds the oesophagus. More highly-developed systems have two large ganglia, the sub- and supraoesophagal ganglia, whereas in less-developed systems, the brain is a single combined mass.

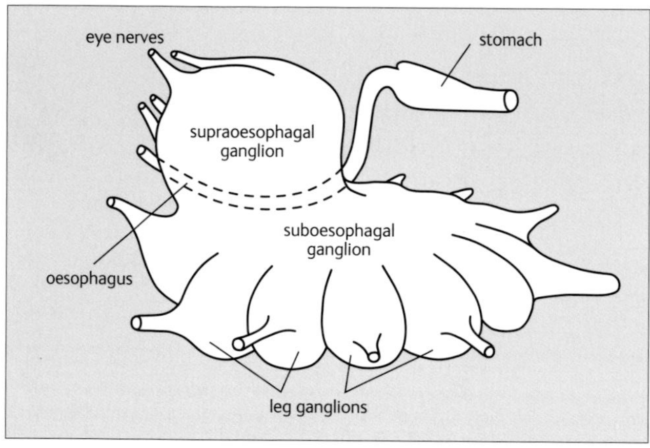

Circulatory system

The arachnid equivalent of mammalian blood is haemolymph. Like mammalian blood, it contains a number of different types of blood cells or haemocytes, involved in wound healing, fighting off infection, blood clotting and other immunological responses. Haemolymph is also responsible for the transport of oxygen.

RESPIRATORY PIGMENTS Unlike mammalian blood, instead of being red, fresh haemolymph is a slightly blue colour, because the respiratory pigment in it, called haemocyanin, contains copper. The oxygen molecules bind to these respiratory pigments and are transported around in the blood system. Haemoglobin in mammals is found in specialized cells, whereas haemocyanin is freely dissolved in the haemolymph. Haemocyanin is not nearly as efficient as haemoglobin, but this is not important because arachnids are inactive most of the time.

OPEN AND CLOSED SYSTEMS The circulatory system of an arachnid (as with all arthropods and molluscs) is known as an 'open' system, as opposed to the 'closed' circulatory system of vertebrates and a few invertebrates, where the blood is enclosed within blood vessels. Contrary to the term 'open', arachnids do actually have quite distinct arteries that branch from the heart through the entire body, arteries even lead right down into the tarsi. All arteries have an open end, from which the haemolymph seeps between the tissues. There are no veins however, so the haemolymph has to flow back to the heart along specific pathways, following the gradient of decreasing pressure. The haemolymph collects in lacunae or empty spaces

RIGHT The basic tubular heart is suspended by ligaments in a cavity or sinus. The haemolymph enters the heart through valves or ostia. More primitive arachnids tend to have longer hearts with several ostia, whereas more specialized arachnids have shorter hearts and a few ostia.

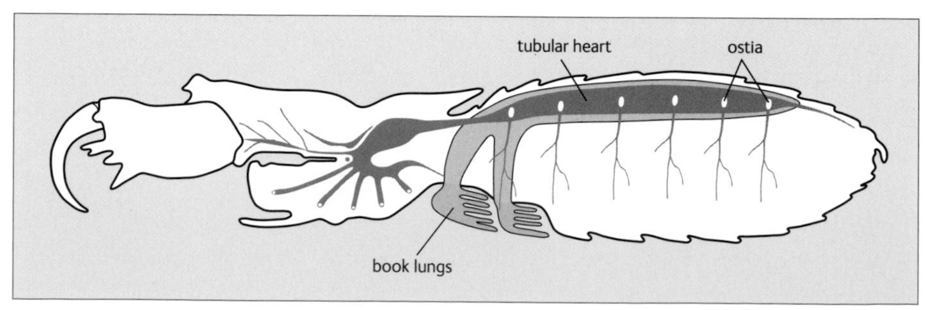

and then passes to the book lungs. Once oxygenated, the haemolymph returns to the heart. Strictly speaking, the arachnid circulatory system should be termed 'partially closed'.

RESPIRATORY ORGANS Arachnids have two types of respiratory organs: book lungs and tracheae. Some arachnids may have just tracheae or book lungs, some may have both, e.g. some araneomorph spiders have one pair of book lungs supplemented with tracheal tubes, and some may have neither. Those without (such as some parasitic mites and palpigrades) absorb oxygen through their very thin exoskeleton.

In terrestrial arthropods, tracheation is a much more efficient system of gas exchange. Complete tracheation is found in araneomorph spiders, Acari, Pseudoscorpiones, Opiliones, Solifugae and Ricinulei. Tubular tracheae are narrow tubes, reinforced by chitin, which run from a porthole-like spiracle opening on the surface of the opisthosoma, and branch throughout the body, allowing oxygen to permeate the tissues. Unlike the opening to book lungs, spiracles are difficult to see. Some spiders also have sieve tracheae, which are delicately branched tracheae, with a slightly different configuration to tubular tracheae. Unlike the tracheae in insects, an arachnid trachea always has an open end that isn't in contact with a cell. Oxygen isn't delivered directly, instead, the haemolymph has to transport the oxygen the last tiny distance to its final destination. However, tracheae still result in a much higher aerobic capacity, which allows running speeds to be much greater and longer.

Book lungs

Book lungs, or lung books, are found in scorpions, spiders, Amblypygi, Uropygi and Palpigradi. They are always situated on the ventral surface of the anterior part of the opisthosoma and open on segments 12–15. Under the flap of the external part of a book lung is a slit-like channel that leads to the internal book lung structure. The lung slits can be enlarged by muscles, which helps to slightly increase gas exchange, but is not a true ventilation of the book lungs. Gas exchange is still mainly a result of diffusion.

The lamellae range from five to 150. In cross-section, the lamellae look like pages of a book end-on, hence the name book lung. They are hollow, which allows the haemolymph to flow through them. There are air spaces between the lamellae, which open to the outside of the body through the channel. As the air passes passively from outside through the channel, it fills the air spaces between the lamellae. As the haemolymph flows through the lamellae, gas exchange takes place, and oxygen is taken up across the very thin cuticle. However, this primitive system is so inefficient, that the capability to deliver oxygen is very quickly overwhelmed by even the minimum of demands required by low-speed movements.

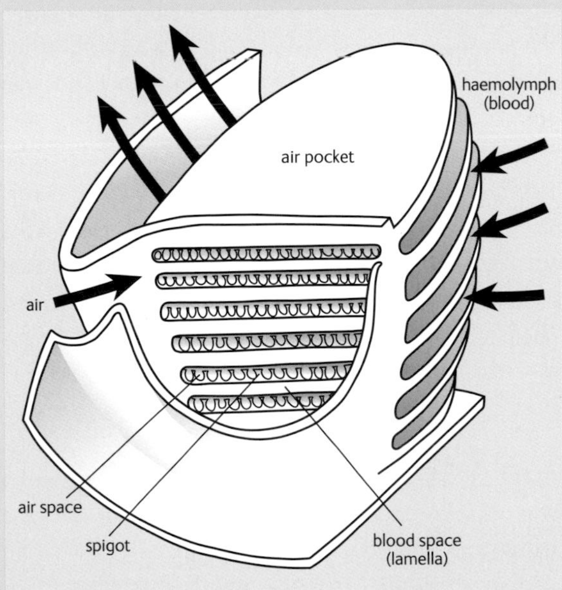

LEFT The book lungs consist of a chamber containing a stack of thin plates called lamellae, made from very fine cuticle and separated from each other by chitinous pillars.

Digestive system

Most arachnids predigest their food outside of their bodies using digestive fluids regurgitated from the mid-gut. This is called 'extra-oral' digestion. In most arachnids, the stomach is attached to the carapace and endosternite (a lightly sclerotized plate located between the alimentary canal and nerve ganglia) by muscles. When these stomach muscles contract, producing a rhythmical pumping action, the stomach cavity greatly increases and acts as a strong pump, sucking the food through the narrow mouth, through the pharynx and the oesophagus, and into the stomach. The gut consists of the foregut, midgut and hindgut. The foregut and hindgut are lined with chitin, so the only section for absorption is the midgut, which has many finger-like projections called diverticula, branching off it at its outer edges. Secretory cells in the epithelium of the diverticula produce the digestive enzymes that are disgorged to break down the tissues of prey. Absorption also occurs in the diverticula, so they act as storage organs, which enable arachnids to go without food for long periods of time.

Excretory system

Excretion is achieved by the explusion of excretory products from the body, and the uptake of excretory products by specialist cells. These are carried out by the coxal glands and nephrocytes in the prosoma, and by the Malpighian tubules in the opisthosoma. Coxal glands are the most primitive of the three excretory systems. A basic coxal gland is a large excretory sac lined with epithelium, which takes up nitrogenous waste and expels it from small orifices behind the coxae. The coxal glands open near the base of the palpal coxae in solifugids, at the base of the coxae of the walking legs in spiders, and behind the fifth prosomal segment in scorpions, pseudoscorpions and opilionids.

Nephrocytes are specialized cells that absorb the products of metabolism from the haemolymph. Arachnid Malpighian tubes or tubules are similar to insect Malpighian tubes, and are the main excretory organs of arachnids. These are slender, tube-shaped muscular glands which extend inbetween the midgut glands and mesodermal tissue, and open into the posterior part of the midgut. Their epithelial cells are concerned with concentrating and storing waste products, transforming them to guanin that is then emptied into the hindgut, from where it is ejected.

Internal genitalia

The male internal genitalia basically consist of one or two testes that produce spermatocytes – male reproductive cells that become spermatozoa. Once the spermatozoa are produced, they pass along sperm ducts called vasa deferentia and out of the gonopore. The female is inseminated by either liquid sperm (as in spiders) or the spermatozoa may be neatly contained in little packages called spermatophores (as in ticks and scorpions), which can be rather complex in design. Opilionids and some mites have an intermittent organ, a penis, used to inseminate the female directly.

The female internal genitalia consist of a single or paired ovary, which produces oocytes – the eggs produced by the ovary before fertilization. Once the oocytes have developed, they pass along oviducts, through the uterus and vagina, and out of the gonopore. Many arachnids have seminal receptacles, which are little pouches just inside the gonopore that store the sperm until the female is ready to fertilize her eggs.

Distribution

As a class arachnids are found throughout the world, and it is only when their distribution is observed at the order level that patterns begin to emerge. However, generalizations can be made, and they can be assigned to four general groups: 1) Arachnids that are confined to the hot tropical and very warm sub-tropical regions, and are more or less continuously distributed there, e.g. Ricinulei, mygalomorph spiders, Amblypygi and Uropygi. They are represented in the tropics by some of their largest species, e.g. scorpions and theraphosid spiders. 2) Arachnids that are also found in tropical and sub-tropical regions but are sporadically and discontinuously distributed there, e.g. Schizomida, liphistiid spiders and Palpigradi. 3) Arachnids that are distributed right to the limits of the temperate zones e.g. Pseudoscorpiones, Acari, Opiliones and most spiders. 4) Arachnids that have made it into polar regions, e.g. a few spider and mite species. There is a sharp distinction between the Antarctic and Arctic arachnid fauna because of the isolation of the Antarctic continent.

Habitats

Arachnids have colonized almost every terrestrial habitat, and secondarily invaded aquatic habitats (both freshwater and marine), although to a much smaller extent. A high proportion of the world's arachnids are found in the tropics, with rainforest habitats being the richest of all. Other habitats include: caves, the tops of mountains, in the desert, up trees, in soil and leaf litter and in gardens. Any one species is often limited to a strictly defined environment. Physical limitations include the requirements for specific humidity and temperatures, and biological limitations include the requirements for specific food and type of vegetation. Each habitat, e.g. woodland, has microhabitats like leaf litter with specific characteristics, and therefore there is a corresponding partitioning of arachnid species relating to these microhabitats. Some species however, can move between different microhabitats.

Dispersal

Arachnids have several different ways of dispersing, including ballooning, walking and rafting, phoresy and incidental dispersal as a result of a parasitic lifestyle. When small to medium-sized spiders and certain mites want to disperse, usually triggered when the environment becomes hostile, they wait until the wind conditions are right, let out a thread of silk and when the air currents catch the silk they are lifted into the air. Of course, many arachnids do not disperse by ballooning, either because the group never evolved ballooning or because they are just too large! They therefore have no other option but to walk, and may not walk far from the maternal nest before setting up home. Although rafting on floating trees and other vegetation is probably a minor method of dispersal for arachnids, it certainly cannot be ruled out in places that are close to a continental mass.

When you are only a few millimetres long, with short legs and no wings, dispersal could be somewhat of a problem. So how do you disperse? Many species of mites and pseudoscorpions have overcome this very large hurdle by hitch-hiking rides from other larger and much more mobile animals, such as flies and beetles. This is particularly effective because these insects fly and can carry their tiny passengers some distance. This association is called phoresy. Ticks are not phoretic, though as a consequence of their parasitic lifestyle, they may be carried around by their hosts.

Biology and behaviour

Venom

It's not just the few spider and scorpion species that produce venom – pseudoscorpions do too. Venom is a term for the poisonous secretions produced by several groups of animals, mainly used for the purpose of subduing prey and for defence. It is usually administered through a bite or a sting, but it can also be sprayed, or may be inhaled. However, it is not harmful if ingested. Arachnid venom actually contains many different substances, such as small amino acids, large, neurotoxic polypeptides (which are the toxins that cause the symptoms of envenomation), and sometimes proteolytic enzymes, which help start the pre-digestion of the prey's tissues.

Silk

Silk production is very characteristic of some of the arachnid orders such as spiders. However, it's not just the spiders that have the skill, pseudoscorpions and mites do too. Not surprisingly, their methods of production are as varied as are their uses. The silk glands of spiders are located in their opisthosoma, and they use silk in many different ways – for prey capture, wrapping and immobilising prey, lining of burrows, retreat construction, sensory enhancement, safety lines, drag lines, dispersal, courtship, sperm webs, making egg sacs, and as an aid in moulting.

The silk glands of pseudoscorpions normally open on their chelicerae, with the glands situated in the prosoma. This silk is not used to trap prey, but to build silken chambers for hibernation, moulting and egg laying.

The silk glands of mites may open on the palps. In several groups of Prostigmata, the silk comes from within the mouth, opening from modified salivary gland secretions. Mites are like spiders in that they use silk in a large variety of ways, including: protection for eggs, chambers for moulting, aerial dispersal, as droplines, silken webs as shelter, and even for prey capture.

Food and prey capture

Nearly all arachnids are predatory. However, mites are the only arachnids with vegetarian and parasitic forms, as well as predatory ones. Mites, along with Opiliones, are also scavengers. As a class, arachnids feed on many different types of food in a wide size range including: other arachnids, insects, Collembola, molluscs, chilopods, crustaceans such as woodlice, nematodes and annelid worms. Vertebrate food includes tadpoles and small fish, snakes and lizards, mice and even small birds! They also obtain food in different ways: ambush predators that lie-in-wait for their prey and then grab it; active foragers for live prey; food stealing; web building (spiders only); parasitism; scavenging for decomposing food (plant and animal), fungi, algae, fruit, microbes. Arachnids use their complex sensory systems (i.e. trichobothria, setae, chemosensory hairs etc.) to locate their prey, before it is grabbed by the palps and chelicerae.

Defence

For many animals, a nice big juicy arachnid makes a decent meal. Apart from such predators, arachnids also need to defend themselves against conspecifics (others of the same species) during e.g. fighting for mates. Arachnids have therefore evolved a whole host of defence strategies, both morphological and behavioural. Sometimes certain behaviour can reinforce the morphological strategy. For example, an arachnid that is camouflaged (morphological) stays perfectly still for hours on end in order to further enhance the defence system (behavioural).

Some orders only demonstrate a few of the following defensive strategies, whereas others (such as spiders) demonstrate them all: escape behaviour; camouflage; warning colouration; threat displays; stridulation; use of venom; defensive secretions; leg autotomy – the deliberate shedding of a body part; dropping and feigning death; burrow structures – these camouflage the burrow entrance from would-be predators.

Adaptations

All arthropods are unable to maintain a constant body temperature, and are therefore known as poikilotherms. They have several ways, both physiological and behavioural, in which they can regulate their body temperature within acceptable limits, and these allow them to live in extremely hot and cold conditions.

DESICCATION TOLERANCE Within a species, desiccation tolerance varies, depending on the sex or the developmental stage. Males are more susceptible to desiccation than females because they have a smaller body volume, and possibly because they spend a lot of their time out-and-about looking for females. Juvenile arachnids are more susceptible to desiccation than adults.

Arachnids have various physiological adaptations against desiccation including guanine crystals. Some species of spiders have guanine crystals just under the epidermis of the cuticle, which actually reflect sunlight, thereby reducing its intensity. Another adaptation is a thick and waxy cuticle, which provides an impermeable quality to the exoskeleton. Arachnids are vulnerable to water loss, particularly through their often lightly sclerotized opisthosoma. Resistance to desiccation is much higher when the cuticle is highly sclerotized and certain quantities and compositions of lipids are present. If, however, critically high temperatures are reached, it is possible that certain cuticular lipids actually hit melting point and with the breakdown of its important barrier, the cuticle is left open to desiccation. Arachnids are much less tolerant to desiccation just after moulting, which is not surprising because their new soft cuticle has not yet developed its waterproof properties. Evaporative water loss is a problem for arachnids that have thin cuticles, and up to 98% of total water loss can occur in this way.

Other adaptations include excreting very dry waste and almost insoluble waste products, such as guanine and uric acid. These compounds can be excreted in crystalline form without having to be dissolved in water, because they are non-toxic (and they are non-soluble anyway). Some species of arachnids are able to control the size of the aperture of the spiracles through muscle control and this might be used to further restrict water loss. Scorpions have narrow slits to their book lungs that are set into the body cavity, which means that water loss due to evaporation, is kept to a minimum. And finally, arachnids that live in dry regions tend to have lower transpiration rates (loss of water through evaporation).

THERMOREGULATORY BEHAVIOUR Arachnids might exhibit one or several different behaviours in order to regulate their body temperature and this is called thermoregulation. Various behaviours include: active avoidance, avoiding extremes of temperature; orientation, moving to more favourable conditions; burrowing, a great way to avoid extremes of temperature on the ground surface (also, those arachnids with book lungs are susceptible to dehydrating quickly, so living in a deep burrow helps to prevent water loss); posturing, e.g. raising the body off the hot ground surface; being nocturnal; drinking, which appears to be a major way that arachnids maintain their water balance and avoid dehydration.

COLD TOLERANCE To survive cold climates and winter seasons, animals need to have special adaptations. Many arachnids spend the winter in a habitat that is insulated from the coldest temperatures such as leaf litter, in soil or under snow. Many overwinter by going into torpor. For example, both juveniles and adults of the aquatic spider *Argyroneta aquatica* move into deeper water, construct a diving bell, and remain inside it throughout the winter. Their metabolic rate drops to less than half its usual rate, and it does not feed. Arachnids may overwinter in different developmental stages, so spiders may remain as spiderlings inside their egg sacs during the winter months, they might overwinter as adults, or they may brave the cold as juveniles. Some arachnids are actually freeze-resistant. They are able to lower the

Moulting

All creatures with an exoskeleton have to moult in order to grow. Therefore, growth is not continuous, but staggered, because increase in growth is only possible during and directly after ecdysis. Arachnids need to moult several times to reach their adult size, and the number of moults varies, depending on the species. An 'instar' is the stage of an arachnid's life in between moults. Usually, moulting ceases once adulthood is reached though adult tarantulas continue to moult once they are adult, because of their great longevity compared to other arachnids.

Moulting is actually a two-part process, triggered and controlled by a hormone called ecdysone. The first internal part, apolysis, is where the old cuticle separates from the new cuticle, which is developing beneath; the old cuticle separates from the underlying hypodermis. A few days before ecdysis, new cuticle begins to develop on the hypodermis, whilst the internal layer of the old cuticle is dissolved by enzymes. As the endocuticle is dissolved, it becomes much thinner. Once secretion of the cuticle is complete and the old cuticle is very thin, ecdysis is ready to begin. This separation is in preparation for the actual shedding, or ecdysis, of the old cuticle. In general, moulting is an independent process though communal moulting has been observed in some arachnids. This synchronization is no doubt related to a chemical cue.

BELOW Once the old cuticle has been shed in moulting, it takes a day or two for the new cuticle to harden. A newly moulted arachnid looks ghostly pale, and the new cuticle is still soft, so the arachnid is extremely vulnerable to predation and dehydration.

temperature of their body at which they would freeze by supercooling, which is the synthesis of antifreeze chemicals called cryoprotectants. These can protect against internal freezing down to -7°C (45°F). Interestingly, there is geographical variation in limits of tolerance to temperature. Cold tolerance is also affected by the degree of dehydration of the body, the contents of the gut, and the relationship between body size and ability to supercool. Not surprisingly, high-latitude and high-altitude species at e.g. 5,500 m (18,000 ft) in the Andes, tend to demonstrate this ability.

Autotomy

Many arachnids are able to shed their legs in a process called autotomy - the deliberate shedding of a body part. This can be extremely useful to an arachnid if it is attacked, or if it were caught in its moult where it would otherwise die. Arachnids that are stung between the leg joints by wasps and bees, also quickly shed their legs, to prevent themselves from succumbing to the venom. Arachnids can't be induced to part company with their legs if they are anaesthetized, so it is certainly a voluntary process. They have a plane of autotomy where the appendage most easily comes away. In spiders, this is usually between the coxa and the trochanter, although many long-legged spiders also have a patella-tibia plane. In Amblypygi and Uropygi, the plane of autotomy is at the patella–tibia articulation, and in Opiliones at the trochanter–femur articulation. At the plane of autotomy, the muscles work together with increased haemolymph pressure in the joint membrane to close the wound. This prevents the arachnid from bleeding to death. In other orders, such as scorpions, there appears to be no particular plane. Many arachnids often feast on their own autotomized leg in a bizarre act of recycling.

Regeneration

Most arachnids are fortunate in that they are often able to regenerate missing limbs or other structures such as chelicerae, spinnerets, and palps. This is controlled by hormones, and is an all-or-nothing affair. If an arachnid has lost a leg, for example, a new leg grows within the remaining coxa. At the next moult, the new leg is revealed. It is thinner and shorter than the other legs because it has had to develop in a restricted space, but it has all the appropriate segments. After a few more moults, it attains the same size as the others. If a leg is lost near to a moult, then a new one won't be generated at that time, but at the one following.

Life history

Reproduction

Arachnids have evolved several mechanisms for transferring sperm. The majority of arachnids, i.e. the Amblypygi, Uropygi, Schizomida, Scorpiones, Pseudoscorpiones, spiders and the Solifugae, do this indirectly. It is transferred either in sperm packages called spermatophores or dropped onto the female's genital opening and pushed in (solifugids). In spiders, males transfer liquid semen using specialized structures on their palps. Only Opiliones and some Acari transfer sperm directly by the use of a penis. Direct transfer (copulation) involves the greatest contact between the sexes. With indirect contact, the male may interact with the female to ensure she picks up the spermatophore, involving medium contact. The transfer that involves the least contact between the sexes is spermatophore transfer when the male isn't even there.

SPERM TRANSFER The spermatophore has a head containing a pair of sperm masses or sperm packages that are usually on a stalk. Inside both the sperm masses and packages, are coiled spermatozoa. In sperm masses, the spermatozoa are held together by a glue-like substance, whereas in sperm packages, the spermatozoa are surrounded by a secretion. This probably helps to embed them in the female's genital atrium. Although indirect sperm transfer is primitive, the spermatophores can actually be complicated in structure. In fact, it is a method not only used by invertebrates, but primitive terrestrial vertebrates, such as newts.

The spermatophores are transferred in different ways. 1) Paired-indirect (without substrate) – spermatophores are transferred to the female without touching the substrate. Male ricinuleids have specialized structures on their third pair of legs that are used to transfer a spermatophore to the female's gonopore. Some solifugid species use their chelicerae to place a spermatophore onto the female's gonopore and then push it in with chewing motions of the chelicerae. Additionally, some mites use their chelicerae to transfer the spermatophore. Because the spermatophore is being transferred without touching the ground, or being left exposed for some time, it is nothing more than a gelatinous sperm-mass and doesn't require any special adaptations. 2) Paired-indirect (via substrate) – the male deposits a spermatophore on the ground, and lures the female to step over it and pick it up. She then gathers the spermatozoa off the top into her gonopore. This is found in scorpions, some Pseudoscorpiones, Amblypygi, Schizomida, Uropygi and some mites. These spermatophores are much more complex in structure. The spermatozoa has to be some distance above the ground, so the spermatophore has a stalk, and often some rather complicated mechanisms so the female can pick up the spermatozoa. A male uropygid uses his specially adapted palpal fingers to push a spermatophore into the female's gonopore. Some pseudoscorpions produce silk threads in a path towards the spermatophore to help the female find it. 3) Disassociated – the male produces spermatophores and leaves them haphazardly around for the female to find at a later time – he won't be there to assist! This mechanism is done by some mites, some pseudoscorpions and palpigrades. These

BELOW Spermatophores can be simple as those of *Mastigoproctus* (Uropygi) (below) or complex as in the ones from *Damon gracilis* (bottom left) and *Euphrynichus bacillifer* (Amblypygi) (bottom right).

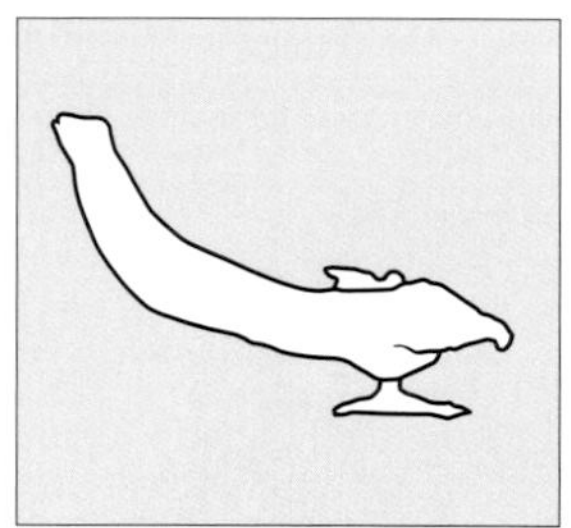

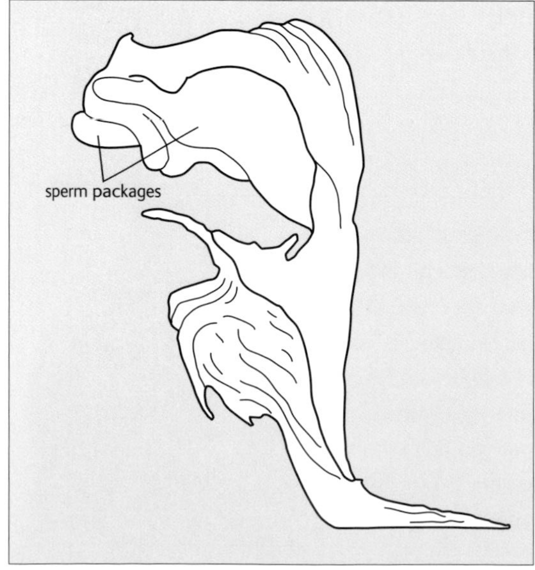

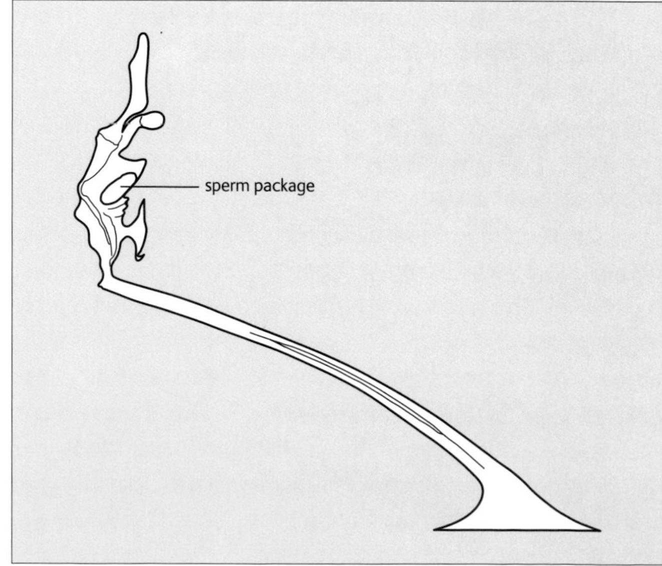

spermatophores must be resistant to desiccation (as with spermatophores in the paired-indirect via the substrate method). Pheromones may also be used to attract females to the spermatophores.

An indirect method of sperm transfer without the use of spermatophores is found in spiders and some solifugids. Adult male spiders produce a droplet of sperm from their gonopore that is drawn up into the specialized structures on their palps. The palps are then inserted into the female's gonopore. Some solifugids deposit a droplet of sperm onto the female's gonopore and then chew it in with their chelicerae

Direct sperm transfer is actual copulation. The only arachnids that have direct sperm transfer are many Acari (some ticks and some mites), and the Opiliones.

COURTSHIP As courtship comes before mating, why is it not discussed before the section on sperm transfer? Well – this is because the type of courtship is related to the sperm-transfer mechanism, so it is necessary to know about sperm transfer first. Why have courtship? It might seem unlikely to us, with our complex human social interactions, that arachnids could indulge in courtship. There are however a few very good reasons in doing so: to reduce female aggression; to make very certain that the male won't be mistaken for prey (especially important in spiders); to make sure that the female is sufficiently stimulated for copulation and to take up a spermatophore, and will therefore be receptive.

In general, it is the male that initiates courtship, which is often stimulated by female pheromones. In scorpions, females leave behind pheromones on the substrate, which are detected by males using their pectines, allowing them to identify and locate potential mates. In spiders, females in the Lycosidae and Araneidae leave pheromones on silk draglines or on webs. So powerful is the stimulus of pheromones, that males may often perform their courtship when the female isn't even there! Courtship is sometimes initiated by male recognition of female morphological characters due to physical contact.

Courtship may be rather basic, or range up to the very complex. It may be similar across several arachnid orders, it may be specific to spermatophore transfer (which involves intense courtship), or it may be specific to one order only. Some types of courtship are similar across most arachnid orders. For example, the stroking of legs and palps, and cheliceral massage are thought to help reduce female aggression. Nuptial gifts are given by males to females in Opiliones and spiders. Some opilionid males may offer a glandular secretion from their chelicerae to their mate before copulation. Male spiders in the family Pisauridae also use nuptial gifts to woo potential mates – these may take the form of a fly wrapped in silk. Some of the courtship behaviours are specific to spermatphore transfer, e.g. the promenade à deux in scorpions is the main part of courtship that enables the male to find a suitable substrate on which to deposit his spermatophore. There are other mating dances performed by other arachnids that produce spermatophores, but they're not referred to as the promenade à deux.

Egg laying and parental care

Eggs are laid by all of the arachnid orders except scorpions and a few mite species, which are viviparous, i.e. they give birth to live young. This is extremely uncommon in terrestrial arthropods. Live birth helps to increase the survival of the offspring. The number of eggs that are laid varies greatly between orders e.g. ricinuleids lay one or two eggs, whereas at the

opposite extreme certain spiders can lay up to 2,000! Eggs may be laid in all manner of places, depending on the group, and may be enclosed in a silken cocoon constructed by the mother, wrapped loosely in threads of silk, or not covered at all.

Although parental care isn't conspicuous amongst the Arachnida as a whole, where it does occur, there are different levels that take several different forms. It can occur just after the eggs are laid until the young are able to disperse and care for themselves. It is thought that the intense predation on eggs by generalist predators actually promotes the selection of parental care. In other words, those arachnids which protect their eggs, will have a greater number of offspring survive, which will go on to produce more eggs than non-guarding arachnids, and so on.

Different forms of parental care in arachnids include: egg hiding; egg covering; chemical protection of eggs; standing guard over the eggs; building a nest for the eggs; carrying around the egg sac; assisting the young when hatching; carrying the babies on the back of the mother; providing food for the young; providing fluids for the young; burrow sharing; food sharing; communal foraging; matriphagy – the eating of their own mothers by young and paternal investment (unique to opilionids).

Post-embryonic development

In general, the different arachnid instars between moults are often known as: larva, nymph and adult. Depending on the order (and indeed family), there may be extra stages. For example, there may be several nymphal stages, known as protonymph, deutonymph and tritonymph. In groups that have more than three nymphal stages, the stages are known as 1st instar, 2nd instar, 3rd instar, and so on. There may be up to 16 instars, so that requires 17 moults in total. The time duration of each instar is very variable, even within an order. For example, the first instar in scorpions can take anywhere between one to 51 days, and there can be 11 months difference in the length of an instar in the scorpion species *Pandinus gambiensis*. Developmental time can vary depending on the amount of food that an arachnid youngster gets over a period of time - a well-fed arachnid has a shorter instar duration than an unfed one.

The age to maturity isn't always known for each arachnid order. When it *is* known, it has been discovered that there can be a great variation within a family, e.g. between 11.5 and 83 months in the scorpion family Scorpionidae. However, scorpions are unusual in the length of time that they spend as juveniles, which is much more than many arachnids. Long generation times result in slow population growth.

In general, arachnid nymphs resemble the adults, apart from the obvious lack of genitalia. There are however, many subtle differences between the various developmental stages. During post-embryonic development, many structures develop in complexity and/or increase in size. These structures can be very indicative of a stage within a particular group, and can therefore help identification. For example, in ricinuleids, the larva always has a two-segmented tarsus on its second leg, as opposed to that of the deutonymph, which always has a five-segmented tarsus on its second leg. The differences in development between the stages include: size of structures; shape of structures; the amount of sclerotization of the cuticle – it becomes more sclerotized as the stages develop and additional structures, such as granulation makes an appearance in later stages; sensory organs – these are underdeveloped in earlier stages and number of setae – these increase in number through the stages.

Longevity

The longevity of arachnids is rather variable. As a class, they live approximately from one year to over 25 years. It is normally sex-dependent, as females are usually the longer-lived sex. For example, within the same species, male tarantulas live for around seven years, whereas females can live for over 25 years. Longevity does not relate to size of the arachnid, so although a ricinuleid is so much smaller than a lycosid spider, the ricinuleid can live up to 10 years, whereas the lycosid will only reach 18 months. The most long-lived groups are tarantulas (Theraphosidae) and scorpions such as *Pandinus* and *Heterometrus* (Scorpionidae).

2 Araneae
Spiders

Spiders are certainly the most famous of all the arachnids. Myths and legends have perpetuated about them through the centuries, many of which are to do with good health or good fortune. In fact, this positive association has actually given a group of spiders within the family Linyphiidae their common name of 'money' spiders! Spiders are notorious for their ability to inject their prey with venom. Perhaps the most infamous is the black widow spider, *Latrodectus mactans*, which can be lethally venomous to humans, though it is very rare for people to die from its bite. However, the vast majority of spiders aren't harmful at all. Although many people suffer from arachnophobia, contrary to popular perception, spiders can be extremely beneficial to human society. They often play a critical role in the ecosystem by feeding on disease vectors and crop pests, and are therefore effective biological control agents.

Spiders are also well known for their ability to spin silk, often used to construct webs, and which is uniquely produced in glands in the opisthosoma, exiting the body through finger-like spinnerets. The amazing properties of spider silk have great commercial potential for products such as bullet-proof vests. The smallest recorded spider by far is the miniscule *Patu digua* (from Colombia), which measures a minute 0.37 mm (1⁄100 in). At the other end of the scale, is a huntsman spider, *Heteropoda maxima*, in the family Sparassidae. This has a leg span of 300 mm (11 ¾ in), the size of a dinner plate.

OPPOSITE Jumping spiders have a square-fronted carapace, with four large eyes that face forward. Their superior eyesight can focus on objects, enabling them to judge distances and so jump accurately. They are strongly dependant on their eyesight for prey capture, orientation and courtship (Guanacaste National Park, Costa Rica).

LEFT The heaviest spider is the massive Goliath bird-eating spider, *Theraphosa blondi* from French Guiana, which weighs up to 155 g (5 ½ oz) the equivalent of five house mice.

Classification and diversity

At the time of writing, there are over 40,000 described valid spider species worldwide, although there are undoubtedly many more that remain undiscovered. The order Araneae is usually split into two suborders, the Mesothelae and the Opisthothelae. The Opisthothelae is then divided into two suborders, the Mygalomorphae and the Araneomorphae. In this arrangement, the suborder Mesothelae contains one family, the Liphistiidae, with 87 species; Mygalomorphae (Orthognatha) contains 15 families with around 2,500 species; and Araneomorphae (Labidognatha), contains 92 families with over 36,000 species.

The Mesothelae and Mygalomorphae used to be lumped together and were often referred to as either mygalomorphs, or Orthognatha. 'Gnathos' means jaw and 'orthos' means straight, so the name Orthognatha referred to the way that the fangs are parallel to each other (see p.43) and articulate downwards in a snake-like strike. However, these two groupings are not closely related. The Araneomorphae and Mygalomorphae are actually more closely related, as signified by their similar arrangement of spinnerets on the tip of the opisthosoma (a maximum of six spinnerets in these groups, whereas Mesothelae spiders have seven to eight), and their lack of body segmentation.

The name Araneomorphae refers to all so-called 'true' spiders, considered to be more advanced than the Mesothelae and Mygalomorphae. Araneomorphs are more widely distributed than mygalomorphs. Araneomorphs used to be called Labidognatha. 'Labidos' means forceps, and referred to the way that the fangs articulate sideways (in an opposing way), like a pair of forceps. Mesothelae spiders have chelicerae which are not parallel, or opposing – these are called plagionaths, 'plagios' means slanting. This may represent the older arrangement from which the Orthognatha and Labidognatha arose. It is beyond the scope of this book to go into detail about all 108 spider families. However, the diversity of spiders is illustrated by introducing a selection of different families, genera and particular species.

Mesothelae

BELOW Mesothelae are effectively living 'fossils', retaining primitive features. The segmented abdomen can be clearly seen on this *Liphistius* (Malaysia).

The suborder Mesothelae contains two extinct families and one extant family – the Liphistiidae. This comprises the most primitive spiders as they retain the primitive feature of a segmented abdomen and have spinnerets located centrally on the ventral side of the opisthosoma , unlike all other spiders, which have them on the tip of the opisthosoma. This rather unusual arrangement has given the suborder its name, because 'meso' means middle and 'thele' means nipple. There are two genera in the family, found only in Japan, China and Southeast Asia. They construct rudimentary trapdoors on their burrows, and are sit-and-wait predators.

Mygalomorphae

Mygalomorphs are currently classified into 15 families. Spiders in this suborder range in size from the small *Homeomma uruguayense* to the heaviest of all spiders *Theraphosa blondi*, and have a great diversity of form.

Tarantulas – Theraphosidae

The most well-known mygalomorph spiders are those in the family Theraphosidae, which contains 903 species. These large, bulky and hairy spiders are most commonly known as 'tarantulas' or 'bird-eating' spiders, as well as 'baboon' spiders, in Africa. Bizarrely, the last two leg segments apparently resemble the finger of a baboon, hence the name. The name 'tarantula' is actually a misnomer. The real tarantula spider is *Lycosa tarantula*, an araneomorph in the family Lycosidae, found in southern Europe. (This spider attained notoriety in medieval Italy because of its venomous bite, which allegedly sent people crazy.) Tarantulas live in most tropical regions of the world. They do not build webs, but live underground in burrows.

BELOW Large, bulky and hairy tarantulas such as this must have been the inspiration behind the scientific name, mygalomorph spiders, because 'mygale' means field mouse!

Trapdoor spiders

This grouping of mygalomorphs has several families, including the Ctenizidae, Actinipodidae and Migidae. The name 'trapdoor' is rather misleading in truth, because many species don't actually build a trapdoor for their burrow. However, those that do so build trapdoors that are specific to their genus and even species. There are 121 species in the family Ctenizidae. They are usually dark in colour, medium to large sized and quite chunky in stature. However, there are a few brightly coloured species too.

Funnel-web spiders – Hexathelidae

Among the 86 species in the family Hexathelidae, known as funnel-web spiders, there are a number of species that are highly venomous to humans. However, there are also araneomorphs known as funnel-web spiders too, e.g. in the Agelenidae, and these are *not* highly venomous, which can cause confusion. One of the most infamous venomous species is the Sydney funnel-web, *Atrax robustus*; it is blue-black in colour, has a glossy velvety texture, large fangs, and stocky spiny legs. It is found in Victoria and New South Wales in Australia. Interestingly, it is the male of this species that is more venomous to humans. *Hadronyche* is another genus of funnel-webs that is potentially highly venomous to humans; *H. formidabilis*, found in northern New South Wales and southeast Queensland in Australia, is very dangerous, particularly to small children.

BELOW Female Sydney funnel-web spider, *Atrax robustus*, feeding on a small kink lizard, Sydney, Australia.

Purse-web spiders – Atypidae

These spiders produce webs that look more like a sock than a purse. The 'foot' part is partially buried in a hole, with the rest above ground, camouflaged by soil particles. The classic purse-web spider, *Atypus affinis*, is the only mygalomorph found in the UK, although it is also found in Europe. This spider is small (up to 18 mm or ¾ in) compared to other mygalomorphs, but the female in particular looks like a typical mygalomorph, with short, stocky legs and large, forward-facing jaws. As with other atypids, *A. affinis* spends almost its entire time in its silken sock. When a potential prey walks onto the sock, it is grabbed by the occupying spider and dragged in. There are 43 species in this family.

Araneomorphae

This suborder contains the majority of all spiders, 94 families, and a selection of families is introduced here. Commonly found spiders are mentioned, along with those species that are particularly notable because of their unusual appearance, behaviour or notoriety in being highly venomous to humans. They are grouped according to how they capture food – by making webs, or as free-living hunters.

Web weavers

ORB-WEB WEAVERS There are several families of spiders that produce orb webs. These include the Araneidae, Nephilidae and Tetragnathidae. Most orb-web builders have spherical opisthosomas, but some are characterized by triangular, spiny bodies, lobed edges, or elongate opisthosomas. The family Araneidae comprises 2,841 species. One of the most recognized orb weavers is the common garden spider *Araneus diadematus*, which is found in Europe, North America and much of Asia and Japan.

The tropical golden orb weavers in the family Nephilidae (which includes 73 species) live up to their name by producing enormous orb webs from silk with the rich colour of gold. Spiders in the genus *Nephila* often have stunningly patterned gold-striped opisthosomas too

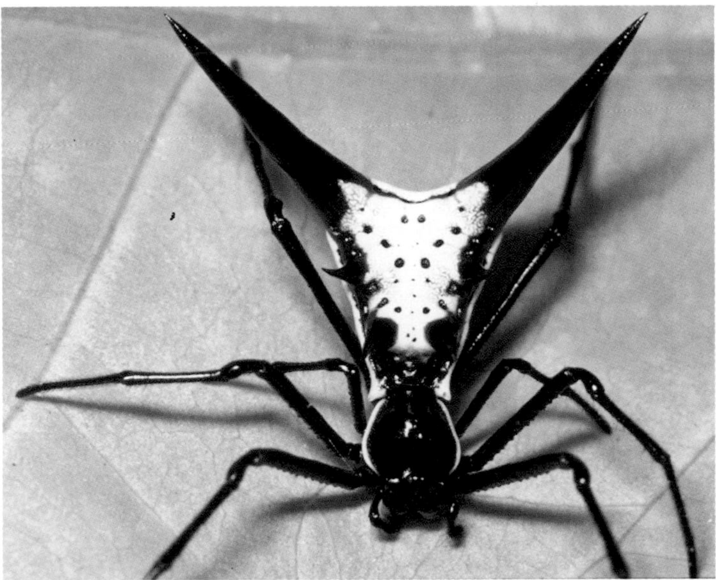

BELOW LEFT The family Araneidae also includes tropical spiders with the most spectacular spiny opisthosomas (Jatun Sacha, Ecuador).

BELOW RIGHT *Araneus quadratus* feeding on a dragonfly. It is so-called because of the distinctive four-spotted pattern on its opisthosoma; found in Europe and Asia.

(see p.48 top). They have a distinctive elongate, tapered opisthosoma and long, thick legs that often possess 'chimney-sweep-style' brushes.

The family Tetragnathidae (958 species) apparently takes its name from the unusual mating position of males and females, which connect their jaws together into a four-sided shape (hence 'tetra'). Males of spiders in the genus *Tetragnatha*, such as *T. extensa*, have incredibly long jaws, with long, slim opisthosomas and legs. They produce orb webs like the araneids, and often hang like sticks at the hub.

SHEET-WEB WEAVERS E.G. LINYPHIIDAE AND AGELENIDAE The family Linyphiidae is the second most diverse, with 4,343 species. Most species in this family are tiny, and so the common name 'dwarf' spider is rather appropriate. In the UK, the majority of species are known as 'money' spiders. They produce sheet webs with no retreat, and they run upside down on the underside of the web.

There are 508 species in the family Agelenidae. These spiders produce sheet webs with funnel retreats. The common northern European house spiders, (which are also found in caves and hollow trees) *Tegenaria*, belong to this family. They produce the infamous cobwebs that are the scourge of every house-proud person. They are hairy, with long hairy legs and can grow to a leg span of around 13 cm (5 in). *Agelena labyrinthica*, also from northern Europe, produces an impressive sheet web on foliage, with a large funnel retreat that extends into the undergrowth (see p.58 top). This spider sits at the entrance to its tube and waits for a passing insect to blunder on its sheet before it dashes out to claim its victim.

TANGLE-WEB WEAVERS E.G. THERIDIIDAE AND PHOLCIDAE The family Theridiidae is large and very diverse, with 2,267 species, and includes the black widow spider. The common name of 'widow' is also used for the whole genus of *Latrodectus*, which includes other infamous species such as the Australian red back, *L. hasselti*, with its distinctive red opisthosoma.

As well as being known as tangle-web weavers, spiders in the Pholcidae family are often called 'daddy long legs' too. This is because they have very long thin legs, and small bodies. The classic 'daddy long legs' spider is *Pholcus phalangioides*, which is found in the UK and Europe, often lurking in the corners of ceilings. This species has an elongate body and very long legs. They often look like they have knobbly 'knees' too, because their patellae can be darker coloured than the rest of their legs. If touched, the spider vibrates so rapidly that it becomes a blur (this is a form of defence). Although these spiders are very much smaller than *Tegenaria* (Agelenidae) spiders, they have been seen to kill and eat their larger relatives. There are 969 species in the family.

BELOW LEFT Spiders in the family Theridiidae are known as comb-footed spiders because the tarsi of the fourth pair of legs have a structure made from setae, which is used to comb out silk from the spinnerets.

BELOW RIGHT The family Theridiidae includes the black widow spider *Latrodectus mactans*, with its red hour-glass marking on the underside of its opisthosoma.

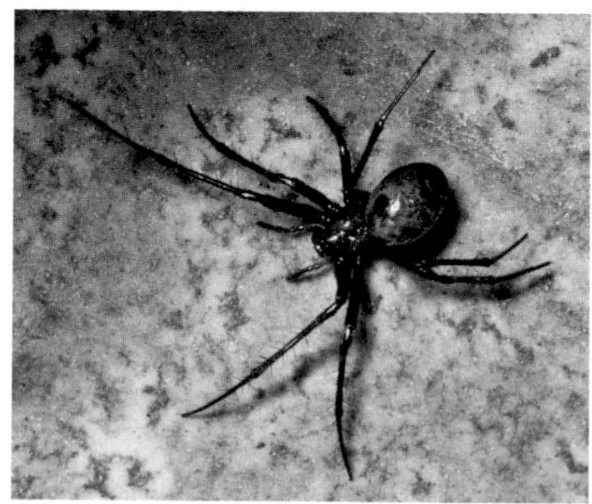

LEFT Deinopids are commonly called ogre-faced spiders because of their very large distinctive posterior median eyes.

LACE-WEB WEAVERS E.G. DEINOPIDAE, ERESIDAE AND AMAUROBIIDAE Lace webs are produced by several different families, including the Deinopidae, Eresidae and Amaurobiidae, and are made from cribellate silk. Amaurobiids are fairly large spiders (5–15 mm or ¼–¾ in) with usually a typical abdominal pattern and a distinctive comb or calamistrum consisting of a row of bristles in females, which is reduced or even absent in males. The calamistrum is located on both metatarsi of the fourth pair of legs, and is used to comb the very fine silk out of the cribellum, which is an extra 'spinning' organ located just in front of the spinnerets. Most species build a typical lace web around holes in walls, fence posts and bark. *Amaurobius fenestralis* is common and widespread in the UK and northern Europe. There are 692 species in the family.

The family Deinopidae, commonly found throughout the tropics, contains 57 species. They are large, slow-moving spiders, with long opisthosomas and stick-like legs. They are grey or light brown in colour. A deinopid produces a small net of highly elastic sticky silk threads on a frame of dry silk. It then suspends itself upside down from a twig or branch, holding the net in the front legs. When an insect passes by, the spider very quickly stretches the net and wraps it around the prey. This has earned the group their other common name - net casting spiders.

There are 101 species in the family Eresidae. They are known as velvet spiders, because of the fine hairs on their exoskeleton. The classic eresid has to be *Eresus sandaliatus*. The small adult male has a bright red livery with black spots, which has earned the species the common name 'ladybird' spider. The female is in total contrast, being much larger, black and velvety. This species is protected by law in the UK, as it is very rare there, as well as in Europe.

RIGHT The Thomisidae look more crab-like than the Philodromidae, because their legs are held out to the sides. They are sit-and-wait predators.

Ambushers e.g. Thomisidae and Philodromidae

This group includes the crab spiders. Named after their crab-like sideways movement, crab spiders are in the Thomisidae (2,062 species) and the Philodromidae (522 species).

Hunting spiders

There are several families of hunting spiders, which include the Lycosidae, Pisauridae, Sparassidae, Oxyopidae, Ctenidae, Scytodidae, Salticidae and Dysderidae. Some families of hunting spiders have wonderfully evocative common names, such as wolf spiders (Lycosidae) and lynx spiders (Oxyopidae), which aptly infer their hunting capabilities!

RIGHT Although usually brown in colour, wolf spiders often have attractive markings, due to pigment in the cuticle and coloured hairs.

WOLF SPIDERS – LYCOSIDAE A large family of 2,320 species. They are covered in dense hairs, and have a great ability to run at high speed. They usually hunt on the ground, but sometimes on low vegetation, and can often be seen running on the ground in large numbers on warm days because many species are diurnal. They were originally given their name wolf spiders as some species are found in large groups. though we now know they are solitary hunters.

HUNTSMAN SPIDERS – SPARASSIDAE There are 1,018 species in this family. Sparassids are known as huntsman spiders because of their prey-hunting habits. They are also known as giant crab spiders, because of their rapid sideways gait. One of the most easily recognized sparassids is the banana spider, *Heteropoda venatoria*, so called because it has occasionally been found among bananas in supermarkets.

LYNX SPIDERS – OXYOPIDAE There are 422 species in this family of diurnal hunting spiders. They hunt on low vegetation, and jump on their prey. This method of prey capture has given them the name of lynx spiders. They have good eyesight (on par with lycosids), using it both in prey capture and courtship, where the male waves his legs and palps at the female.

ABOVE Large nocturnal sparassid spiders generally have a flattened body, enabling them to squeeze into narrow cracks. They have long hairy legs, which only seem to move in the horizontal plane, and are held with the front two pairs bending forwards, and the back two pairs bending to the rear (Ifaty Forest, Madagascar).

NURSERY WEB, FISHING AND RAFT SPIDERS – PISAURIDAE There are 331 species in the family Pisauridae. These are fairly large spiders with long and robust legs, and often have lateral stripes on their carapace and opisthosomas. They can be seen hunting on vegetation or on the surface of ponds. Adult females carry large white egg sacs underneath their bodies, held in place with the chelicerae and palps. Species in the genus *Dolomedes* like moist habitats, *D. fimbriatus* and the rarer *D. plantarius* are found near ponds and can often be seen skimming along on the water's surface. They are able to catch and consume aquatic insects and fish around 4–5 times their own body size, and because of this are also known as raft or fishing spiders (see p.36).

LEFT In general, lynx spiders have an oval prosoma and a tapering opisthosoma, long thin legs with distinctive black spines, and often intricate patterning on their carapace and thorax. Some have beautiful green colouration (Berenty Reserve, Madagascar).

ABOVE Pisaurids attach their egg sac onto low vegetation and cover it with a tent of silk just before the spiderlings are due to hatch. Female pisaurids construct a particularly large 'nursery web' and stand guard over their young (Thursley Common, Surrey, UK).

ABOVE RIGHT *Dolomedes fimbriatus* feeding on a damselfly (Chobham Common, Surrey, UK).

OPPOSITE The Brazilian wandering spider, *Phoneutria nigriventer* (or Brazilian armed spider), from South America, is one of the most venomous spiders of all (Machilla National Park, Ecuador).

BELOW *Dysdera crocata* is a shiny spider, with reddish prosoma and legs, and a greyish cream opisthosoma. It lives in a silk retreat and comes out at night to hunt woodlice, which it can pierce with its powerful fangs. It is often found living among woodlice (Minsmere Reserve, Norfolk, UK).

SPITTING SPIDERS – SCYTODIDAE There are 192 species of spitting spiders. The most well known is *Scytodes thoracica*, which is found in Europe and the USA. It is a distinctive yellow and black creature, with a peculiar domed prosoma, which houses the venom gland (see p.68).

JUMPING SPIDERS – SALTICIDAE The Salticidae is the most diverse family of spiders with 5,088 species. Salticids have great jumping skills, which make them extremely good hunters. Other spiders, such as lycosids and oxyopids, can also jump, but they only manage short distances. These small diurnal spiders are visually very characteristic. Salticids can be extremely beautiful, as they often have coloured and iridescent hairs. They are capable of seeing colour, so their courtship is a very visual affair. Males will perform species-specific movements in front of the female by waving their legs and palps in a kind of semaphore and twitching their opisthomas. Usually, the parts of the body used in courtship are brightly coloured.

WOODLOUSE HUNTERS – DYSDERIDAE Dysderids are nocturnal hunters with powerful jaws, which feed on well-armoured prey, including isopods, beetles and millipedes. *Dysdera* is the most common genus in this family of 494 species. *Dysdera crocata* is widespread in the UK, although it is also found in Europe, New Zealand, Australia, South Africa, Japan and North and South America.

WANDERING SPIDERS – CTENIDAE There are 472 species in the family Ctenidae. These rather large-bodied spiders are often seen wandering over foliage or on the ground in search of prey, which is how they got their common name. Ctenids are some of the most venomous of all spiders.

Colouration

It's easy to think that spiders only come in shades of brown, but there are some amazingly colourful ones. Apart from brown, spiders can be black, green, red, yellow, pink, white, blue, and even metallic gold! Many spiders exhibit a combination of colours, such as adults of *Steatoda paykulliana* (from southern Europe, North Africa and western Asia), which are black and orange, and adult males of *Missulena insignis* (from Australia and New Guinea), in which the chelicerae and part of the prosoma are red, and the opisthosoma is blue. Not only do spiders vary in colour and pattern between species, there is also variation within species, such as with the Hawaiian happy face spider *Theridion grallator*, in the family Theridiidae. Colour and different patterning are used in camouflage, mimicry, warning colouration, courtship and thermoregulation.

ABOVE This salticid spider (Mexico) exhibits a combination of colours with an orange opisthosoma, and yellow prosoma.

RIGHT The colourful ventral surface of an orb-web weaver spider, *Leucage* sp. (family Tetragnathidae) (Poring Hot Springs, Sabah, Borneo).

Colour changes

Some spiders are able to change colour to blend in with their environment. Such colouring may be cryptic to some organisms but not to others. So, although it may be possible to see thomisids on their flower heads because they don't appear to match the colour very closely, they may in fact be invisible to their potential prey. It's not just the thomisids that have this reversible skill. Spiders in the families Oxyopidae (e.g. *Peucetia viridans*), Linyphiidae, Araneidae (e.g. *Araneus quadratus*), and Sparassidae are also able to do this. Some spiders (e.g. *Araneus diadematus*) are able to adjust the brightness of their opisthosomas depending on the light intensity in which they find themselves. This is very useful as they are able to drop from their webs if predators approach, and then become dull coloured and cryptic, so that they blend in with the ground.

Spiders with pigmented cuticle or hairs are unlikely to be able to undergo colour changes, because their colouring is fixed. However, spiders that are coloured by excretory products, such as the araneids *Cyrtophora citricola* and *Argiope* species, can change by retracting guanocytes located in the diverticula, away from the surface of the exoskeleton, thereby changing their body colour from white to brown. However, not all colour changes are as instantaneous as in these chameleon-like spiders. It usually takes the crab spider *M. vatia* 10–25 days to change from one colour to another.

ABOVE Females of the crab spider *Misumena vatia* (Thomisidae) are able to change their colour from white to pink to yellow and back again, to match the colours of the flowers in which they are concealed. This makes them less visible to potential prey, and also means that they are less visible to their own predators (Slapton Ley, Devon, UK).

Anatomy

External anatomy

A spider's prosoma and opisthosoma are joined by a thin waist called a pedicel, which is a narrowed seventh segment. There is a great deal of movement at this joint, as the spider can move its opisthosoma from side to side.

EXOSKELETON People often find spiders repellent because of their hairiness. But why are spiders so hairy? Do these hairs have a function, or are they merely decoration on the exoskeleton? Interestingly, there are several different types of setae found on spiders, depending on the species, and these have various functions: as sensory structures; for defence; for provision of colour; for attachment of spiderlings; to enable spiders to breathe under water, or to live on its surface; and for adhesion.

The most common sensory hair on a spider's cuticle is the tactile hair and is found on both body and legs. Trichobothria are only found on certain leg segments and palps, and are much less numerous than tactile hairs. Chemoreceptive hairs are found mainly on the tarsi of the

RIGHT Dorsal view (left) and ventral view (right) of generalized spider.

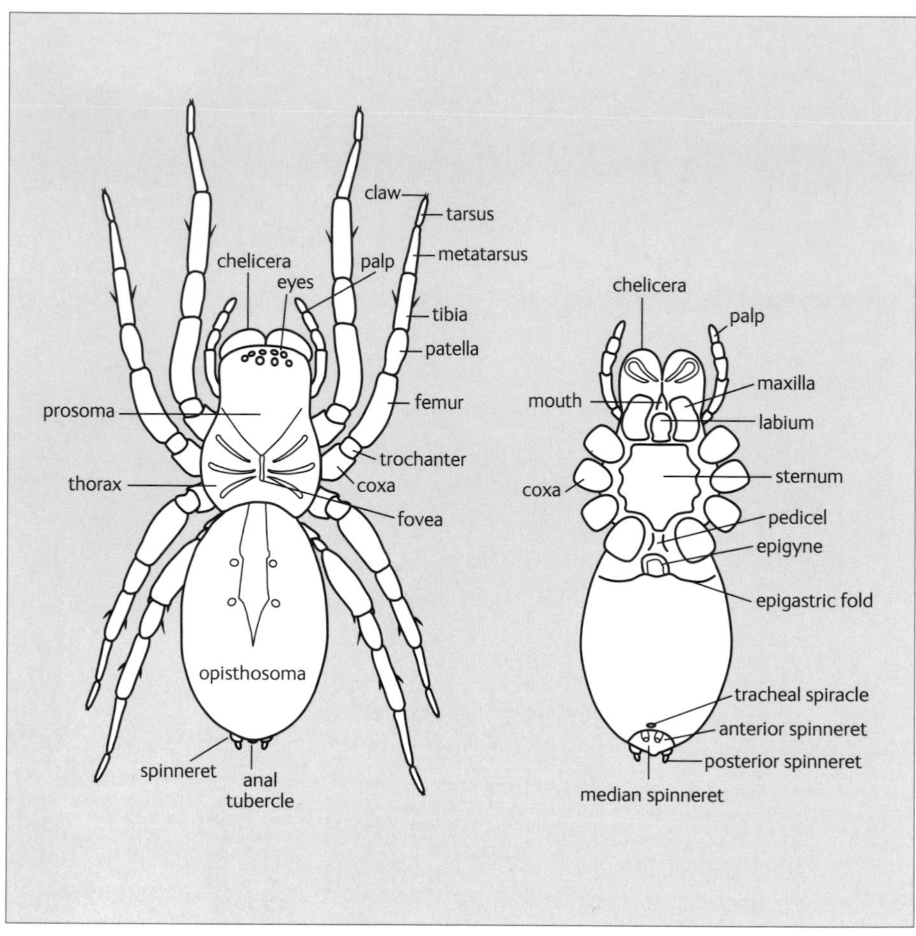

BELOW The junction between the head and thorax, the cephalothoracic junction, often forms a slight V-shaped depression, at the point of which is a deeper depression called the fovea, a site of internal muscle attachment for the sucking stomach.

first legs and on the palps, but some are also found on the mouthparts. As well as hairs, spiders have other sensory structures including slit sensilla, which are found all over the body surface, especially on the legs and particularly near the joints.

Some spiders are able to modify the chemical profile of their exoskeleton, to effect a kind of sensory camouflage. For example, if a spider enters an ant nest, the ants recognize it as 'foreign' by the chemical compounds on its cuticle. To gain safe access, some intruders have evolved chemical mimicry. *Cosmophasis bitaeniata* is myrmecophilous (i.e. it habitually shares the nest of an ant colony), and is able to acquire the cuticular hydrocarbons of the particular colony of its host ant, *Oecophylla smaragdina*, by feasting on ant larvae! Because of its deception, the spider is treated like a nest mate, and is not attacked.

PROSOMA The carapace is slightly domed in most spiders. In some, however, the domed shape is modified into strange extended structures as in males in the linyphiid genera *Walckenaeria* and *Peponocranium*. Ventrally, the prosoma has a large sternum. The coxae of the legs are located laterally.

LEFT AND BELOW The arrangement and size of the eyes is important in classifying different spider families. They are given specific names in pairs, depending on whether they are in the anterior or posterior row, and whether they are in the middle or on the edge (Danum Valley, Southeast Asia and Mantadea National Park, Madagascar).

EYES Most spiders have eight eyes, but some have six (e.g. Oonopidae, Dysderidae and Segestriidae), some four, some two, and cave-dwelling species often have none at all. The eyes are usually arranged in two rows, or occasionally three, at the front edge of the carapace. However, not all eyes are arranged in this way. They may be grouped closely together on top of raised areas, as in tarantulas, or they may be on the tips of extended cephalothoracic structures, as in *Walckenaeria* and *Peponocranium* males.

For the most part, vision is a minor sense for spiders, because many are nocturnal so are more reliant on touch and taste. Web weavers especially can produce an intricate web when it is pitch dark, so they don't require good eyesight. However, the eyes of such spiders *are* able to detect subtle changes in light intensity, so the spider knows if it is night or day. Additionally, the eyes are able to detect movement, alerting the spider to danger and allowing it to drop out of its web for protection.

For some hunting spiders, such as Salticidae and Lycosidae, however, vision is particularly important, for both prey capture and courtship. Their eyes are arranged so that they have good all-round vision simultaneously. Lycosids and agelenids additionally use polarized light for orientation, as we might navigate using a compass. Of all the spiders, salticids have the best vision. Not only can their eyes discriminate light from dark, but they can also focus on objects because they have muscles at their bases, which is unique to the family. Their main eyes detect detailed images, whereas their secondary eyes are primarily movement detectors. They also have binocular vision, which is crucial to establishing distance – a vital attribute for a jumping spider. Salticids also differentiate between different colours, and even perceive their own image in a mirror. Their excellent vision gives these little spiders an almost human-like quality; if you approach, they will turn around to look at you!

Eight-eyed spiders have two different types of eyes – the main eyes and the secondary eyes, which vary slightly in their structure, and have different functions. Both types work together to provide an image for the spider. The anterior median eyes are always the main eyes, and are particularly distinctive because they appear dark (due a lack of a tapetum – see below). The main eyes have a large depth of field to deal with objects at close range. The main eyes of most spiders are small and have few visual cells, although those of Salticidae are specialized and allow the spider to perceive high-resolution images.

The secondary eyes, i.e. all those except the anterior median eyes, are thought to be particularly good for seeing in low light levels. Additionally, in some families such as the Lycosidae, they also provide a sharp image. The secondary eyes look light in colour because of reflected light from the tapetum. This is a layer of small crystals that reflects light back onto the retina, possibly allowing nocturnal spiders to see more effectively at low light levels. The structure of the tapetum varies between different spider species and may not be present at all. The posterior median eyes in the nocturnal spider *Denopis* (Deinopidae) are extremely large, and are around 3,000 times more sensitive than the main eyes of the diurnal *Portia* (Salticidae).

Salticid vision is particularly effective, compared with that of other spider families. The main eyes each have a large lens, a retina with four layers of visual cells and a large vitreous body. Although the visual field is small, muscles at the base partially compensate for this because they can move the retina in different directions. The four-layered retina has 1,000 visual cells, and sensitivities to different spectral wavelengths. The secondary eyes possess relatively large visual fields, and these overlap. As a result, a salticid has binocular vision. Moving prey is usually detected by the binocular vision of the secondary eyes. If the prey is closer than 200 mm (8 in), the spider turns to face it. The main eyes fix the object onto the centre of the retinas. If the prey moves, so do the retinas, keeping the image locked. The shape of the object is quickly 'scanned' (the main eyes are rotated about the optical axis), and if it is recognized as prey, the spider will stalk it and finally jump on it. Salticid eyes are on a par with the compound eyes of insects, and the resolution is superior.

FORMIDABLE JAWS The chelicerae are sited at the front of the prosoma. Each chelicera consists of two hinged parts – the large basal part, and the curved fang, which has a finely serrated edge. The fangs are hinged outwards away from the cheliceral grooves when the spider bites. Near the tip of each fang is a tiny opening to a fine duct (tube), which leads through the length of the fang to a venom gland.

Depending on the spider family, the venom gland may be within the basal part of the chelicera, or extend into the prosoma. When the spider bites, venom is ejected from the venom gland, along the duct, through the fang and into the prey. Spiders in several families have tooth-like processes either side of the cheliceral grooves, which are used to mash up their prey. Males in the jumping spider genus *Myrmarachne* have enormous chelicerae, but they actually lack ducts in their fangs, so they can't envenomate prey. Instead, these well-endowed males skewer their prey with their fangs, like kebabs!

Chelicerae have many uses apart from prey capture and feeding. They are used to protect the owner from would-be predators, and for digging burrows – such spiders often have a group of short, stout spines (called a rastellum) at the base of the chelicerae. Chelicerae are also used as weapons in fighting, to carry egg sacs, and even to produce sounds. They may be larger in males than in females in the same species or they may be large in both sexes, as in the Tetragnathidae.

OPPOSITE When the fangs are not in use, they are usually hinged inwards and rest in a groove along the top edge of each basal segment, much like a blade in a Swiss Army knife.

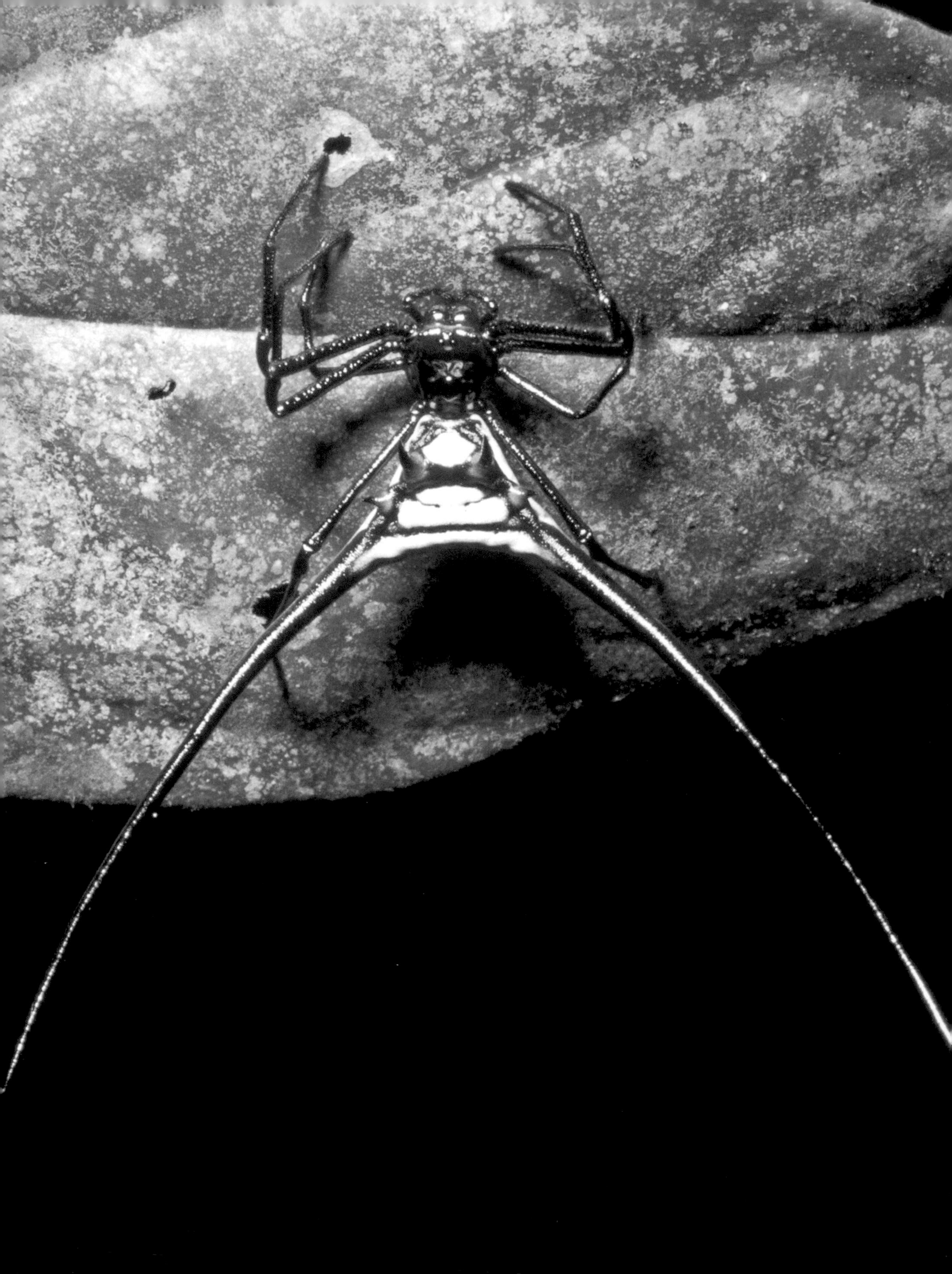

PALPS Spider palps look like a fifth pair of legs (see p.40 bottom). In some male tarantulas, they are almost the same length as the walking legs, but are usually quite a bit shorter. The palps are not used for locomotion, however. They are used to manipulate prey, and spiders often touch prey with the palps and front legs in order to 'taste' their potential meal with chemoreceptive hairs. Adult males have three times as many chemoreceptive hairs as females, because they use them to follow female pheromones. Palps are also used in courtship.

An adult male spider doesn't have a copulatory organ (as do Opiliones, for example) to transfer sperm to the female. It is transferred indirectly using structures on the palpal tarsi, which have developed during a succession of moults. These can be clearly seen by the naked eye and look like little boxing gloves. Such structures are not seen in any other arachnid.

Adult male palps are unbelievably varied in complexity and structure. In its simplest form, the palpal structure consists of a teardrop-shaped bulb. This contains a spiral, blind duct which leads from the semen reservoir and out through the long pointed tip called the embolus. Such palps are found in the Theraphosidae. The majority of palps, however, are more complex than this basic design. Complex male palps may additionally consist of hard, sclerotized parts (sclerites), soft elastic areas (haematodochae), and protrusions (apophyses). When they are not in use the palps are collapsed to protect the embolus, and are partially housed in a hollowed-out sclerotized structure called the cymbium. However, during use, haemolymph is pumped into the haematodochae, expanding and pushing out the sclerites, which may lock in position, depending on the species.

To be ready for copulation, a male needs to transfer his sperm from the genital opening at the epigastric fold, to his palps. To do this, he produces a little web (or even one thread) onto which he deposits a droplet of sperm – he then dips the tips of his palps in and the sperm is drawn up. It is then stored in the blind duct, until it can be transferred to the female. During mating with a simple palp, the embolus is merely placed into the female's genitalia. With the embolus of complex palps however, the erect sclerites and apophyses couple with the female's intricate genitalia. In some spiders, the embolus is unfurled by increased haemolymph pressure, and can extend out into a long thread, which is much longer than the male's body! All complex palps are species specific, and are used for identification.

OPPOSITE AND BELOW In a few spiders, the dorsal surface of the opisthosoma is actually very hard and spiny, so the opisthosoma can only expand ventrally (Guanacaste National Park, Costa Rica and Jatun Sacha, Ecuador).

OPISTHOSOMA The opisthosoma is a soft, bag-like structure that is covered by thin exoskeleton, so that it can expand after feeding, or to allow the development of eggs, in females. In male spiders, it is generally quite tapered and slim, but in females, the opisthosoma is often larger and more rounded. In females of *Theraphosa blondi*, the opisthosoma is almost spherical (see p.27). The opisthosoma is often hairy, and may have patterns and colours.

LEGS Spider legs can be very spiny (such as in the lynx spiders, see p.35 bottom) and may possibly be used to capture prey. Spiders in the families Theridiidae and Nesticidae have serrated bristles on the tarsi of the fourth pair of legs, and the Mimetidae have curved spines. Cribellate spiders are unique in having a calamistrum on both metatarsi of the fourth pair of legs, which is used to comb the very fine silk out of the cribellum. Adult male theraphosids have specialized spines and spurs on the front legs, which are used to force open the fangs of the female and prevent her from attacking the male during mating.

There are claws at the end of each tarsus, which can be moved by muscle action. Hunting spiders have two claws that help the spiders to mechanically interlock with rough substrates. Web-weaving spiders have three claws. The central claw is used to grasp a silk thread and hold it against V-shaped cuts on serrated bristles opposite. This enables the spider to climb a vertical thread without slipping.

Some wandering spiders also have tufts of hairs called scopulae on each tarsus. The scopulae have thousands of branched ends or 'end feet', which enable the spider to walk on slippery surfaces, both horizontal and vertical. Attachment is not by suction, but by weak electrostatic forces, called van der Waals forces, creating a kind of 'molecular Velcro'. A fine film of water enhances attachment. Geckos use the same kind of attachment mechanism. These scopulae also give the spider a firm grip on its prey. Another attachment method has recently been discovered in *Aphonopelma seemanni*. These Costa Rican zebra tarantulas have nozzle-like structures on their feet, which exude silk-like secretions enabling them to adhere to smooth surfaces during locomotion.

Spiders use muscles to flex their legs inward during walking and running, but hydraulics to extend them; their haemolymph acts as the hydraulic fluid. When the legs are flexed, the hydraulic fluid is pushed back upwards. Spiders walk in a specific pattern, always moving two sets of legs alternately. Even if some legs are missing, the spider can adapt its gait to achieve a smooth rhythm. Spiders only lift one or two legs at a time; to speed up, they just go faster, but they don't lift up any more legs. Spiders quickly become exhausted because the elevated heart rate needed for hydraulic extension of the legs can only be maintained over relatively brief periods. The resulting pattern of movement in short, sudden bursts is one feature of spiders that often unnerves people.

The majority of spiders move forward only, and they have two pairs of legs directed to the front, with the other two pairs to the back. This arrangement is known as 'prograde'. Spiders such as crab spiders and huntsmen have all four pairs of legs fanned out to the sides, which bend backwards but face to the front. This arrangement is known as 'laterigrade'. These spiders run sideways and backwards, like miniature crabs. Spiders with legs of equal length run faster than spiders that have legs of different lengths, not surprisingly. Anyone who has tried to catch a *Tegenaria* (in the family Agelenidae) will know that they can run rather fast – in fact, they can reach a speed of 40–50 cm (15¾–19¾ in) per second! Certain spiders can jump. However, unlike fleas or grasshoppers, these spiders do not have enlarged hind legs for jumping. Instead, the jumps are achieved by a sudden extension of the fourth and third pairs of legs, caused largely by an increase in haemolymph pressure, as well as to relaxation of the muscles.

EXTERNAL GENITALIA In the female, on the underside of the opisthosoma near the pedicel, there is a fold called the epigastric furrow. Within the Araneomorphae, if the spider is an adult entelegyne then the external opening of the genitalia is visible just above the fold.

(Entelegynae is a group of spiders where females have complex external genital structures in the form of epigynes, which are sclerotized plates with various infoldings.) Males have correspondingly complex palps. In some spiders, e.g. Araneidae, the epigyne extends outwards into an elongate structure called a scape. This locks onto the male's palp when the embolus is inserted. Because the epigyne is specific to each species, it provides a useful identification aid. Haplogynae spiders have a primitive form of genitalia where the female has the copulatory opening internally within the gonopore. They therefore don't have a sclerotized epigyne. Males of these species have rather simple palps. Adult female mygalomorphs also don't have epigynes. The male's genital opening lies within the epigastric fold, and so cannot be seen. Interestingly, adult males have spigots just in front of the fold, which produce silk for the sperm web.

RESPIRATORY STRUCTURES Depending on the family, a spider may have book lungs, tracheae, or a combination of the two. If book lungs are present, they can be seen externally as hairless cuticular flaps, which look a bit like gill flaps on fish, one on each side of the epigyne. Primitive spiders (Mesothelae, Mygalomorphae and Hypochilidae) have two pairs, which are found on the second and third abdominal segments. In spiders that have tracheae, the single spiracle is located just in front of the spinnerets. Sometimes there are two spiracles, depending on the species.

Spiders are able to regulate the opening of their spiracles and book lung slits by specific muscles. Closing of the slits is due to relaxation of the opening muscles, so is not 100% complete. This means that they are able to control (to a certain extent) moisture loss as a result of respiration. Unfortunately, they lose control of these muscles when the atmosphere contains 10% carbon dioxide or above. This results in unregulated water loss and the animal can fatally dehydrate. Fruit importers use elevated carbon dioxide levels for transporting fruit in order to delay ripening – the side effects of this can be fatal to spiders, as well as other arthropods.

SPINNERETS At the tip of a spider's opisthosoma, there are usually three pairs of finger-like structures called spinnerets (see p.40 left). Abdominal spinnerets are unique to spiders. Silk produced in internal glands passes through up to 600 fine ducts, through the spinnerets and out of tiny openings called spigots. These operate as valves that control the flow of silk. Spigots evolved from hollow setae.

In the primitive Mesothelae, there are four pairs of spinnerets and they are located in the centre of the opisthosoma near the book lungs. The anterior median pair is visually absent in mygalomorphs, so that they often have two pairs. Spinnerets are extremely mobile. They can move both independently and together in a co-ordinated way during any silk-spinning activities, because they are controlled by muscles and haemolymph pressure. Their arrangement on the opisthosoma is sometimes very specific to the family level, e.g. Hahniidae.

CRIBELLUM Cribellate spiders also have an extra spinning organ called a cribellum, located just in front of the spinnerets. Made of one or two plates, the cribellum has hundreds of spigots, which connect to the cribellate silk gland via ducts. From these fine spigots, very thin silk threads are produced, which are combed out by the calamistrum. Originally, all

BELOW: The difference in size between the sexes is taken to the extreme in the genus *Nephila*. The tiny male can be seen to the rear of the large female's opisthosoma (Jatun Sacha, Ecuador).

BOTTOM As with all arachnids, females are usually larger than the males. This can be seen in this pair of *Dolomedes fimbriatus* (UK).

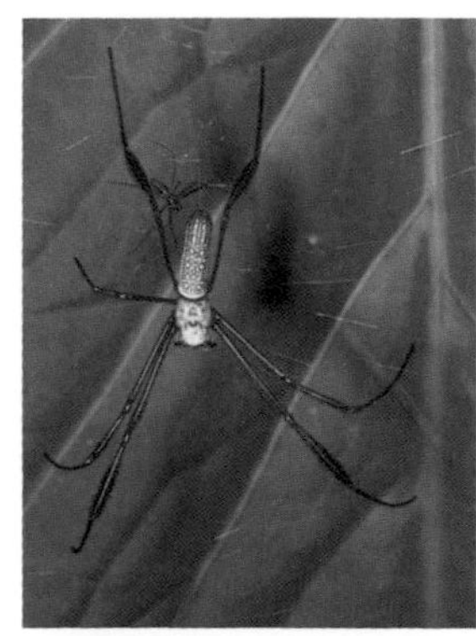

araneomorph spiders would have had a cribellum, but in many species it is now lost. In most araneomorphs, the cribellum has been reduced to a small, vestigial bump called a colulus, which has no function, or may even be absent.

SEXUAL DIMORPHISM Some spider species demonstrate quite pronounced sexual dimorphism in terms of size, colour and different structures. There is quite often a difference in colour between the sexes, with males having much more brightly coloured livery than females. One of the best examples is E. cinnaberinus, where the adult male's bright red livery with black spots has earned the species its common name, ladybird spider. The female is black and velvety. Sex-related differences in colour are especially evident in the Salticidae and Lycosidae, most likely because spiders in these families have good eyesight and are able to see colour.

Sexual dimorphism can be prevalent in several different structures. The most obvious is the difference in the palps. Males often have larger chelicerae than females, e.g. in the common jumping spider, *Salticus scenicus*. In another salticid genus, *Myrmarachne*, the male's chelicerae are five times larger than the female's! The prosoma can also vary between the sexes; e.g. some male linyphiids have enlarged prosoma domes, e.g. *Walckenaeria* and *Peponocranium*. Males of several species often have mating spurs on their front legs used to hold open the female's fangs during copulation. Sexual dimorphism can also be seen on a smaller scale too, e.g. male palps have around three times as many chemoreceptive hairs as those of a female.

Internal anatomy

CENTRAL NERVOUS SYSTEM Spiders have a rather sophisticated and compact central nervous system, which integrates a variety of sensory information. The central nervous system consists of two large interconnected ganglia, located in the prosoma – the supraoesophageal ganglion, and the suboesophageal ganglion. The supraoesophageal ganglion consists of the 'brain' and fused cheliceral ganglia. The brain receives information from the optic nerves of all eight eyes, and so has well-developed areas that convert visual signals into physical responses. For example, a jumping spider receives visual input about the position of its prey, and can adjust its own position of attack in response. It is the most complex part of the central nervous system, and is much larger in wolf spiders and jumping spiders than in others, because they have much better eyesight. The cheliceral ganglia are connected by nerves to the chelicerae, pharynx and venom glands.

The larger suboesophageal ganglion links non-visual sensory input (from tactile hairs, slit sensilla and so on) to motor output. It co-ordinates the appropriate response e.g. running away, to a stimulus e.g. air movements caused by a human with a slipper. Nerve bundles lead from the central nervous system to the extremities and to organs in the opisthosoma, and constitute the peripheral nervous system. As a result of this sophisticated arrangement of central nervous system and peripheral nervous system, external sensory input can be stored in the brain and recalled later. In other words, a spider has a memory e.g. an orb-web weaver is able to relocate prey previously captured and left wrapped up in its web.

CIRCULATORY SYSTEM A spider's heart consists of a tube, which runs from front to back in the upper part of the opisthosoma beneath the exoskeleton. The beating heart can therefore be seen if a live specimen is studied under a hand lens or microscope. On the walls of the heart are several ostia, which open into the cavity, and can be closed by valves. There may be between one and nine pairs of ostia.

RESPIRATORY SYSTEM In spiders, there are different combinations of respiratory organs. The primitive spiders (Mesothelae and Mygalomorphae) have two pairs of book lungs – the anterior pair serves the prosoma, and the posterior pair serves the opisthosoma. Some families, e.g. Pholcidae, have only one pair of book lungs. In other spiders, there is a combination of one pair of book lungs and one pair of tubular tracheae, e.g. Araneidae and Lycosidae. Tubular tracheae are generally considered to be the more advanced respiratory system, and a combination of book lungs and tracheae delivers oxygen more efficiently to the body than with book lungs alone.

The tracheal system varies in distribution and size between different spider families. In some, e.g. Filistatidae, the tubes are very short. In others (e.g. salticids), the tubes are very long and branched, and reach right into the prosoma and extremities. In this system, gas exchange is more efficient because of the extra tubes, so they have a smaller heart, which beats slower. As well as tubular tracheae, in some families, e.g. Caponiidae, there are also 'sieve' tracheae. These are delicately branched, with a slightly different configuration to tubular tracheae.

DIGESTIVE SYSTEM Below the chelicerae are the mouthparts, which consist of the mouth, rostrum (upper lip), labium (lower lip) and endites (chewing mouthparts). In araneomorph spiders, the front edge of each maxilla is serrated, and is used to cut into prey. The inner

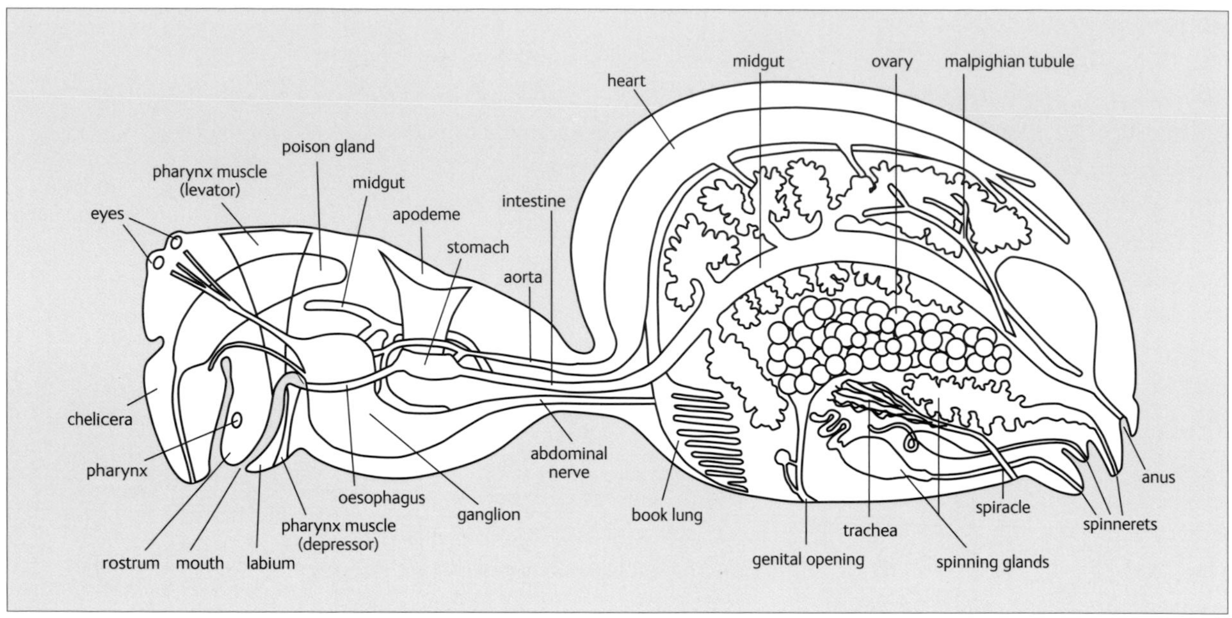

ABOVE Generalized section through a female spider showing the digestive system and the book lungs and tracheae; some spiders have both respiratory systems as shown here.

surface of the endites is covered in fine hairs, which filter the incoming liquid (see p.43). The hinged and movable rostrum is to the front of the mouth, and to the back of the mouth is the labium. Between the rostrum and labium, and behind the mouth, is the pharynx, which is a short tube leading to the oesophagus. The pharynx is lined with microscopic teeth that act as a micro-filter for the liquefied food.

Liquefied food is sucked through the narrow mouth by the muscle action of the pharynx, and the wave-like contractions of the main pump – the sucking stomach. Only very small particles are sucked in, because the rest are filtered out by bristles around the mouth and in the pharynx. The food passes through the oesophagus, into the stomach, and then into the midgut, which is a large, widely branching structure, ending in diverticula. Unlike the pharynx and sucking stomach, the midgut lacks a thin cuticular lining, so it's here that the absorption of minerals takes place. In some spider families, the gut branches occupy most of the opisthosoma, and part of the prosoma, and they surround many internal organs. The diverticula may extend into the coxae of the legs, and in jumping spiders may even penetrate between the eyes.

EXCRETORY SYSTEM The Malpighian tubules empty into a sac called the cloacal chamber, which is a widening of the posterior end of the midgut. The chamber is connected to the hindgut, which is attached to the anus. Inside the Malpighian tubules, products of metabolism are stored and concentrated. These products are then expelled into the cloacal chamber. The main excretory products include uric acid and guanine, which are nearly insoluble in water, and so tend to crystallize. The excretion of uric acid helps to reduce water loss, because it doesn't need diluting with water like urea, the chief nitrogenous waste of mammals. When passed from the cloacal chamber, through the hindgut and out of the anus, they appear as small white deposits. Guanocytes in the opisthosoma may also take up excretory products. In fact, there is a mechanism by which the excretion of guanine can be blocked, so that more can be stored. There are two pairs of coxal glands in the Mygalomorphae and Liphistiidae, which may play an important part in water balance. In the Araneomorphae, there is just one pair.

Silk glands

The silk glands are located in the ventral half of the opisthosoma. There are seven different types of gland, and each produces a different type of silk. No species possesses all seven, but orb-web weavers have five. Depending on the type of gland, they may be shaped like chilli peppers, greatly lobed, or have long and twisted tubular ends. Each type of gland is associated with particular spinnerets, and secretes a specific kind of silk with its own characteristics, as outlined in the table.

Type of gland	Type of silk
aciniform	swathing silk for wrapping prey, sperm web or egg sac silk, stablimenta silk
aggregate	sticky coating to go on the spiral thread
ampullate	safety line, frame thread silk, ballooning silk
cribellate	highly woolly silk – this silk passes through the cribellum and is combed out by the calamistrum
flagelliform	thread of the sticky spiral (only found in orb-web weavers)
pyriform	silk for attachment discs, to anchor silk threads

VENOM GLANDS In tarantulas, the venom glands are located just within the base of the chelicerae, whereas in most other spiders they extend out into the prosoma. Some spiders even have lobed glands, e.g. Filistatidae. Venom glands are tubular in shape, with a narrow duct leading to the chelicerae. Spiralled around the glands are strong muscles, which expel the venom. Venom glands also have their own nerve supply, to ensure rapid triggering of venom release.

INTERNAL GENITALIA In the female, there are a pair of ovaries, which look like elongated bunches of grapes, located in the ventral part of the opisthosoma. The oviducts extend from the ovaries – the end part of the oviduct is called the uterus externus. In the Mygalomorphae and the Haplogynae, the uterus externus extends into the gonopore (external genital opening). The spermathecae are directly connected to the uterus externus, where fertilization occurs, hence the name 'external uterus'. In the Entelegynae, the epigyne has several infoldings, which act as the sperm ducts and spermathecae.

During copulation, the male inserts his embolus into the external genital opening. It is therefore imperative that the male and female structures match! The embolus may reach right up to the seminal receptacles, and this is where the sperm is stored until the female starts to lay her eggs. In this way, the sperm from one copulation can fertilize several egg batches, in the uterus externus. The genitalia of the male are less complex than those of the female. The male's paired testes, which have a great number of coiled loops, are located in the opisthosoma. The loops are lined by epithelial tissue, in which sperm production occurs. The testes both feed into a duct that is joined to the outside through the epigastric furrow; here sperm is released onto the sperm web, where it is taken up by the palps.

Distribution and habitats

Spiders are truly cosmopolitan arachnids, found throughout the world except in Antarctica. They occur at all elevations from sea level up to 5,000 m (16,000 ft). The distribution of some spider families is almost worldwide, e.g. Salticidae, while that of others is rather restricted e.g. the Liphistiidae (Mesothelae), which are only found in Japan, China and Southeast Asia. Many distribution patterns are the result of geological history and past climatic change. Other patterns are the product of aerial dispersal. For example, the islands of the Great Barrier Reef in Australia have been colonized by spiders from seven families, but not by large spiders such as mygalomorphs mainly because of aerial dispersal.

Other spiders are occasionally dispersed by humans, usually in fruit; e,g, *Latrodectus mactans* into the UK in grapes from the USA; *L. hesperus* in grapes from Mexico and *Heteropoda venatoria* in bananas from the tropics and subtropics. Fortunately, the vast majority of these species don't naturalize because they are not adapted to the climate. The distribution of other species seems to be widening, perhaps due to global warming e.g. *Argiope bruennichi* is moving northwards in the UK.

From forests to fresh water

Spiders are found in almost all habitats – the tops of mountains, deserts, rainforests, up trees, in soil and leaf litter, in gardens, and even in fresh water. The richest of all habitats for spiders is forest. However, any one species is often limited to a strictly defined environment. Physical limitations include the requirement for specific humidity and temperatures, and biological limitations include the need for specific food and type of vegetation. Each habitat has microhabitats with specific characteristics, so there is a corresponding partitioning of spider species relating to these microhabitats. Some species, such as jumping spiders, however, can move between different microhabitats. A high proportion of the world's 40,000 spider species are found in the tropics, with rainforest habitats being the richest of all. To minimize competition in these species-rich habitats, spiders have evolved to be more specialized than species in less diverse environments.

Aquatic Araneae

The freshwater spider, *Argyroneta aquatica* (family Cybaeidae), is adapted to an unusual habitat. Found in the UK, as well as northern Europe and Asia, it is the only completely aquatic spider in the world. *Argyroneta* has a dense mat of hairs on its opisthosoma, which traps air. When the spider goes to the surface, an air bubble clings to the hairs, called a plastron. Oxygen in the air bubble is taken up in the usual way though the spider's book lung slits and spiracle (the spiracle tends to be near the centre of the opisthosoma in this family), thereby allowing the spider to breathe underwater. Gas exchange also takes place through the spider's thin abdominal exoskeleton; carbon dioxide passes through the exoskeleton and into the air bubble. Occasionally, the spider goes back up to the surface to renew the air bubble thereby preventing carbon dioxide poisoning. The female spider lays its eggs and tends its young in the 'diving bell' of air bubbles under the water, which is held in place by a mesh of silk threads attached to aquatic plants.

The aptly named fishing spiders, in the family Pisauridae, live around the edges of ponds, and on the surface of water. They move around totally unhindered as they hunt for food, and search for potential mates; the female leaves a tantalising pheromone-impregnated silk dragline on the surface of the water for the male to slavishly follow.

OPPOSITE The freshwater spider, *Argyroneta aquatica*, is a completely aquatic spider. An air bubble on the plastron can be clearly seen (Wicken Fen, Cambridgeshire, UK).

RIGHT With larger spiders, the metatarsi as well as the tarsi are sometimes in contact with the water surface, while large spiders such as *Dolomedes* also rest their body on the surface.

Spiders from families that are normally considered to be terrestrial, such as *Pirata piscatorius* (Lycosidae), also hunt on the surface of water. In fact, these spiders can actually walk as well on water as they do on land. Smaller species, such as *Pardosa amentata*, can raise their bodies above the water. *Dolomedes plantarius* from Europe is actually known as the Great Fen raft spider, such is its ability to live on the water's surface. Sadly, *D. plantarius* is now listed as vulnerable on The World Conservation Union's Red List of Threatened Species.

Such spiders have certain adaptations allowing them to lead an aquatic lifestyle. Pisaurids have dense setae on their cuticle that are hydrophobic so they remain totally dry, even when submerged. These hairs act in the same way as feathers on aquatic birds. It has been demonstrated that pisaurids are 50 times more resistant to wetting than the land-bound Pholcidae. These setae also allow them to keep well supported by the water's surface tension.

Spiders aren't just found at the margins of fresh water they are also found at the seashore, between the low and high tide marks, where they live in rock crevices and hunt for crabs and sand hoppers. To survive being submerged by every tide, one species, the aptly-named *Desis marina*, produces a silken tube which is used as a retreat. The spider hides inside and seals the entrance with a thick sheet of silk. This keeps it nice and dry, even though the retreat is totally submerged.

BELOW The desert spider, *Carparachne aureoflava* (Sparassidae) lives in a burrow on the slopes of sand dunes in the Namibian deserts. To avoid predatory wasps, it flips its body sideways and cartwheels on its bent legs down the sand dunes, moving up to a staggering 1 m/s, and appears as a blurred ball because it is travelling so fast (Skeleton Coast, Namibia.)

Amongst humans

As human populations encroach upon natural habitat, it becomes inevitable that spiders come into contact with humans. Around the world, there are several species of spiders, which happily live among us. House spiders in Europe include the 'daddy long legs' spider *Pholcus phalangioides* (Pholcidae) and species in the genus *Tegenaria* (Agelenidae). *Tegeneria*

spiders especially seem to be found in bathrooms! It's a common experience to go to run a bath and, as you reach over to put in the plug, find a *Tegenaria* of 'vast' proportions sitting in the bottom. But how did the animal get there? Did it come up the plughole, or through the overflow? The answer is more basic – it has fallen in and can't get out. The bath is like a big 'pitfall' trap, with slippery sides. The poor stranded creature is most likely to be a male that had been roaming around to find a female. Attempts to flush it back down the plughole only result in an unfortunate drowning incident, or else the spider remains on the water surface because of surface tension, and evades disposal! People in the southern USA often encounter black widows in their homes or outbuildings, living alongside them. In tropical regions, large wolf spiders often come into houses, but can be very useful, because they will eat pest insects, such as cockroaches.

Dispersal

Ballooning

Spiders can travel several hundred kilometres by ballooning, as part of the 'aerial plankton', and can reach high altitudes. Spiders have been caught from aeroplanes at several thousand metres! Because of their ability to balloon, spiders are often one of the first forms of life to colonise, or re-colonise land, e.g. after a volcanic eruption. During mass dispersal, the surrounding ground and foliage can be covered by vast amounts of silk thread, known as 'gossamer'. *Missulena insignis* (from Australia) is one of the few mygalomorphs that can disperse by ballooning. Of course, many spiders cannot disperse by ballooning so they walk. Baboon spiderlings walk away from their mother's burrow, but won't go very far if there is a good patch of ground close by. In fact, there may be a large cluster of youngsters' burrows around that of the adult female. A count of over 100 burrows in just 80 m^2 (861 ft^2) of ground has been discovered in Africa. Spiders also occasionally disperse on floating rafts of vegetation.

General biology and behaviour

Silk

Spider silk is a truly amazing substance. It can out-perform most industrial materials, especially in terms of toughness. For example, weight for weight, it is five times as strong as steel, yet it is extremely elastic. Silk is a fibrous composite protein, made from protein crystals embedded in a protein matrix. The crystals are responsible for the strength of the silk, while the matrix has properties like rubber. The water content of silk also determines its elasticity. A dry thread breaks if it is extended beyond 30% of its length, whereas a wet thread can extend up to 300% before breaking.

The difference between silk that has come through the cribellum and that which comes through the spinnerets is basically in its structure. In cribellate silk, there are one or two straight threads surrounded by a dense network of dry, woolly 'catching' threads, which are brushed onto the straight threads by the calamistrum. These catching threads (also known as 'hackle bands'), act like Velcro on the feet of insects, and so don't require any glue. As a result, this silk doesn't need renewing daily, like the wet, sticky non-cribellate (ecribellate) silk.

Interestingly, spiders with a cribellum often use a combination of cribellate and ecribellate silk e.g. in the orb web of *Uloborus*. Lace webs, produced by several different families

including the Deinopidae, Eresidae and Uloboridae, are made from cribellate silk. *Amaurobius* (Amaurobiidae) builds a typical lace web around holes in walls and fence posts. This spider comes out and investigates if a tuning fork vibrating at a certain pitch is held against its web! Sticky ecribellate silk consists of a core fibre, with added viscous silk laid on top. The viscous silk separates into droplets on the core fibre. When it dries out, it loses its elasticity and stickiness.

What determines the colour of silk is uncertain. However, it does come in variety of different colours, depending on the spider species. Most silk is white, but the golden orb weaver *Nephila* spins beautiful gold silk orb webs, as its name suggests. In addition, some freshly produced cribellate silk is bluish, or it may be pale green, pink or silver.

BELOW We think of a spider squeezing silk out of its opisthosoma as we might squeeze the last blobs of toothpaste out of a tube. This is not the case, because the spinning glands don't possess muscles to expel the liquid silk. The silk is usually drawn out of the spinnerets by the spider's hind legs or by the weight of the spider. This individual of *Araneus diadematus* is in the process of producing an orb web, UK.

SILK PRODUCTION The silk glands are located in the opisthosoma, and each type of gland produces a specific type of silk. The liquid silk hardens just inside the spinneret ducts, and so leaves the body in a hardened state as a solid fibre. This hardening occurs because of the re-alignment of the molecules in the silk strand, and is irreversible. The silk is kept as a liquid until required.

An exciting new observation about spider silk was reported in *Nature* magazine in 2006 by the Max Planck Institute in Germany. They discovered that Costa Rican zebra tarantulas (*Aphonopelma seemanni*) produce a silk-like secretion, from tiny nozzles on their tarsi, which enables them to adhere to smooth vertical surfaces. This secretion isn't the same as the silk used to build webs and line burrows. Rather, it resembles the attachment silk used to cement draglines to surfaces, and the adhesive agent produced by *Antrodiaetus unicolor* (Antrodiaetidae; sub-order Mygalomorphae). The secretion is laid down as a liquid, which then solidifies, thereby gluing the silk threads to the substrate. This discovery has a great impact on how the evolution of spider silks is viewed. It could be that tarsal silk production evolved independently. On the other hand, it may be that silk production to increase traction evolved first, with silk produced from abdominal spinnerets evolving later.

USES FOR SILK All spiders produce silk, but they don't all use it in the same way. Mygalomorphs, for example, produce only two or three types of silk, which they use to line their burrows, to construct egg sacs and to make a mat on which to moult. Hunting spiders use silk to construct semi-permanent shelters, which enable them to move to where prey is available. Orb-web weavers can produce up to five different types of silk, which they use to produce spectacular prey-catching webs. (Some uses for silk are summarized opposite.)

Uses for silk	Examples
wrapping and immobilizing prey	Many spiders use silk to wrap their prey, in order to immobilize it in preparation for feeding.
lining of burrows and trapdoors	Mygalomorphs and some araneomorphs live in burrows, and line them with silk. This insulates the burrow, and helps it keep its shape. Trapdoor spiders, construct a door for the burrow, made from silk and soil, with a hinge of silk.
sensory enhancement	A spider's web acts as an extension of the spider's sensory faculties. For example, a spider sitting in the hub of its web will feel the vibrations of a fly stuck in the catching thread, some distance away.
safety line and dragline	A safety line is a silk thread attached to the substrate (such as a leaf) from the spinnerets, enabling the spider to retain contact with the substrate at all times. This is particularly important in spiders like salticids that are always on the move. As the spider descends on a safety line, it can stop or slow itself like an abseiler, by pulling the thread out to the side with one of its fourth legs. The spider can then climb back up the thread. Draglines are produced almost constantly when a spider is walking around. Also, many spiders use pheromones in courtship and these are often distributed on silk threads that drag behind the female spider.
construction of retreats	Free-ranging spiders that don't live in burrows or on webs produce silken cells, under leaves for example, which they use as retreats e.g. Thomisidae and Clubionidae.
egg sacs and brood chambers	Females produce egg sacs (cocoons), brood chambers and nursery webs out of silk, in which to protect their developing eggs.
moulting	When large spiders moult, they must lie on their backs. Tarantulas therefore produce a small web on the ground, like a mat on which to lie.
dispersal	Some small spider species, a few mygalomorphs and some medium to large-sized species such as *Araneus* spp., *Clubiona* spp., *Pardosa* spp. and *Tetragnatha* spp. , use silk threads to disperse by ballooning.
courtship	Some spiders use silk during courtship. For example, male thomisids loosely truss-up females in silk threads before they mate, and male araneids pluck the webs of females to alert them to the fact that they are potential partners, and not dinner!
sperm webs	Adult male spiders have to transfer sperm from their genital opening to their palps. To do this, they produce a small sperm web on which to deposit the sperm (possibly so that it doesn't get contaminated by particles), and then draw the sperm up in their palps.

ABOVE A summary of the main uses of spider silk and examples of their uses.

Evolution of webs

Considering that the order Araneae has over 40,000 species, it is not a surprise that spiders display an amazing variety of web designs. But not all spiders produce webs. It is estimated that around 45% of all species do, the rest are hunting spiders, and that hunting is more advanced than web building. Web evolution is certainly not straightforward. It is unlikely that there is a single evolutionary line leading to each of the main types of web. Development is more likely to have recurred many times in many groups, although it is possible that the orb web originated only once, from a cribellate common ancestor. It is likely that, when insects evolved the power of flight, there was selection pressure for spiders to evolve aerial webs. Web evolution can be simplified into the following steps, but note that the stages did not lead from one to the next.

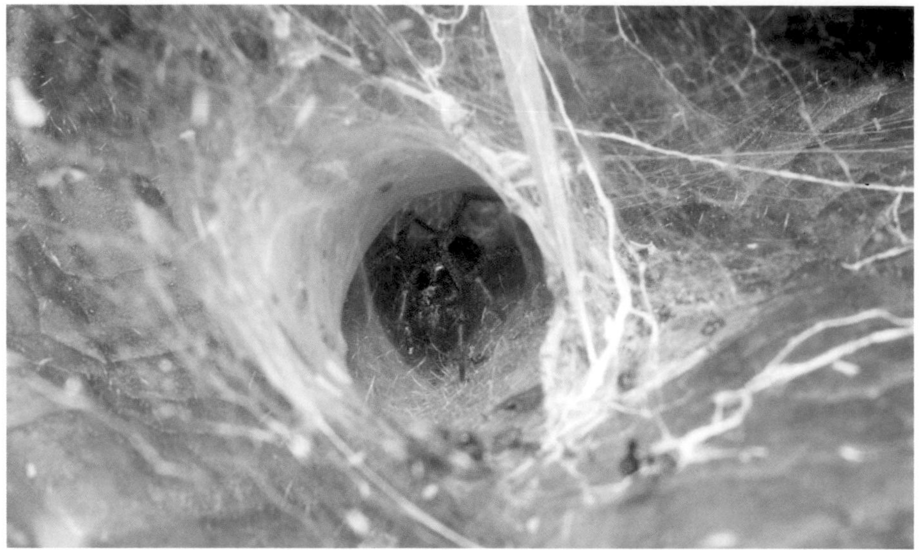

RIGHT Webs of *Agelena labyrinthica* are very common, and can be seen draped on vegetation, reaching a size of over 30 cm (11¾ in) (Wicken Fen, Cambridgeshire, UK).

SILK-LINED RETREAT WITH A COLLAR *Segestria* (Segestriidae) produces a tubular retreat that extends into holes within brickwork, fencing, trees and so on. Signal threads radiate out from the opening, and inform the spider inside about potential prey in the neighbourhood. It is easy to see how the production of webs like this could have gradually evolved into the making of sheet webs. Liphistiidae also construct tube webs. The common name of *Atypus* (Atypidae), the purse web spider, is a bit of a misnomer, because it produces a silken tube more like a sock. The spider digs a tunnel up to 50 cm (19¾ in) into the ground, which it lines in silk. This silken tube extends up to 8 cm (3¼ in) out of the burrow and along the ground, normally under vegetation. It is anchored at the end by a few silk threads. The spider, upside down within the tube, bites prey walking on the upper part of the 'sock', and then drags it into the tube and eats it.

SILK-LINED RETREAT WITH A SHEET WEB Agelenids build impressive sheet webs with tubular retreats, which may lead into natural holes in trees, or disappear into thick vegetation. The large and hairy *Tegenaria* frequently builds sheet webs with retreats inside buildings, especially sheds, which are called cobwebs. Agelenids hide in their retreats and only come out onto the sheet when they detect prey. Other spiders that build silk-lined retreats with sheet webs are found in the families Dipluridae, Eresidae, and Amaurobidae.

BELOW Spiders in the family Linyphiidae produce horizontal sheet or hammock webs. These petite arachnids can festoon an entire area of grassland with their modest webs measuring just a few centimetres each.

AERIAL SHEET WEB WITH RETREAT The families of spiders that construct aerial sheet webs with retreats include the Agelenidae, Theridiidae and Linyphiidae. Sheet or hammock webs consist of a fine silk mesh sheet with threads rising up from the sheet surface. These threads might pull the sheet upwards, creating a slight dome. The silk doesn't need to be sticky, because an insect will get trapped among the thread maze, drop onto the sheet, and be bitten by the occupant of the web, hanging beneath.

SCAFFOLD OR TANGLE WEBS These are constructed by the Theridiidae, Pholcidae and Nesticidae. The web is basically an intersecting mass of scaffold threads, often with droplets of glue. They look like a tangle, hence the name, but actually have a definite structure. They are primarily designed to trap flying insects. When their prey is trapped in the tangle web, certain theridiids fling silk threads to stop it struggling. Although produced by species within the same family and even genus, gum-footed webs are functionally quite distinct from scaffold webs. Gum-footed webs are designed to trap insects crawling on the ground (particularly ants). In certain species e.g. *Steatoda castanea*, the web is anchored to the ground by threads with glue on their lower ends. When an insect walks into the tightly stretched trap threads, they break and spring up, trapping the prey. Such webs are also constructed over the surface of water.

AERIAL SHEET WEB WITH SOME REGULAR THREADS This can be classed as a 'pseudo' orb web. There are roughly convergent radial threads in the same plane, across which zigzags of cribellate silk are arranged. The web can appear to have more than one hub. This type of web is produced by families such as the Eresidae and Dictynidae.

THE ORB WEB The orb or 'wheel' web is the classic spider's web. They are constructed in a huge range of sizes, from less than 10 cm (4 in) (many *Cyclosa* spp.) up to at least 1 m (40 in) in diameter (*Nephila* spp.)! Built mainly by the Araneidae, Tetragnathidae and Uloboridae, and the cribellate Uloboridae, orb webs usually consist of a two-dimensional polygonal frame, with between 3 and 100 convergent radial threads and a spiral of silk laid over them, which may be sticky (ecribillate silk), or woolly (cribellate silk). But what about the mechanics of it – how is it constructed? And why don't spiders get stuck on their orb webs? To answer these questions, it is necessary to look at the basic construction of a typical orb web – that of the European garden spider *Araneus diadematus*.

First, the spider must produce a bridging thread, which is the starting point of the whole web. For this, the spider either walks across vegetation, letting out a line of silk behind it, or it drifts a line of silk on air currents across a gap, until the silk attaches itself to a rigid structure. The spider pulls the line taut, then walks across it letting out a much thicker line of silk, which is the actual bridging line. Next the spider walks back along the bridging line, trailing a sagging line, which it attaches to a rigid structure. The spider then walks along this sagging line until it reaches the centre, where it attaches a second thread. It drops down on this thread until it reaches an anchor point to attach it. The spider has now produced a Y-shaped structure – the point at which the lines on the Y-shape intersect will be the hub. The spider then fills in radial threads to the hub, and constructs framing threads to which the radial threads will be attached on the outer edge. The spider uses its legs to 'measure' the required distance between each radial thread.

With work on the radial threads finished, the spider now lays down the spiral thread. First, it lays down a widely spaced temporary spiral thread, working from the hub out to the edge of the web. The spider then turns around, and starts laying the more closely spaced permanent spiral thread, eating the temporary thread as it spirals back towards the hub. This thread is the key to the web's success, because it is coated in a sticky substance, and acts as the 'catching' thread. The sticky coating is hygroscopic (attracts water), making the catching thread very elastic. This is important, because it means the web absorbs the kinetic energy of insect impacts without breaking.

It is truly amazing to think that orb webs are constructed by touch alone, as orb web weavers have poor eyesight, and that all this work takes less than 30 minutes! The spider's swiftness is just as well, because the 'catching' spiral's stickiness soon becomes ineffective, making it necessary to remake the web each day. The spider retains the bridging thread, and normally eats the rest of the web (pre-digesting it first, of course, before sucking it up).

Although the ability to produce an orb web is an inherited trait, spiders do show a reasonable amount of flexibility during construction. If, for example, some radial threads are experimentally removed several times, while the spider is laying them down, it keeps replacing them, rather than just continuing in a 'programmed' way.

How the orb web is used

Web-weaving spiders have poor eyesight, so they rely on a sense of touch. A spider's web might be described as its 'sensorium' – that is, it acts as an extension of the spider's sensory faculties. For example, a spider sitting in the hub of its web is able to feel the vibrations of a prey insect struggling in the catching thread, even though the insect might be some distance away. At the hub, is the 'free zone', which is an area free of the sticky catching spiral. When a spider feels a vibration in its web, it plucks the threads in the free zone to discover the direction from which the vibration is coming, so that it can orientate itself correctly before rushing out onto the web. Orb webs are often built asymmetrically, with the lower part larger than the upper part. This is because the spider can run faster into the lower section (because of gravity) than the upper section, so prey capture is more efficient there. This is why many spiders sit facing downwards in the hub of their web.

So, how *does* a spider avoid getting stuck on its own web? There are several ways it does this. First, it normally moves around on the non-sticky radial threads. It also minimizes contact with the web by walking on 'tiptoe' on its tiny tarsal claws, and it tends to hang beneath the web (which is slightly off the vertical) when it crawls so its weight isn't pushing it down onto the silk threads. If the worst happens, the spider knows how to get unstuck – it pulls other strands of the web together, sticking them to each other to free it. Not only do spiders avoid sticking to their own webs but, in general, if placed on another spider's web (even if it is of a different species, although in the same genus) they don't stick to that either!

BELOW An orb web with well-developed tubular retreat, plus occupant (Ranomafana National Park, Madagascar).

Orb web variations

The typical orb web described above is built with ecribellate silk. Webs built by spiders of the family Uloboridae use their woolly cribellate catching silk, rather than sticky silk, for the catching spiral and their webs are orientated horizontally, rather than vertically. There are around thirty described variations on the orb web. One of the major variations is that not all orb-web spiders sit at the hub, because certain species have evolved the use of a signal thread leading from the hub to a retreat. The spider only ventures onto the web in daylight if it receives prey vibrations through the signal thread. The spider *Cyrtophora* (Araneidae) produces a horizontal orb-like web structure, with a maze of threads above and below. The threads above the hub are taut, and so pull it into a raised arch, giving it the appearance of a domed tent.

Stablimenta

Orb webs of certain species have an additional feature – the stablimentum. This is a zigzag band of silk which may bisect the web. In fact, they are highly variable between and within species, and even vary between different webs of the same spider! Several possible functions for the stablimentum have been suggested. The stabilizing function hinted at by the name is misleading, however. In some species, such as *Cyclosa*, the stablimentum certainly acts as camouflage for the spider. The spider adorns itself with old shed skins and remnants of prey, and then sits in the hub with legs pressed into its body. In this way, the spider blends in perfectly with the debris on the stablimentum and can't be seen. In the webs of conspicuous spiders, it probably acts as a warning device to alert birds to the web's presence, so they don't fly into it. Other suggested functions include the attraction of insects, as the silk reflects UV light, or protection against the Sun's radiation. Alternatively, it might serve as a moulting platform for large orb weavers such as *Nephila*.

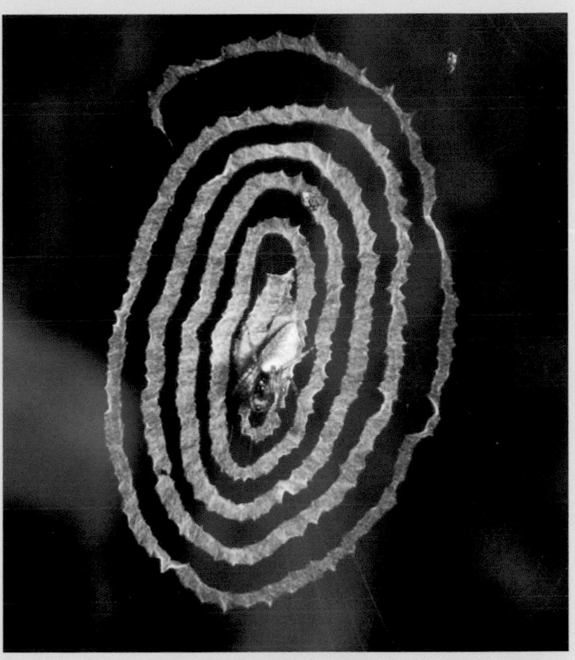

ABOVE AND BELOW A stablimentum has different structures; it may be a small spiral at the hub (*Cyclosa* sp.) or consist of two bands creating a cross (*Argiope* sp.) (Costa Rica and Danum Valley, Southeast Asia).

Orb web derivatives

Because of its complexity, the orb web was regarded until fairly recently as the pinnacle of web construction. However, the designs of certain webs are considered to be 'derived' from (descended from) orbs webs, making them more 'advanced' (though only in the sense that they presumably came later in evolutionary time – they are not necessarily 'better', or more efficient, than orb webs). Modifications of the orb web can range from a basic reduction of a few of the elements, to an almost total transformation, which is likely to be due to prey specialization.

TRIANGULAR WEBS The triangle spider, *Hyptiotes triangularis*, constructs a web that resembles a triangular segment from an orb web. The mooring thread that leads from the tip of the triangular web is held by the spider, which has attached itself to a rigid structure. The spider thus forms a 'living link'. When an insect flies into the web, the spider releases the tension on the thread, and the web collapses on the prey, thus ensnaring it.

LADDER WEBS An unusual derived orb web is the ladder web of the spider *Tylorida* (Tetragnathidae) from New Guinea. This web is about 1 m (40 in) high, but only 15 cm (6 in) wide, with the hub near the top and horizontal 'spiral' threads. The construction of this kind of web has evolved to catch moths. When they blunder into the web near the top, they fall, loosing many scales, and are caught at the bottom.

BOLAS SNARES The most extreme derived orb web has to be that of Bolas spiders. Several genera in the Araneidae are found as far a field as Costa Rica and Australia, and use silk to capture prey in a very ingenious way. These canny spiders produce a silk thread with a globule of sticky silk on the end. This silken snare strongly resembles a bolas, which is a rope with weights attached. The spider attaches itself to a branch with one of its first pair of legs, and swings the bolas in small circles. Bolas spiders prey on male moths in the family Noctuidae, which they attract by mimicking the odour of female moth sex pheromones. When a male moth flies close, the spider swings the sticky globule, the moth gets stuck, and the spider reels it in. Bolas spiders don't always use a bolas for hunting. For example, *Mastophora dizzydeani* sometimes hangs from a silk line with spread front legs, but with no sticky 'death trap'. Bolas spiders are not unique in using chemical attractants – spiders in the genus *Kaira* also attract prey in this way.

Manipulated web building

Spiders demonstrate typical, species-specific web-construction behaviour. However, this can be manipulated by outside factors. Amazingly enough, spiders can be drugged to produce irregular orb webs. By giving a spider strychnine or caffeine, for example, webs of a certain irregularity can be produced. It is possible to determine the imbibed chemical simply from the appearance of the web structure.

It has also been discovered that parasites of spiders can manipulate their host's web-building behaviour. For example, the larva of the ichneumonid wasp *Hymenoepimecis argyraphaga* induces its host, an orb-web weaver called *Plesiometa argyra*, to produce a 'cocoon' web, which the larva uses to pupate after it has killed its host. The cocoon web is constructed by continued repetition of the early steps in normal orb-web building. It is thought that the larva induces these changes in the spider's behaviour by a fast-acting chemical, which at the same time suppresses the rest of the typical orb-web building behaviour.

OPPOSITE Net-casting spiders or ogre-faced spiders (Deinopidae) produce a small net of highly elastic sticky, silk threads on a frame of dry silk. It then suspends itself from a twig or branch, holding the net in the front legs. When an insect passes by, the spider very quickly stretches the net and wraps it around the prey (St Luce, Madagscar).

Venom

Spider venom is, not surprisingly, a subject that fascinates and terrifies people. This section gives a basic background about venom.

WHY DO SPIDERS PRODUCE VENOM? A spider produces venom primarily to subdue its prey. Quick immobilization is the key function of venom – the lethal effect is secondary. With prey that might fight back, such as an insect with a sting, it is particularly important to knock it out quickly. So, crab spiders (Thomisidae) that prey on bees, for example, have a venom that is faster acting and stronger than that produced by spiders that feed on harmless prey. *Cupiennius salei*, a formidable spider in the family Ctenidae, has been shown to inject more venom into dangerous or aggressive prey, and less into non-problematic prey. In fact, it injects no more venom than is necessary, and uses it in an economic way. Venom may contain many different substances, including proteolytic enzymes (enzymes that split proteins), so might play a minor part in the digestive process, by starting to break down the prey's body tissues. Venom is also useful for protecting the spider from would-be predators, although it is rarely used in self-defence. Interestingly, there is chemical variation in venom between close relatives.

ARE ALL SPIDERS VENOMOUS? All spiders are venomous, apart from one family, the Uloboridae, which doesn't produce any venom at all. In general, hunting spiders have more potent venom than web spinners, with *Latrodectus* (widow spiders) a very notable exception. However, it is important to note that although all spiders are venomous to their *prey*, the effect of their venom on humans is another matter. Although many spiders are capable of biting humans, they are often non-aggressive, and only bite if provoked. Even when spiders *do* bite humans, there is often very little effect. The worst that most people suffer is an experience similar to a bee or wasp sting, with localized pain and perhaps a feeling of being unwell for a few days. It is worth remembering that many more people have allergic reactions to bee and wasp stings than to spider bites, which can be particularly problematic because bees and wasps are encountered more often and are usually in swarms. Fatalities from spider bites are extremely rare. To put things into perspective, there are around 200 species of spider known to be harmfully venomous to humans, out of over 40,000 species in total, so it really is a case of the small minority giving the rest a bad name.

WHICH SPIDERS PRODUCE THE STRONGEST VENOM? Surprisingly, strength of venom is not related to the size of a species. So, for example, a Mexican red knee tarantula, *Brachypelma smithi*, is much less venomous to humans than an Australian red back, *Latrodectus hasselti*, even though the red back is many times smaller. Unfortunately for dogs, tarantula bites are often fatal to them, whereas tarantulas rarely inflict a venomous bite on humans. With most spider species, females tend to be more dangerous than males because they are usually larger, and so produce a greater quantity of venom.

Possibly the most notorious spider of all time is *Lycosa tarantula*. Over the centuries, this spider (which may have been named after the city of Taranto, southern Italy), was blamed for sending people almost to the brink of madness because of its venomous bite. It is possible that the venomous *Latrodectus tredecimguttatus* was in fact to blame, but many more people suffered from the so-called 'tarantism' than could have reasonably been bitten!

Common name(s)	Scientific name	Range
black widow	*Latrodectus mactans* (Theridiidae)	southern USA, Mexico
Australian red back	*Latrodectus hasselti* (Theridiidae)	Australia, Gulf region, introduced to New Zealand and Japan
Sydney funnel web	*Atrax robustus* (Hexathelidae)	Victoria and New South Wales, Australia
Brazilian wandering spider (also known as Brazilian armed spider)	*Phoneutria nigriventer* (Ctenidae)	southern Brazil, Paraguay
brown recluse	*Loxosceles reclusa* (Sicariidae)	Central America, southern USA

LEFT A few of the most well-known spider species that are venomous to humans.

The symptoms of tarantism included vomiting and giddiness, which indicate envenomation by *L. tredecimguttatus*. However, other alleged 'symptoms' included a great sensitivity to music, and to rid their bodies of the venom the 'victims' would dance wildly for hours. This dance became known as the 'tarantella'. It seems likely that the too-regular incidences of tarantism were most likely caused by mass hysteria, or occurred as a reaction against strict laws of behaviour.

WHY DOES VENOM AFFECT HUMANS? Spider bites are common in most parts of the world, but the vast majority do not cause major problems. Unfortunately, the severity of spider bites is often exaggerated and there are conflicting reports about envenomations. However, to put things into perspective, a recent large study in Australia demonstrated that even with the presence of highly venomous spiders, such as the Australian red back (*Latrodectus hasselti*) and Sydney funnel web (*Atrax robustus*), most bites were minor.

Many spiders produce venom that is specific to their preferred prey, which is usually invertebrate. Spiders that are venomous to humans do not produce their unusually strong venom in order to subdue humans – it is purely incidental that their venom has an effect on humans. The venom is potent to humans because it has effects that interfere with essential body functions.

The effects of envenomation vary from person to person. A large person will be less severely affected than a small person, because the greater volume of body tissues has a greater 'dilution' effect. The health and age of a person also have a major impact on the envenomation effects. The effects of a spider bite vary with the spider's state, too. For example, the spider might have eaten recently so its supply of venom is reduced; the spider might be old or in ill health; the angle of strike might be such that it is not sufficient to pierce the skin, and so on. All this means that only a very small percentage of spider bites are effective envenomations.

Classification of venoms

Spider venom can essentially be divided into two different categories – neurotoxic venom, which affects the nervous system, and cytotoxic venom, which mainly affects the tissues. Neurotoxic venom is produced by spiders such as *Latrodectus mactans, L. hasselti, Steatoda* spp., *Atrax robustus* and *Phoneutria nigriventer*. Nowadays, less than 1% of people who are bitten die. After treatment, recovery can be relatively rapid and even without treatment, patients normally recover within a few weeks. It depends on the amount of venom injected, and the age and health of the victim. Interestingly, some mammals are much more susceptible to a *Latrodectus mactans* bite than others, e.g. sheep, cows and horses are even more sensitive than humans, whereas rabbits and goats are much *less* affected. Severe envenomation from funnel-web bites only develops in 10–25% of all cases. Funnel-web neurotoxicity can cause death within 15 minutes, especially in children. Interestingly enough, it hardly affects dogs and rabbits. What may be surprising is that very few deaths have been known to have occurred from the venom. After severe envenomation by *Phoneutria* death can occur within 2–12 hours, but this is very rare. Nearly all victims recover within 24–48 hours.

Cytotoxic venom causes necrosis, which means that the tissues of the skin die and bacterial infections can then set in. Necrotic ulceration of the skin is due to enzymes in the venom attacking the skin. As the enzymes continue to work, a deep wound can occur. Additionally, the auto immune system of the victim affects the skin too. Such wounding is very slow to heal, and may take 1–2 months, with up to 15% chance of major scarring. The cytotoxic venom *of Loxosceles reclusa* also affects the kidneys, which can lead to kidney failure and possibly to death. (The overall condition caused by the venom is called loxoscelism.) However, even in those parts of the world inhabited by *Loxosceles*, spider bites are a rare cause of necrotic lesions.

Prey capture

All spiders are carnivorous. They nearly always eat live food, although some will accept dead foodstuffs. Insects, particularly flies, and arthropods such as Collembola, are very important spider prey. Other arthropod favourites include beetles, cockroaches, grasshoppers, butterflies, and of course other spiders. Vertebrate suppers are unusual, but can include tadpoles and small fish, which are consumed by 'fishing spiders'. Large desert spiders are known to eat geckos. Reports of tarantulas eating birds are clearly the reason they are often called bird-eating spiders, but such reports are difficult to verify. However, *Nephilgenys cruentata* quite regularly catches and feeds on small birds in its web.

Most spiders are really not that fussy about the prey they feed upon, though some spiders are highly specialist, and eat only one type of food e.g. *Ero* and *Palpimanus* eat other spiders. Webs can act as selective filters for potential prey, thereby imposing a certain degree of specialism on their owners. In fact, webs of a certain spider species tend to trap the same size of prey and possibly the same species. For example, they might easily catch plant-sucking insects such as aphids, but not be successful at catching insects with good eyesight and an ability to fly well, such as hoverflies. Other feeding specialists include ant feeders, e.g. *Zodarion* spp. and termite specialists (*Ammoxenus, Microheros, Mashonarus*).

As we have seen, spiders construct a vast array of webs, snares and other web derivatives to catch prey, but they have also evolved many other clever ways of obtaining food. The following sections cover some of the main methods of food capture (apart from the use of silk), along with some more unusual methods, but are in no way exhaustive.

ABOVE A spider eating a large riodenid butterfly (Jatun Sacha, Ecuador).

LEFT A spider eating a cockroach (Taman Negara National Park, Malaysia).

HUNTING SPIDERS The hunting (or wandering) spiders are thought to have secondarily reverted from snare building to a nomadic lifestyle, where they hunt their prey directly. Hunting spiders often use mechanical variation to locate their prey. They are able to distinguish between vibrations caused by a buzzing insect and background noise. Some hunting spiders use sight to locate prey, being guided by movement. Salticid spiders are also hunting spiders, with an amazing ability to jump. The cryptic jumping spider *Portia* is an incredible spider with a particular liking for feeding on other spiders; it often stalks its prey stealthily and grabs it at close range.

ABOVE A hunting spider nestles under leaves ready to start hunting for its prey (Danum Valley, Sabah).

AMBUSHING SPIDERS An ingenious way to capture prey is to lie in wait and grab it when it goes past, and many spiders have adopted this crafty technique. Ambushing spiders often rely heavily on camouflage, both their own cryptic markings and concealment of a burrow. Trapdoor spiders, such as *Anidiops villosus* (Ctenizidae) from Australia, lie in wait for prey in their camouflaged burrows. *Anidiops* attaches grass stems in a radiating pattern around the burrow entrance, in order to transmit prey vibrations into the burrow. Other trapdoor spiders, such as *Cteniza moggridgei*, sit with the door to the burrow slightly open and their legs on the edge of the rim, ready to grab unsuspecting passers-by.

FISHING SPIDERS Amazingly, some hunting spiders are very proficient at hunting and catching food on the surface of water. The raft or fishing spider, *Dolomedes fimbriatus* (Pisauridae), is a proficient diver, catching much of its prey under water, and is capable of catching and consuming fish around 4–5 times its own body size. Fishing spiders row themselves along the water surface holding their second and third pairs of legs quite rigidly, and dragging the fourth pair behind. Their front legs are stretched forward and used as feelers; when prey is detected, spiders such as *Dolomedes* can increase their speed up to five times faster than their normal 'rowing' speed. Pisaurids can also detect the vibrations of prey insects struggling to escape the water's surface tension. The water spider, *Argyroneta aquatica*, (see p.53) catches insect larvae and small crustaceans encountered as it swims around. It also catches insects that have fallen onto the water surface, by swimming up beneath them. Because spiders digest their food *outside* of their bodies, the spider has to return to its submerged 'diving bell' air bubble to consume its meal.

BELOW A fishing spider waiting on the edge of the water to catch prey (Costa Rica).

SPITTING SPIDERS The spitting spider, *Scytodes thoracica*, which is found in Europe and the USA, is yellow and black with a distinctive domed prosoma containing a large gland – the anterior part produces venom and the posterior part produces a kind of glue. To capture its prey, the spider contracts muscles in the prosoma, squirting a mixture of venom and glue out of the chelicerae. The spider 'spits' this substance in a zigzag fashion. The venom in the mixture paralyzes the prey, while the glue sticks it firmly in place, ready for the spider to consume it.

FOOD THIEVES Some spiders steal food from others. For example, the whole lifestyle of *Argyrodes* (family Theridiidae), is based on being a thief. The spider builds its web very close to an orb web, to which it attaches signal threads at the hub. When the owner of the orb web starts to wrap prey just captured at the hub, vibrations are transmitted along the signal threads to the *Argyrodes*. This thieving spider then carefully makes its way into the other web, locates the wrapped prey stored at the hub, cuts it out, and eats it.

BLOOD SUCKER The wonderfully named 'mosquito terminator', *Evarcha culicivora*, is a beautiful red, black and gold jumping spider found in Kenya and Uganda, which has a most unusual choice of food. The little beast feeds on mosquitoes, with a preference for *Anopheles*, targeting females that have just had a blood meal. It behaves like a small, predatory mammal when it is hunting and capturing its prey. Professor Robert Jackson and Dr Simon Pollard in Christchurch, New Zealand, have discovered that it launches into a feeding frenzy, and can kill as many as 20 mosquitoes in quick succession! No other spider is known to feed on blood, directly or indirectly. It is interesting to note that female *Anopheles* mosquitoes are the vectors of human malaria – could these mosquito terminator spiders form the basis of a biological control mechanism to conquer malaria?

NECTAR FEEDERS Although all spiders are carnivorous, some actually supplement their diets with things more vegetarian. Different species of jumping spiders and crab spiders have been observed eating nectar, a foodstuff much more readily associated with bees. In laboratory tests using 90 different species of salticids, all 90 fed from flowers, suggesting that the behaviour is widespread in this family. In an investigation involving spiderlings of the crab spider *Thomisus onustus*, those fed on nectar had a significantly greater survival rate. Nectivory has also been observed in male crab spiders of the species *Misumenoides formosipes*. They are much more prone to dehydration than females, because they have a large surface-to-volume ratio. By drinking nectar, they are able to replace essential fluids in their bodies. Also, those males drinking nectar actually lived longer. So it seems that nectar acts as a good, long-term energy source, especially when there is a lack of insects, and as a source of fluid!

LEFT A vegetarian spider! This spider feeds mainly on the beltian bodies (structures found on the tips of leaves) of *Acacia* trees (Guanacaste National Park, Costa Rica).

Feeding and digestion

Once prey is subdued by venom, or wrapped in silk, the spider regurgitates digestive fluid onto its surface, or through holes left by the fangs when the venom was injected. Some spiders widen the holes by opening and closing their fangs. Certain species feed from several sites on the same prey item. Even in species that touch the prey, there is no seal between the prey surface and the mouthparts. Uloborids are rather extreme in this practice, because their mouthparts never normally touch the prey at all. Instead, they entirely coat the prey with digestive juices, which probably flow through cracks in the prey's exoskeleton. The enzymes in the digestive fluid quickly begin to break down the prey's tissues, including the intersegmental membranes, and they liquefy. The spider then sucks up this 'soup' through the narrow mouth, using the pharynx and strong stomach muscles.

More and more digestive fluid is regurgitated, and liquefied tissue sucked up. The spider sucks in only very small particles, filtering out the rest using bristles around the mouth, and the flattened pharynx. Although some spider venom contains digestive enzymes, it doesn't play a significant part in the digestive process. The spider doesn't poison itself with its own venom as it ingests its prey, as the injected venom is denatured (rendered useless) before it is imbibed.

Spiders such as tarantulas, with tooth-like processes on their chelicerae, mash up their prey while they suck out the bodily juices. In a way, it's a bit like a human diner cracking open lobster claws and sucking out the contents, all that's left is a plate of inedible fragments. In the case of a tarantula's dinner, the corpse ends up as an almost unidentifiable ball of exoskeleton. Spiders without these cheliceral teeth do not masticate their prey. Instead, they suck out the body contents, leaving an undamaged shell that looks like a still-living creature. However, they are still able to access the rather hard to reach areas. Some spiders masticate soft and small prey, but leave larger prey intact.

Although many spiders pump in digestive juices, it has been shown that certain species of Theridiidae, Thomisidae and Uloboridae, do not inject digestive juices into their prey under pressure. Instead, the liquid flows into the prey by capillarity (where a film of liquid is drawn in). As the spider continues to regurgitate, the liquid moves further and further into the prey. Fluid may also flow out of the prey by capillarity too.

OPPOSITE TOP A *Pepsis* wasp dragging a paralyzed tarantula back to its burrow (Los Cedros, Ecuador).

OPPOSITE MIDDLE Certain species of wasp sting a spider to temporarily paralyze it, and then lays an egg onto its opisthosoma, where it can't be dislodged. The spider recovers quickly and continues its usual behaviour. However, it is very soon in trouble, because when the egg hatches, the wasp larva clings onto the spider, and eats it alive (Jatun Sacha, Ecuador).

OPPOSITE BOTTOM A spider killed by a pathogenic fungus (Jatun Sacha, Ecuador).

Spider enemies

Apart from slipper-wielding humans, spiders have several other kinds of enemies in the forms of predators and parasites. Ironically, one of the main enemies of spiders are spiders themselves. Anyone who has inadvertently put two spiders together in a collecting tube will be able to testify that the larger of the two will immediately attack and feed upon the smaller. Pirate spiders, of the family Mimetidae, make a culinary living by feeding upon other hapless spiders. Those in the genus *Ero* will creep into the web of another spider and pluck the threads to lure out the owner, which it then attacks and eats! Spiders of the genus *Cupiennius* (family Ctenidae) have been observed hunting in *Nephila* webs, and jumping spiders of the genus *Portia* also display considerable spider-eating tendencies.

There are many insect predators of spiders, which include ants, flies, beetles, bugs, bush crickets and praying mantids. It is well known that ants are formidable beasts – they swarm over almost anything in their path, including spiders. Such unlucky arachnids are often dismembered by the ants, and transformed into a 'take away' meal. Certain ants feed their

young on spider eggs, which they take back to their nests. Non-insect predators include scorpions and centipedes. The centipede *Scolopendra westwoodi* feasts upon trapdoor spiders – it dashes down into the spider's burrow and rapidly consumes it in its own home. Vertebrate predators include fish, birds, reptiles such as lizards, amphibians such as toads, and mammals including bats, shrews and monkeys, as well as marsupials. Bandicoots, for example, are very keen on trapdoor spiders and funnel web spiders, which they pull out of the ground in their silk-lined burrows. The bandicoot will briefly chomp on the spider, before consuming it, web and all!

Parasites of spiders include insects, nematode worms (roundworms) and fungi. The main insect parasitoids are wasps. Those in the family Pompilidae attack spiders exclusively. Two other wasp families, the Sphecidae and the Ichneumonidae, also feed on spiders or their eggs. Several pompilid species will attack a spider in its burrow, lay an egg on it and then seal it up in its own home, or drag the paralyzed spider off to a specially dug burrow and enclose it there. The spider then becomes a macabre live-food larder for the wasp larva when it hatches. Technically, the wasp larva is a predator since it doesn't live on the body of its paralyzed prey.

Large spiders are not immune to wasp attacks. Indeed, tropical tarantulas are parasitized by the enormous *Pepsis* wasp, ominously named the tarantula hawk wasp, which has a body length of 8 cm (3¼ in). Tarantulas don't usually fight against the wasp's attack, they usually panic and try to run away! Even *Latrodectus mactarus* has its very own enemy, the wasp *Chalybion cyaneum*. Certain flies from the family Acroceridae also attack spiders. A maggot crawls into the spider's opisthosoma through one of the book lungs, and develops there until it reaches the fourth instar. It then gruesomely devours the tissue of the entire opisthosoma, not surprisingly killing the spider in the process. In a final insult to the deceased arachnid, the replete larva then crawls out of the shell of the opisthosoma and onto the spider's web, where it pupates. Other flies lay their eggs on spider egg sacs, and the spider eggs are munched when the fly eggs hatch.

Nematodes also eat the poor spider from the inside out. The female nematode lays great numbers of eggs inside the spider's body, and when these hatch, the tiny worms imbibe the spider's body fluids. The spider's body gradually distends into a vast wriggling mass of worms, which inevitably kill it.

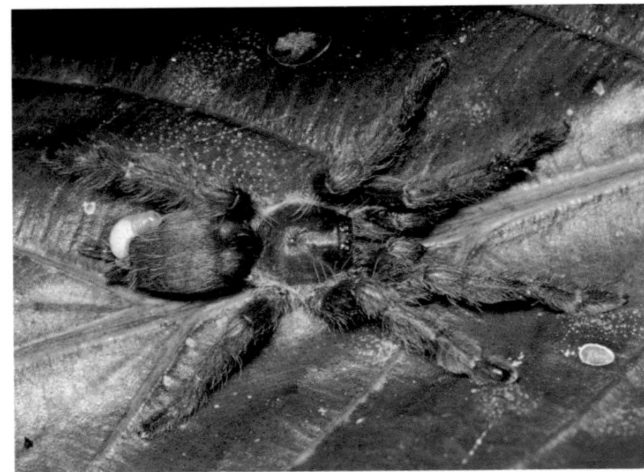

Spider sociality

Social living in adult spiders is extremely rare, occurring in around 40 species out of over 40,000 spider species in total. Nearly all of the 40 species actually live together in colonies, such as *Anelosimus eximius* (Theridiidae) from the tropical rainforests of Peru. Such colonies are able to overpower potential predators and prey by co-operative behaviour. *Stegodyphus* (Eresidae) colonies contain up to several hundred individuals. Their colonies are so-called 'open societies' as there is indiscriminate intra- and interspecific tolerance. New colonies are founded by the fusing of different colonies and by ballooning individuals. African species build their large sheet webs in spiny trees. Within the sheet web, there is a central silk nest that houses the spiders and shelters them against most predators. Additionally, some species of tarantulas are known to be social when kept in close proximity in captivity.

One of the few social species that doesn't live in webs is *Delena cancerides*. Individuals live under the bark of dead trees in massive colonies of up to 300. Such groups may have only a single female with several clutches of offspring or may contain social groups of multiple adults. They share prey but are extremely aggressive to non-colony members.

Defensive behaviour and aggression

Apart from the use of venom, spiders have several defensive adaptations. These include escape behaviour, camouflage, warning colouration, mimicry, threat displays, stridulation, urticating hairs and other defensive structures, defensive secretions, leg autotomy, dropping and feigning death, web detritus and stablimenta, and burrow structures.

OPPOSITE Spiders with disruptive colouration have distinct colour patterns, such as bright contrasting colours, stripes a spots e.g. *Argiope* (Araneidae) (Namibia).

ESCAPE BEHAVIOUR Running away from a would-be predator is obviously a good move, and this is a common strategy for spiders. However, there are alternative escape methods. For example, pisaurids avoid predators by moving away over the surfaces of streams and ponds, and may disappear underwater if threatened. One of the most unusual 'get away' tactics is the 'cartwheeling' behaviour of *Carparachne aureoflava* from Namibia, which escapes its predators by flipping its body sideways and rolling at great speed on its bent legs down the sand dunes.

BELOW A highly camouflaged diplurid spider on a lichen-covered tree trunk (Guanacaste National Park, Costa Rica).

CAMOUFLAGE If an animal is the same colour or texture as its surroundings, it is difficult for predators, and potential prey, to locate. Such concealing colouration is known as camouflage or crypsis. Physical attributes are often enhanced by behaviour – spiders may stay still for long periods, or press their bodies very close to the substrate to avoid shadows.

The lichen huntsman, *Pandercetes gracilis* (Sparassidae), from Australia and New Guinea, all but disappears when it rests on lichen-covered tree trunks. The body and legs are very hairy, which breaks up its outline. Additionally, it presses itself very close to the trunk. The bark spider, *Caerostris sexcuspidata* (Araneidae), from Africa and Madagascar, really doesn't resemble a spider at all when it rests on a branch. It has a pair of protuberances on the opisthosoma, and the opisthosoma overhangs the eyes on the front of the carapace, which helps to camouflage it against the tree bark. *Mastophora caesariata* resembles the tattered cocoon of a large caterpillar or detritus covered with fungal growth, and it is extremely difficult to see when it is on the bark of a tree. Disruptive colouration is another form of camouflage. This colouration helps to disguise the spider's shape, and to render it non spider-like from a distance. Zebra and giraffe are examples of mammals with disruptive colouration.

WARNING COLOURATION Spiders don't always blend in with their backgrounds. They may possess warning colouration (aposematic colouration) – conspicuous markings which indicate to a predator that the animal is distasteful or venomous. It is thought that *Latrodectus* species, with their distinctive red markings, may be aposematic although the effects of these colours on potential predators have yet to be researched.

MIMICRY This occurs where a species has evolved to look like another. In spiders, there are two types – Batesian mimicry and aggressive mimicry. Batesian mimicry is where, in order to avoid predators, an 'unprotected' species (the Batesian mimic) resembles a 'protected' species – the model, which has a noxious quality of some kind. *Myrmarachne plataleoides* (from southern Asia) lives on trees and shrubs that are also inhabited by red ants. Although the ants aren't fooled by this arachnid subterfuge, the mimicry is thought to fool other predators that avoid ants, because of their vicious bites and formic acid defence. Spiders in the family Gnaphosidae are also good ant mimics, as are the Corinnidae.

The main purpose of aggressive mimicry is to conceal a predator from its prey item. Some spiders mimic inanimate objects such as bird droppings – *Celaenia excavata* is a bird-dropping mimic from Australia. It looks just like an unappetising blob of avian faeces, with its dirty white, grey and black lumpy opisthosoma. It emits pheromones to attract males of the moth species *Spodoptera mauritia* and grabs them when they come to investigate. Spider mimics often display associated behaviour that reinforces the mimicry, e.g. *Phrynarachne rugosa* from Madagascar and sub-Saharan Africa sits motionless for hours on foliage, and often spins a messy white patch of silk alongside, to help its subterfuge.

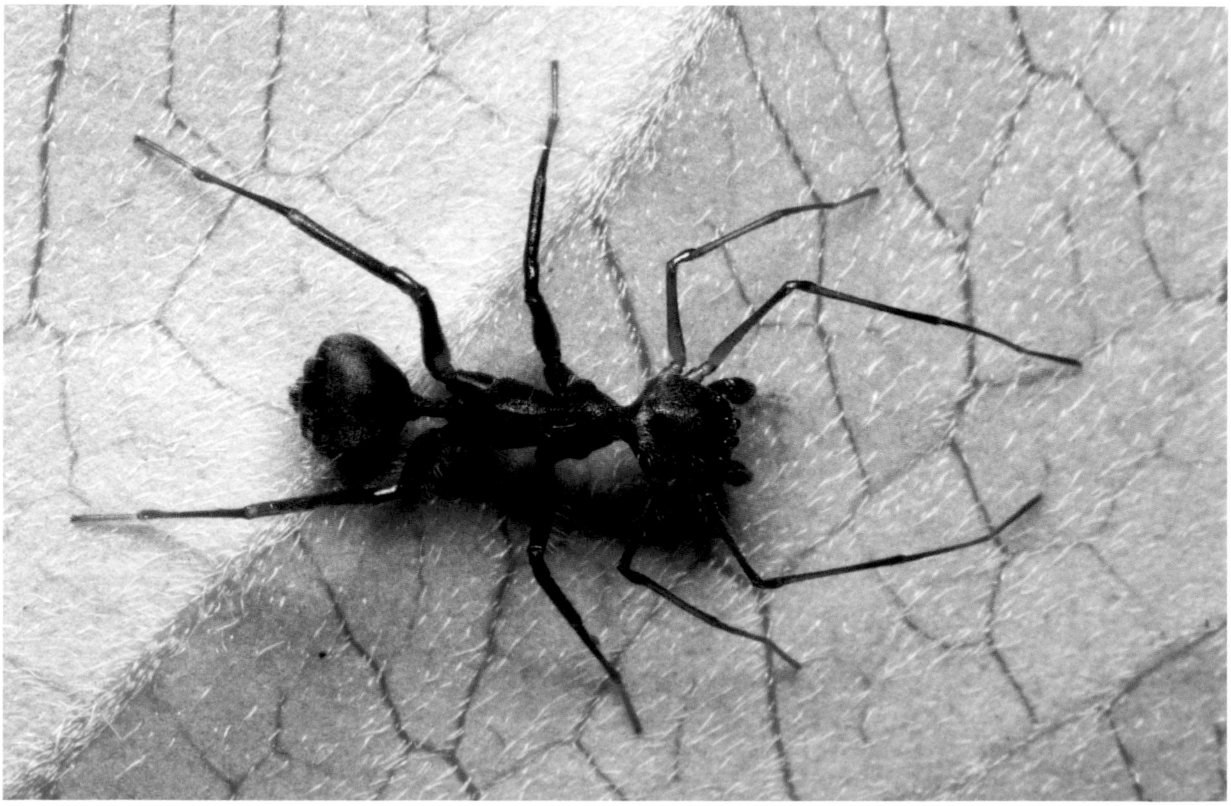

BELOW Ant-mimicking spiders (Batesian mimics) really are masters of disguise. These spiders have long, slim legs and a constricted pedicel, and they often move their first pair of legs as if they are antennae. They may extend their palps or chelicerae forward, so that they look like the head of an ant. Salticids are particularly good ant mimics (Jatun Sacha, Ecuador).

LEFT A crab spider mimicking a fallen rainforest flower (Jatun Sacha, Ecuador).

THREAT DISPLAYS AND STRIDULATION Tarantulas often throw their legs back and open their chelicerae when threatened. They are then in position to strike. Spiders, scorpions and and often use threat displays like this to frighten would-be predators, thereby allowing the creature to make its escape unharmed. In some species, such threat displays are supplemented by colour patches in conspicuous places. Some theraphosids are able to produce an audible hissing sound when threatened, by rubbing together setae on the chelicerae and on the palps. This is known as stridulation.

BELOW Tarantulas from the Americas possess urticating hairs that grow on their opisthosomas and are used in defence. This is almost unique to the Araneae, and is only found outside the arachnids in other invertebrates such as the Lepidoptera (Guanacaste National Park, Costa Rica).

URTICATING HAIRS AND OTHER DEFENSIVE STRUCTURES There are four basic types of urticating (irritating) hairs, all of which have a penetrating end and barbs along the shaft and point, making them look like tribal weapons. Type I hairs cannot penetrate the skin deeply, but can cause ophthalmia nodosa (inflammation of the eye). Type II hairs are incorporated into the silk lining of tunnels. Indeed, both *Theraphosa blondi* and *Megaphobema* incorporate such hairs in their egg sacs. Type III hairs can penetrate up to 3 mm (1/10 in) into skin, and type IV hairs cause irritation in the upper respiratory tract of predators.

If disturbed, a tarantula quickly rubs off a cloud of these hairs with its hind legs. There is a constriction at the base of each hair, which enables it to snap off easily. Small rodents have been known to die from suffocation when they get these hairs in their throats! Tarantulas from Africa and Asia don't have these hairs, and tend to be more venomous and aggressive. Spiders also have other defensive structures. For example, adults of *Gasteracantha cancriformis* have spines to protect the normally soft and vulnerable opisthosoma against predators.

DEFENSIVE SECRETIONS Several spider families also use chemicals as part of their defensive armoury. They may release nasty odours, such as those used by the bolas spider *Mastophora caesariata* from Costa Rica. Some discharge the odour over quite a distance. For example, *Phoneutria rufibarbis* (Ctenidae) apparently discharges from its anus a fluid that smells of ammonia over 0.5 m (1½ ft)! The spitting spiders, *Scytodes*, also use their venom-squirting skills to repel would-be predators.

DROPPING, FEIGNING DEATH AND LEG AUTOTOMY Many web-weaving spiders drop to the ground when approached by a potential predator and this proves to be very effective. Some also 'play dead' (thanatosis) when they hit the ground as part of this subterfuge, which is particularly useful against other spiders, insects and birds, though thanatosis is more common in insects than spiders. Spiders are often able to escape predators by autotomizing a leg.

WEB DETRITUS AND STABLIMENTA Some webs contain the leftovers of old meals, remnants of egg sacs or pieces of plant material. These can be extremely good at camouflaging the occupant, if it rests among this rubbish. For example, *Uloborus plumipes* is a small spider that builds an orb web. Its first pair of legs is long and plumed, and it has a humped opisthosoma. It sits in its web with the first pair of legs extended, among a line of insect husks on a band of silk across the hub, and is perfectly camouflaged. It is found in Europe, Africa and Asia, and has recently been imported into the UK from the Netherlands, and can be found in vast numbers in greenhouses. Stablimenta are sometimes used to camouflage the occupant (see p.61).

RIGHT In some species, different parts of the web are used to hang detritus such as grass and leaves. It's been observed that spiders with such decorations are attacked much less frequently by predatory paper wasps (Thursley Common, Surrey, UK).

BURROW STRUCTURES Spiders that live in burrows often protect themselves against predators by constructing structures such as trapdoors, cryptic turrets of soil around the entrance, and subsidiary tunnels in which to hide.

Adaptations to extreme physical conditions

Spiders, like all arthropods, are unable to maintain a constant body temperature. They are known as poikilotherms. However, spiders have several ways (both physiological and behavioural) of regulating their body temperature within acceptable limits, which vary with species. Some spiders are tolerant to desiccation or the cold, and many demonstrate thermoregulation (which is physiological). Temperature preference and water loss rates depend very much on the habitat in which the species is found, as well as whether the species is nocturnal or diurnal, summer or winter active. As an order, spiders cope well with temperatures in the broad range between 3.5°C (38°F) and 30°C (86°F), although many desert spiders will tolerate well above 30°C (86°F), at least for moderate periods of time. At the other extreme, *Pityohyphantes phrygianus*, *Lepthyphantes minutus* and *Lycosa agrestis* have been recorded as active on snow between –3°C (27°F) and 9°C (48°F) in Poland. In fact, spiders can maintain winter activity under snow, because it creates an insulating blanket.

DESICCATION TOLERANCE Spiders are vulnerable to water loss, particularly through their often lightly sclerotized opisthosomas. Resistance to desiccation is much greater when the exoskeleton is highly sclerotized and certain quantities and compositions of lipids are present. If, however, critically high temperatures are reached, it is possible that certain cuticular lipids actually hit melting point, and with the breakdown of this important barrier the exoskeleton is left open to desiccation. Unsurprisingly, spiders are much less tolerant to desiccation just after moulting, because their new soft exoskeleton has not yet developed its waterproof properties. Evaporative water loss is a problem for spiders that have thin exoskeletons, up to 98% of total water loss can occur in this way. Book lungs, large surface area and small size also add to the problems. Drinking appears to be a major method by which spiders maintain their water balance and avoid dehydration.

Spiders that live above ground are much more tolerant to desiccation than those living below ground. Because mygalomorphs have book lungs, this makes them susceptible to dehydrating quickly, so living in a deep burrow helps to prevent water loss. Within a species, desiccation tolerance varies, depending on the sex and developmental stage. Males are more susceptible to desiccation than females because they have a smaller body volume, and possibly because they spend a lot of time out in the open looking for females. Juvenile spiders are more susceptible to desiccation than adults. Females carrying egg sacs are more at risk of desiccation than those without.

COLD TOLERANCE To survive cold climates and winter seasons, animals need to have special behavioural, morphological and physiological adaptations. Spiders that are able to survive cold conditions deal with them in several ways. First, many species over-winter by going into torpor (a state of inactivity). For example, both juveniles and adults of the aquatic spider *Argyroneta aquatica* move into deeper water, construct a diving bell, and remain inside throughout the winter. Their metabolic rate drops to less than half its usual rate, and it does not feed. Spiders may over-winter in different developmental stages. So, depending on the

species, they may remain as spiderlings inside their egg sacs during the winter months, they may over-winter as adults, or they may brave the cold as juveniles. Spiders that over-winter in juvenile stages often remain active throughout the whole cold season, because immatures are much more tolerant of low temperatures than adults. Some species are freeze-resistant. They are able to lower the temperature at which their body would freeze by 'supercooling', which is the synthesis of antifreeze chemicals called cryoprotectants. These can protect against internal freezing down to −7°C (19°F). There is geographical variation in limits of tolerance to temperature. Cold tolerance is also affected by the degree of dehydration of the body and the contents of the gut.

THERMOREGULATION Like reptiles, spiders are known to regulate their body temperatures through particular physiological adaptations and behaviour. Several species of spiders utilize guanine crystals as part of their colouration. These crystals, located just under the epidermis of the cuticle, reflect light, which helps to reduce the heating effects of intense sunlight. Many araneids, even those without guanine crystals, are a pale reflective colour on the surface of their body that is orientated towards the sun (usually the dorsal surface), but dark on the ventral surface.

Spiders might exhibit one or several different behaviours in order to regulate their body temperature – active avoidance, orientation, burrowing and posturing. If it gets too hot, the spider gets going. Not surprisingly, active avoidance of this kind is one of the most common behaviours.

A spider will move itself to more favourable conditions through orientation. This might involve large movements such as moving in and out of a burrow, or small movements such as aligning its body to catch the rays of the sun. For example, the lycosid *Pirata piraticus* moves deeper into the lower layer of the sphagnum bogs of northwest Europe where it lives, when the temperature increases on the surface in the summer months. A spider might avoid the constant to-ing and fro-ing between the sunlight and the shade by orientating its web strategically. For example, *Micrathena gracilis* orientates its web at an angle to the sun that allows it enough heat, without it being scorched by high intensity solar radiation.

BELOW A common spider orientation is seen on log piles, where an adult female lycosid is sunning itself and its egg sac. As a result of raised temperature, the rate of embryonic development increases in the egg sac (Epping Forest, Essex, UK).

Burrowing is a great way to avoid extremes of temperature. When it gets too hot in the burrow, the spider may plug the entrance, which raises the humidity inside. When spiders construct burrows in arid areas, they make the burrows deeper than those in non-arid regions, in order protect them from high temperatures.

Spiders may adopt particular postures in order to regulate their body temperatures. For example, the orb-web weaver *Nephila clavipes* will follow the sun's movements by altering the angle at which it holds its opisthosoma. As with many animals that live in desert regions such as lizards, diurnal spiders raise their bodies as far as they can off the desert substrate in order to reduce its heating effect. This is called stilting. Such spiders often have long legs.

Burrows – a life underground

Many non-web-spinning spider species produce burrows. These are used to avoid predators, to escape extremes of temperatures, and to raise young. Ground-dwelling tarantulas, such as baboon spiders, use their chelicerae as digging tools and their palps to excavate the dug-out substrate. The burrow is comfortably lined with silk, which extends out to form a rim that is sometimes adorned with plant material. In some species, threads leading from the rim warn the burrow's occupant of approaching prey. During the day, the spider stays at home and covers the entrance of the burrow with silk. It is not just tarantulas that produce burrows. Araneomorphs, such as lycosids, like to live underground too. Females in particular spend most of their time in their silk-lined burrows, when they are not hunting.

ANTI-PREDATOR DEVICES Burrows aren't just a hole in the ground. Some have 'doors' such as those constructed by the trapdoor spiders. The door and hinge are constructed from silk, and make the burrow almost invisible.

Burrows may also have added structures around the entrance. For example, *Lycosa tarantula* constructs little turrets of soil, which decrease the rate of burrow invasion by the scorpion *Buthus occitanus*, probably because they make the burrow entrance cryptic – the larger the turrets, the smaller the chance of scorpion invasion.

BELOW LEFT AND RIGHT
The entrance to a trap door spider's burrow. When the door is closed, the burrow becomes invisible. The hinge of the door is made from silk (Ecuador).

Some spider species have evolved a very cunning strategy of digging a second, shallow subsidiary tunnel (in some cases complete with trapdoor) that leads off the main burrow. If a predator invades the burrow, the spider quickly dashes into the shallow tunnel and shuts the door behind itself. It holds the door closed and remains very still. The predator is fooled into thinking the spider is no longer at home and abandons its search.

The pellet spider, *Stanwellia nebulosa*, from Australia, uses a variation on this theme. It also constructs a secondary tunnel – in this case, it is more of a pocket – off the main burrow. The spider lives in the bottom of the burrow below the pocket, in a silk sock-like burrow lining. If a predator invades the burrow, the spider tugs the silk sock, which is attached to a compact pellet of silk and soil grains positioned in the pocket. The pellet falls down into the burrow, blocking the predator's way. Maybe the most quirky of all burrow defensive mechanisms is that of species in the genus *Galeosoma* (Idiopidae). These spiders have a specially hardened spherical shield on the rear of the opisthosoma, which they use to block the burrow entrance – their own backsides form the trapdoors to their burrows!

SPIDER SQUATTERS Digging a burrow is obviously costly in terms of energy, and the substrate may not be easy for an invertebrate to dig if it is dry, hard and compacted. So, if the opportunity presents itself, taking over an already constructed but empty burrow would certainly seem to be a good strategy, and it appears that's exactly what spiders do, given the chance. In the Nylsvley Nature Reserve in South Africa spiders from three different families – Agelenidae, Pisauridae and Pholcidae – all coexist in the same abandoned mammal burrows. In fact, the larger the burrow, the greater the number of individuals and number of species that coexist. Indeed, it was recently discovered that a tarantula (*Sericopelma*) and a toad were sharing a burrow in Panama! In studies where artificial burrows were dug, an increase in spider settlement resulted, showing that spiders certainly are opportunists.

Moulting

All spiders need to moult several times to reach their adult size. Small spiders go through five moults, while large spiders go through around ten moults. The process of moulting in tarantulas continues after maturity has been reached. Perhaps this is because of their great longevity compared to other spiders. Although moulting is usually an independent process, it has been observed in some species that live close together (e.g. the colonial species *Eriophora bistriata*) that the spiders moult at the same time.

At the start of moulting, or ecdysis, the weight of the prosoma increases by as much as 80%, as the heart rate increases and pumps in extra haemolymph from the opisthosoma. Usually, the carapace is very rigid, but most of the endocuticle has dissolved away from inside, so it is quite flexible. The combination of a thin carapace and an enlarged prosoma results in the exoskeleton tearing along the edges of the prosoma. This is aided by the chelicerae, which move back and forth. The carapace eventually lifts off on a flap of exoskeleton.

Tearing continues along each side of the opisthosoma, which has shrivelled because haemolymph has moved into the prosoma. Muscular contractions of the opisthosoma finally separate it from the old exoskeleton, and the opisthosoma is lifted out. As the opisthosoma is removed, the legs and extremities are freed. This is the point in the process where the spider is most likely to get trapped. Because legs that have newly emerged from a moult are not rigid enough to support the weight of a spider, smallish spiders suspend themselves from a

silk thread in order to moult. However, larger spiders such as tarantulas and some wandering spiders produce a mat of silk on the ground, and lie on their backs to moult. Once the legs are freed, the spider stretches and contracts them as they harden, in order to maintain flexibility. Amazingly, the entire process of moulting takes only 10–15 minutes in small spiders, although it might take hours in tarantulas. The sign of a good moult is when the moulted exoskeleton is intact, and looks just like a live spider.

Life history

Reproduction

When a male is sexually mature and ready to mate, he usually stops catching food and feeding – all his energies are focused on reproducing. Before he goes on the hunt for a mate, he must fill his palps with sperm. For this, he either produces a little horizontal triangular web, or a single thread of silk, onto which he deposits a droplet of sperm from his genital opening. He then alternately dips his palpal bulbs into the droplet and draws up the fluid.

COURTSHIP When it comes to courtship, spiders are strangely reminiscent of humans in some ways, because they are known to use perfume (pheromones), bright colouration and 'personal' touching (stroking of legs and palps) to attract the ideal partner. Spiders such as tarantulas stroke each other's palps and legs before copulation. In some species, e.g. *Atrax robustus*, the male lulls the female into a trance-like state by repeatedly tapping her with his palps and first pair of legs. A few may even take part in loosely tying with silk threads! In other spiders, such as *Nephila* (Nephilidae) and *Xysticus* (Thomisidae) the male loosely ties the female with silk thread before copulation. Such involved courtship rituals might seem wasteful of energy and effort, but there are very good reasons for indulging. First, apart from a very few exceptions, female spiders are larger than males and have voracious appetites, so males are at risk of being eaten – they need to make *very* certain that they won't be mistaken for prey, and courtship rituals help to achieve this. Second, the male needs to make sure that the female is sufficiently stimulated for copulation. There are many varied courtship rituals. In fact, almost every spider species has its own style. However, they can be loosely grouped into three categories, as described below.

PERFUME AND VIBRATIONS In the Lycosidae and Araneidae, for example, female pheromones on silk draglines or on webs bring on courtship behaviour in males. So powerful is the stimulus of the pheromone, that males often perform their courtship when the female isn't even there! A male *Dolomedes* (Pisauridae) can even track pheromones on a female's dragline across water. Courtship in araneids often involves both female pheromones and the plucking of the web by the male. The vibrations produced are specific to each species, so the female will only respond to a male of her own species. In some species, the females respond in a species-specific frequency. These vibrations differ to those of a trapped fly, so the female won't rush out and consume the male in a case of mistaken identity! The use of vibrations for courtship is not confined to webs – certain species will use the surrounding substrate or foliage. For example, lycosid males drum on leaves with their palps, and some make vibrations on the substrate by beating it with their opisthosoma.

BRIGHT COLOURATION AND MOVEMENT Salticids have exceptionally good eyesight, and are capable of seeing colour, so their courtship is a very visual affair. In certain species, when the male waves and moves his legs and palps, the female signals back by vibrating her palps, or imitating his movements. Lynx spiders (Oxyopidae) and wolf spiders such as *Lycosa rabida* (Lycosidae) also use visual courtship, and some lycosid males use palpal semaphore too. However, this is largely confined to those genera, such as *Pardosa*, that have ornamented palps or front legs.

RIGHT The male salticid performs species-specific movements in front of the female, by waving his legs and palps in a kind of semaphore, and twitching his opisthosoma. Usually, the body parts involved in courtship are brightly coloured (Taman Negara National Park, Malaysia).

BELOW A pair of courting salticid spiders (southern Mexico).

NUPTIAL GIFTS Some spiders have taken the mating game one step further – male spiders of *Pisaura* and *Dolomedes* (Pisauridae) use nuptial gifts to woo potential mates. These gifts are actually captured insects wrapped in silk. It is thought that such courtship tactics have evolved to allow the male precious time to escape after copulation before being eaten, because the female is so involved in consuming her gift that she leaves the male alone. However, it has recently been discovered that in trials using males of *Pisaura mirabilis*, 90% of males that offered a gift managed to copulate, compared to a meagre 40% of males who didn't offer a gift, and *none* of the participating males were eaten, even those who didn't present a gift. The duration of copulation increased with the size of the gift, and the longer the copulation time, the more eggs were fertilized. In other words, if a male offered a large gift, he got to fertilize a lot more eggs than a male who didn't come so well prepared.

SIZE DOESN'T MATTER Some males are able to mate with females without them knowing! This happens in the genus *Nephila* where the males are so small that the female is unaware of his presence (see p.48 top). He is able to creep up and mate while she is preoccupied.

SPIDER SEX During copulation, the male inserts his palpal organ into the female's genital opening and deposits sperm in her spermathecae. Males with simple palps, e.g. tarantulas insert the whole bulb, whereas those with complex palps, e.g. *Araneus* only insert the tip of the embolus. When the haematodochae are inflated with pumped-in haemolymph, the tip of the embolus is pushed into the female's sperm duct. The complexity of male and female sexual organs is matched. You would think that such species-specific 'equipment' would prevent different species from mating and producing hybrid offspring. However, it has been shown that it is largely species-specific courtship that prevents interspecific mating, rather than the complexity of the male and female genitalia. Slightly different habitats and seasonal variation in courtship also reinforce the separation. Closely related species that reach the copulatory stage do in fact produce hybrid offspring, e.g. *Tegenaria*. After copulation, some species live together (although this is extremely uncommon), others part never to meet again and sometimes repeated mating occurs between the same individuals.

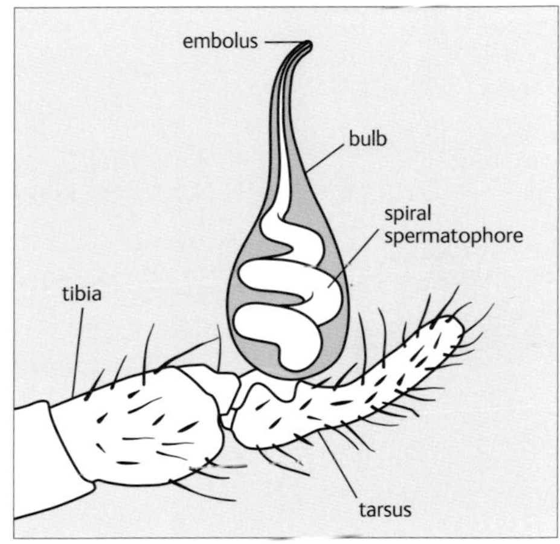

ABOVE Simple male palp of the spider, *Segestria florentina*. The tarsus of the palp carries a pear-shaped bulb, which acts as a reservoir for the spiral spermatophore.

SEXUAL CANNIBALISM Female spiders are renowned for eating their partners after copulation. However, such post-coital consumption is really quite rare. Where it does happen, it is often solely because the male has been mistaken for food! In most species, the males are able to make their escape unharmed. In fact, males of many species have adaptations to hold the female's jaws away, so that she can't attack. For example, males in the families Theraphosidae and Hexathelidae have spurs on their legs to hold the female's fangs. Males in the Tetragnathidae have very large jaws, which interlock with those of the female. Yet others have lumps and bumps on their prosoma that the female grabs with her fangs during mating.

Males of certain species have evolved cunning strategies to avoid being eaten. For example, a sparassid male may guard a sub-adult female until she moults into an adult. Just as she emerges and before her new exoskeleton has hardened, the male mates with her. This ensures that she hasn't got the strength to eat him! This is also very common among salticid species.

Unfortunately for males in a few species (e.g. *Latrodectus hasselti*), they must place their opisthosomas directly in front of the female's chelicerae to be in the right position for copulation. The male thus sacrifices himself to ensure a successful mating. This is no doubt how the name 'black widow' came about for the genus *Latrodectus*! However, although *Latrodectus* females are renowned for eating their male partners, males are also known to cohabit with females. Certain species in a few other families, such as the Nephilidae, also indulge in sexual cannibalism. Hungry virgins in poor condition are more likely to be cannibalistic. It might seem rather gruesome to consume one's sexual partner, but there is an evolutionary reason for it. Many males die very shortly after mating anyway, and the males that are consumed confer an advantage to their future offspring by providing the expectant mother with essential nutrients. In this way, the males increase their chances of passing on their genes successfully. Another advantage is that cannibalized males mate for longer, which results in increased fertilisation success. Cannibalism is therefore in the interest of both the male and the female.

PROTECTING PATERNITY Male spiders have different ways of ensuring their own paternity. Some males plug the female's epigyne after copulation to ensure that no other males can mate with her. In some species, such as the golden orb weaver *Nephila fenestrata*, the tip of the embolus breaks off within the epigyne. The male is therefore emasculated in order to protect his sperm. In other species, there is a protruding sclerotized tongue-like structure on the epigyne, which is pulled off by the male making further copulation impossible. Interestingly, in a selection of species from different families (e.g. *Tegenaria* in the family Agelenidae, *Latrodectus* in the family Theridiidae and probably most salticids), the male lives in the same web as the female. This cohabitation enables the male to guard the female from other potential suitors.

PARTHENOGENESIS In some spider species, is would appear that males are dispensable. For example, over 1,000 specimens of *Theotima minutissima* (in the family Ochyroceratidae) have been collected from tropical regions in Asia, the Caribbean and South Africa, but not one male specimen has ever been found. As this gender ratio is clearly not usual, it was strongly suspected that *T. minutissima* must be parthenogenetic (able to reproduce asexually). To test this theory, specimens of this fascinating spider were reared in the laboratory, and it was found that a second generation spiderling had offspring that were viable – thus proving that *T. minutissima* is indeed parthenogenetic.

Egg laying

A female will store the male's sperm until she is ready to lay her eggs. As the eggs pass out of the female's genitalia, the sperm cells are released from the spermathecae and the eggs are fertilized. The number of eggs laid varies enormously from species to species e.g. *Cupiennius* (Ctenidae) lays 1,500 to 2,000, whereas *Heliophanus* (Salticidae) lays 20 to 40. The eggs are nearly always protected by silk, so they are not exposed to the elements and predators. Some spiders, such as *Pholcus* (Pholcidae), wrap only a few silk threads around the eggs, while others lay their eggs in their silken retreat and remain inside to guard them, e.g. *Clubiona*. Most spiders produce an egg sac (cocoon). Egg sacs are important because they maintain a certain level of humidity and temperature, as well as helping to protect the eggs from predators and parasitoids.

Egg sac designs

A typical egg sac consists of a basal plate, with a covering plate on top. However, the designs are specific to particular species, and can sometimes be used to identify the adults. The various designs include ones that are flat, or globular and suspended by a thread, or spherical and woolly, and yet others are papery and shaped like fried eggs. Structural variations include a cylindrical wall between the plates, as produced by *Araneus quadratus*, and an inverted 'wine glass' shape, with separate internal egg and brood chambers, as produced by *Agroeca brunnea*. The 'wine glass' stem suspends the egg sac from a twig or branch to protect it from predators. Some spiders camouflage their egg sacs by gluing on particles of plant, soil or small stones. *Theraphosa blondi* and *Megaphobema* species incorporate urticating hairs into their egg sacs. Spider egg sacs can be found in a variety of places. For example, they can be seen on the upper and lower leaves of vegetation, on the bark of trees, in webs, under stones and attached to ceilings inside buildings.

ABOVE A globular egg sac suspended by a thread, with glued on particles (Madagascar).

BELOW A highly camouflaged spider straddling its very flattened egg sac (Zombitse National Park, Madagascar).

RIGHT Lycosids with their egg sacs attached to their spinnerets are also a common sight. These spiders often sit in the sunshine turning the egg sac frequently, so that all the eggs receive the same amount of heat (Danum Valley, Sabah).

Maternal care

Spiders exhibit different degrees of maternal care. Some lay their eggs in egg sacs, and then abandon them to lay others elsewhere. Some mothers abandon their eggs because they die soon after laying them, e.g. orb-web weavers. Some stand guard over the egg sac, e.g. Thomisidae, whereas others carry them around. For example, pisaurids are frequently seen carrying their eggs sacs on the underside of their bodies, held in place by their chelicerae and palps, which means they are unable to feed during this time. Lycosids valiantly defend their eggs if threatened with their removal, and search for them for hours if they are taken away. However, they can be fooled into accepting a substitute egg sac, such as a small paper ball of the appropriate size and weight.

A female lycosid assists her babies when they are hatching, by tearing open the egg sac with her chelicerae. When the babies emerge, they scurry onto their mother's back, where they hold onto special club-shaped hairs. They may stay there for some time, before they disperse. If they are removed, then the female goes around gathering up her offspring and getting them back on her opisthosoma.

The nursery web spider, *Pisaura mirabilis*, lays her eggs on plant stems, and constructs a tent-like web around them. She then sits guarding her eggs on the outside of the web (see p.36 top left). When the spiderlings hatch, they are contained in the nursery web, still protected by the attendant mother. After moulting, they disperse.

Some spiders actually provide food for their babies, a level of care only found in about 20 species out of over 40,000. Some species catch prey, kill it and leave it for the spiderlings to feed upon. In other species, such as the theridiid *Theridion sisyphium*, the female regurgitates pre-digested food for her babies, much like a bird feeding her chicks! Spiderlings are also known to stroke their mother's legs in order to request more food.

MATRIPHAGY Perhaps the highest form of all maternal care is found in spiders that demonstrate matriphagy, the eating of their mother by young spiders. This gruesome-sounding action appears to go against the very laws of nature – how does it make evolutionary sense?

Like sexual cannibalism, matriphagy confers a great advantage to the offspring, which is good news for the offspring and their mother. This was cleverly demonstrated recently by a simple experiment in Japan. In the species *Cheiracanthium japonicum* (the Japanese foliage spider), youngsters feast upon their mother shortly before dispersal. When they are ready to leave the breeding nest, they have already moulted into the third instar stage, have put on more than three times their pre-matriphagous weight, and have longer legs. They are therefore already well suited to a solitary hunting lifestyle. However, those spiders that had their mothers removed before they could tuck in were severely disadvantaged. Most were unable to moult into the third instar before they dispersed, so they had shorter legs, and they had a much lighter body weight. Their survival rate was reduced.

Post-embryonic development

Unfortunately, arachnologists simply cannot agree on the developmental terminology of spiders. There is a plethora of terms that are often difficult to match up as referring to the same stage. For example, the post-embryo spider – the stage at which the spider hatches out of the egg – has been called, among other things, nymph, larva, prenymph and protonymph. (The most tongue-in-cheek name, used in pet trade literature, is 'egg with legs'!) Other developmental terms are confusing too. 'Spiderling' is very commonly used but does not define a particular stage; it covers instars 1-4 or possibly more.

The developmental stages in a spider include: 1) egg hatches into post-embryo, which is immobile; 2) post-embryo moults into 1st instar larva; 3) 1st instar moults into 2nd instar, and the mobile animal emerges from the egg sac and begins to feed – the nymph; this stage has adult-like characteristics such as functional spinnerets; 4) varying number of moults (depending on species, sex and condition); 5) penultimate instar moults into the adult (also known as the ultimate instar); 6) ultimate instar moults into 1st post-ultimate instar, and so on – only spiders that are long-lived such as female theraphosids, go passed the ultimate instar.

LEFT Young araneomorph spiders in a nursery web (Usk, Wales, UK).

OPPOSITE Freshly hatched (post-embryo) tarantula spiderlings.

The number of moults before adulthood depends on the species, the sex of the spider, and the environmental conditions in which the spider lives. For example, *Dictyna coloradensis* matures after 6–10 moults, with 8–13 days in between. However, there may be over 20 moults in long-lived species such as females in the Theraphosidae. Other variations include *Mastophora hutchinsoni*, where males that have just emerged from their egg sacs are only 1–2 moults away from sexual maturity, and the precocious males of *Mastophora cornigera* that are already sexually mature when they emerge! Males need fewer moults to get to maturity than females, because of their small body size. They therefore mature more quickly.

Longevity

Many spiders have a fairly short lifespan. In many species in temperate regions, the eggs are laid in summer, and will over-winter. The adult females that result from these eggs then lay eggs themselves the following summer, before dying. Lycosids and salticids can live for up to 18 months, but many araneomorphs live for up to two years. Mygalomorphs are known to take up to 12 years to reach maturity, and can live for over 20 years. Within all species, the males draw the short straw, because they never live as long as the female e.g. male mygalomorphs live up to seven years. As harsh as it seems, once they have mated (perhaps several times), they serve no further biological purpose and so die.

3 Amblypygi
Whip spiders

Amblypygids are bizarre-looking arachnids, with highly flattened bodies and spiny, robust raptorial palps. They possess a peculiar-looking first pair of legs, which are extremely long, thin and whip-like. Amblypygids are considered (albeit superficially) to look like spiders, hence the common name, 'whip spiders'. They are distinguished from the rather similar-looking Uropygi by the fact that they don't possess a 'tail' (flagellum), hence the other common name, 'tail-less whip scorpion'. The scientific name reflects this, too, because it means 'blunt rump'! The smallest amblypygids, a petite 5 mm (3⁄16 in) long (from chelicerae to tip of opisthosoma), are in the Neotropical genus *Tricharinus*. The world's largest, at a formidable 45 mm (1¾ in), is *Acanthophrynus coronatus* from Mexico.

OPPOSITE If cornered, an amblypygid extends its legs and rocks backwards and forwards, extending its palps to show its sharp defensive spines. With further harassment, it makes rapid jerking movements and thrusts its palps towards the aggressor.

Classification and diversity

There are currently around 140 valid amblypygid species in 17 genera, within five families – the Paracharontidae, Charinidae, Charontidae, Phrynidae and Phrynichidae. Amblypygids are really rather uniform. Any variation that *is* present is generally found in the palps, the length of the antenniform legs, the overall size of the species, or in the colour. For example, *Euphrynichus amanica* from East Africa has extremely long palps, and can attain a whopping leg span of 40 cm (8 in). *Xerophrynus machadoi* from Namibia and Angola has impressively spiny palps, which are in contrast to the rather bald palps of *E. amanica*. *Stygophrynus* species (from Southeast Asia) have extremely long antenniform legs. Most amblypygids are some shade of brown, but *X. machadoi*, with its spiny palps, is almost black and *Phrynichodamon scullyi* from South Africa has a white carapace.

BELOW Species in the African genus *Damon*, can display vivid markings of contrasting bands and spots, such as this individual in Atewa Forest Reserve, Ghana. The colour variation is due to pigment in the cuticle and a coloured 'powder' produced by glandular pores.

Anatomy

The exoskeleton is usually covered with different sized tubercles, setae and spines, as well as many glandular pores (which can be opened or closed by valves), and epidermal glands. In

OPPOSITE The eyes of an amblypygid (*Phrynus tessellates*, Saba, Netherlands Antilles), are arranged into three groups – a pair on the central tubercle at the front of the carapace, and a group of three either side of the carapace. The eyes can detect varying light intensities, and if suddenly illuminated the amblypygid will quickly find a dark place.

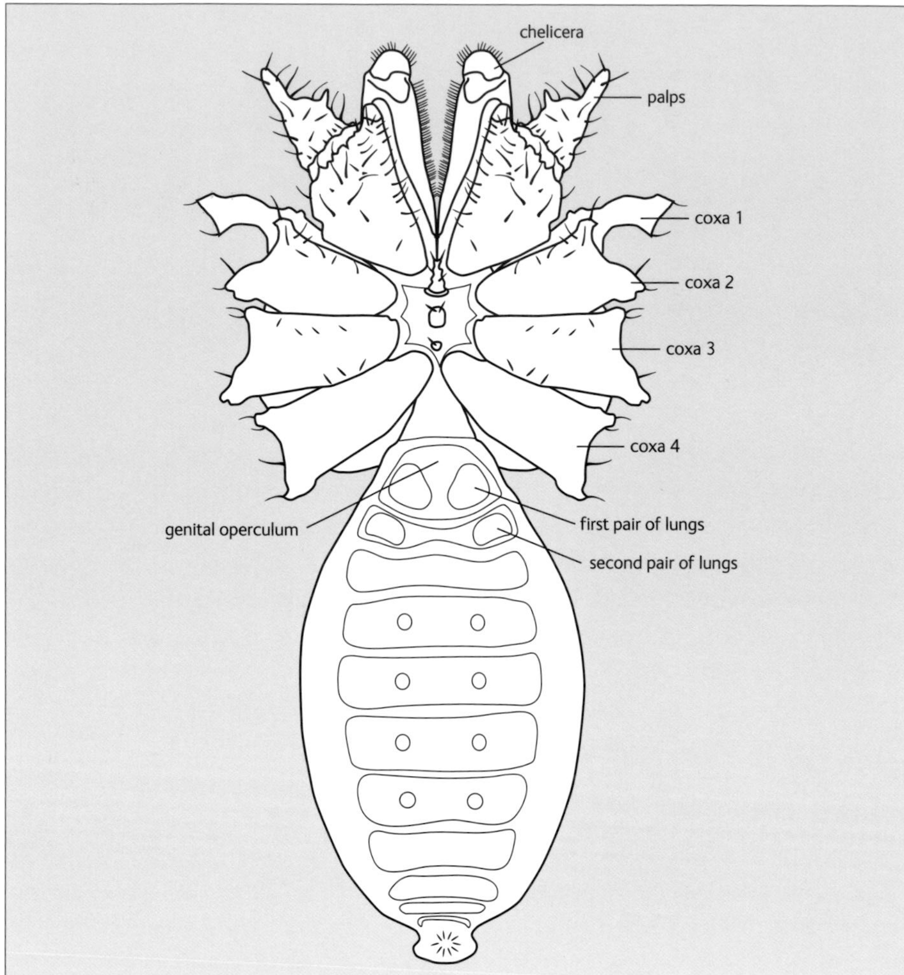

RIGHT Ventral view of an amblypygid, *Charinus milloti*.

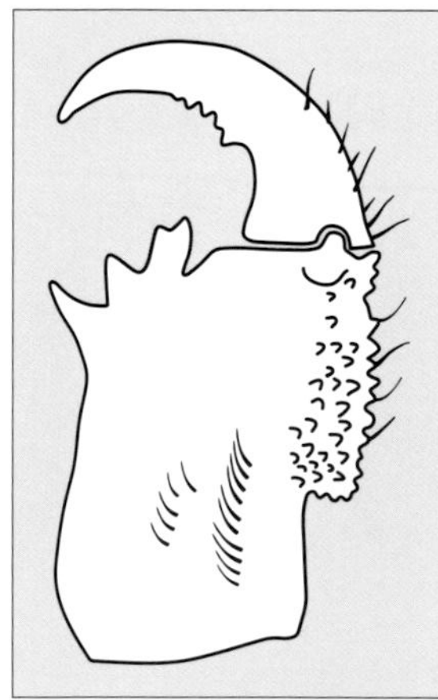

ABOVE The fang on the chelicera often has a row of denticles or teeth that fit between two rows of denticles on the base. The underside of both fang and base are covered in fine dense hairs.

species within the Phrynichidae and Phrynidae, these glands secrete tiny globules of unknown chemical composition in the first few days after moulting, which give the cuticle a brownish-yellow, powdery look, particularly on the dorsal surface of the chelicerae, the palps, the basal part of the legs and the prosoma. The parts of the exoskeleton that are cleaned on a regular basis tend to be quite smooth.

External anatomy

PROSOMA Dorsally, the prosoma is covered by a kidney-shaped carapace, which is always broader than it is long (see opposite). The deep depression along the midline of the carapace is equivalent to the fovea in the spider i.e. an inner attachment point for the muscles of the sucking stomach. In most species, the carapace is rather rough, due to numerous setae and tubercles of different sizes. Ventrally, the coxae of the palps and legs cover most of the area, leaving a small area in the centre. This is divided into three or four small, sclerotized sternites, which may bear setae.

EYES Amblypygids have eight eyes, although they are often totally absent in true cave-dwelling species.

Raptorial palps

Amblypygid palps are *totally* different to those of spiders, but are similar to those of uropygids. They don't have the appearance of walking legs because they are raptorial, i.e. adapted for catching and holding prey, and this is their primary function. In immature and young adults, where food catching is very important, the short and robust palps have many spines. However, in large adults, for which mating and fighting are more important, there are fewer spines. In the Phrynichidae, however, the main spines have shifted to the end of the tibia, and are used together with spines on the tarsus to form a movable pincer. This prehensile 'hand' is ideal for catching smaller prey. In some species, other spines are reduced or totally absent, so that only the spines that form the 'hands' remain.

In young animals, the palps are shorter than the body, but often get a lot longer in older animals, as a result of positive allometric growth (the growth of a particular structure at a constantly greater rate than the rest of the body). Even after reaching sexual maturity, amblypygids still grow and moult. So, long palps with a few spines belong to the old timers, and this means that within a species, different-sized animals may have different forms of palps. Although palps are often about the same length as the walking legs, in some amblypygids, such as in *Heterophrynus* sp., they are much longer than the walking legs. One might expect such long, thin and fragile palps to be a disadvantage to their owner because of a propensity to break. However, they would also allow the animal to be a good ambush predator, because it doesn't have to get close to its prey to catch it.

ABOVE *Heterophrynus* sp. from Koashen District, Guyana. The palps are held tightly against the front of the body when the animal is resting or walking around, but when the amblypygid is approaching a potential prey the palps are extended outwards, and held parallel to the ground, with the joints between the femur and tibia at right angles. In most species, the palps can only be used in this plane.

BELOW The palps are armed with ferocious-looking sharp spines, which vary in number, length and position depending on the species, and also vary with age. As the palps close, the spines act as a catching basket for large prey (Jatun Sacha, Ecuador).

CHELICERAE The most anterior appendages on an amblypygid are the chelicerae, which are tucked under the carapace. These are oriented in the vertical plane, like the fangs of mygalomorph spiders and uropygids. Each chelicera consists of two parts – a large base attached to the first ventral segment of the prosoma, and a movable finger or fang, which closes against the base to form a grasping mechanism.

WHIP-LIKE ANTENNIFORM LEGS An amblypygid antenniform leg has the same divisions as a walking leg, but the femur is 25–50% longer, and the tibia and tarsus are divided into many tiny divisions, called articles. The number of articles is species specific. The coxae of the antenniform legs are positioned slightly higher than those of the walking legs, which enables the antenniform legs to be moved across the dorsal surface of the body while being kept in the same plane. The antenniform leg retains the same joints as the walking leg. However, because of the articles, the tarsus can bend midway, which gives it great mobility. Because there is also great mobility at the trochanter-coxal joint, the legs themselves can rotate in almost any direction.

Amblypygid antenniform legs are extremely sensitive to natural chemical cues and vibrations, because they are covered in many varied sensilla, which are concentrated on the tarsi. These sensilla include bristles, trichobothria and slit sense organs. The bristles are probably chemoreceptors, and the trichobothria and slit sensilla are mechanoreceptors. When the amblypygid is active, the antenniform legs are held out sideways and are almost constantly moving in sweeping horizontal movements, unfolding and extending where necessary. When the tarsi of the legs come into contact with an object, they will lightly tap at shortly spaced intervals. The antenniform legs can even feel around objects and into holes. They move independently of each other, making it possible for the owner to be detecting both in front and behind at the same time! At rest, the antenniform legs are folded neatly across the body, keeping parallel to the substratum.

ROBUST WALKING LEGS The other three pairs of legs are much more robust, and much shorter (see opposite top). They are used in locomotion, and are referred to as walking legs. Claws on the tarsi of the walking legs are very important in allowing amblypygids to attach themselves to rough vertical surfaces, where they may spend their entire time. In adults of several genera, for example *Stygophrynus,* there are also pulvilli (adhesive pads) between the tarsal claws. An amblypygid walks forward slowly and cautiously, with its body held high. If suddenly alarmed, it is able to run very fast sideways, like a crab spider. It turns its legs into a flattened position onto its opisthosoma so that it can squeeze into narrow crevices. This doesn't cramp its style, however, because it is still able to run in this position! The walking legs, like the antenniform legs, are also important as sensory structures. They too have a variety of sensilla, such as bristle sensilla, slit sensilla, lyriform organs and trichobothria. There are between 66 and 300 trichobothria on the walking legs, depending on the species, and they are very important in locating moving prey.

OPISTHOSOMA The most striking thing about an amblypygid's opisthosoma is that it is so distinctly flattened, in stark contrast to a spider's rotundity. This enables the animal to squeeze into very narrow cracks and crevices, protecting it from predators and the elements, especially the heat of the sun. The other striking difference is that the opisthosoma has 12 distinct segments (see opposite top). Each segment consists of a tergite and a corresponding

sternite. The tergites often have many cuticular lumps and bumps, which bear setae. The sternites, however, are normally much smoother, although they still have some setae. The intersegmental membrane between the tergites and sternites is a great 'hunger indicator' – in ravenous beasts, the membrane is highly folded, but in well-fed animals, the membrane is wider and stretched. The opisthosoma is attached to the prosoma by a thin waist called the petiolus, which is joined to the prosoma along its whole width, rather than by a thin pedicel like in spiders.

EXTERNAL GENITALIA Both males and females have a genital operculum (flap), which covers the gonopore (reproductive aperture) (see p.92 top). This is on the ventral surface of the opisthosoma.

RESPIRATORY STRUCTURES In amblypygids, these consist of two pairs of book lungs. These open externally as slits in the cuticle and are located on the second and third segments of the ventral surface of the opisthosoma (see p.92 top). Some amblypygids are capable of breathing through a plastron in the same way as the spider *Argyroneta aquatica* does.

VENTRAL SACS As the name suggests, these are sac-like structures lined by a soft cuticle, which are thought to absorb water from humid air. Not all amblypygids have them, but they are present in a variety of species within three families, e.g. the Charontidae, the Charinidae and the Phrynichidae. If present, the sacs are located on the ventral side of the opisthosoma, on the third sternite. They are not normally visible, because they are drawn inwards when not in use.

SEXUAL DIMORPHISM In amblypygids, sexual dimorphism is not terribly common. Even where it does exist, it is not very extreme. The main difference in amblypygids is longer palps in the male, due to positive allometric growth. This can, however, be identical in both sexes as they develop. Less common differences include the reduction of palpal spine length found in males of some species, and a conspicuous fold that runs around the posterior part of the opisthosoma in females within the genus *Damon*.

Internal anatomy

CENTRAL NERVOUS SYSTEM This consists of a supraoesophageal ganglion (consisting of the brain, along with the cheliceral ganglion) and suboesophageal ganglion located in the prosoma. The brain receives information from the optic nerves, and contains association centres (concerned with memory and learning). The much larger suboesophageal ganglion consists of fused ganglia from the appendages, and links sensory input to motor output. Nerve bundles lead from the central nervous system to the extremities and to organs in the opisthosoma, and constitute the peripheral nervous system.

CIRCULATORY SYSTEM The heart of an amblypygid is located in the upper part of the opisthosoma. A major blood vessel, the aorta anterior, runs from the heart into the prosoma, enlarges to become the aorta crassa, and branches into finer blood vessels, which serve all of the appendages and organs. Finely branched arteries, the arteria laterals, run longitudinally into the opisthosoma, and serve the book lungs, midgut, reproductive organs, and other tissues.

LEFT View of a whip scorpion showing the palps and chelicerae tucked in towards the mouth.

DIGESTIVE SYSTEM Each papal coxa has an endite (this is equivalent to a maxilla in spiders), and these are fused ventrally under the mouth to form a cavity in front of the mouth (the pre-oral cavity). In the 'floor' of this cavity is a trough, which funnels the liquefied food to the mouth. The walls of the endites are covered with hairs, which filter out particles from the amblypygid's food.

Food is taken into the body in a liquefied form (see *Feeding and digestion*, p.100). The prey's liquefied tissues are drawn through the pharynx and oesophagus, and into the stomach by the powerful sucking stomach muscles. From here, the food passes into the midgut, which is directly behind. The midgut branches into two large diverticula, which both divide further into four smaller diverticula called caeca. These caeca reach right into the leg coxae, thereby maximising the amount of gut that can be fitted into the prosoma! The gut doesn't stop there however, as nearly the whole opisthosoma is filled with gut diverticula too. The hind and foregut are lined with cuticle, but the midgut is unlined, and this is where absorption occurs. The midgut diverticula are surrounded by spongy mesodermal tissue that acts as a cellular larder, providing a large food storage area, which can keep an amblypygid going for weeks or even months. Nearly all of the opisthosoma and part of the prosoma are filled with midgut glands and mesodermal tissue.

EXCRETORY SYSTEM Excretion in amblypygids is not well understood. In an amblypygid, the posterior midgut widens to form a rectal pocket, which is connected to the anus by the short hindgut. The anus leads into the anal chamber, which is covered by the anal operculum. Faeces collect in the posterior midgut and rectal pocket, before being jettisoned from the body. The faeces come from two pairs of Malpighian tubules, which open into the posterior part of the midgut and extend inbetween the midgut glands and mesodermal tissue; they may even reach into the prosoma in adult animals. In spiders, Malpighian tubule excretory products include guanine and uric acid; in amblypygids, the excretory products have yet to be studied, though it is likely that they also excrete uric acid.

Other excretory organs include the coxal glands, which are located in the prosoma next to the brain. They are thought to be involved with ultrafiltration of the haemolymph, but their role in excretion has not been studied in depth. Nephrocytes are also involved in excretion.

These are large cells located in the prosoma near the suboesophageal ganglion, which are assumed to take and store waste products from the haemolymph.

INTERNAL GENITALIA The female has a pair of ovaries, a pair of oviducts, a uterus and a genital atrium with gonopods (claspers). The paired ovaries are two long, flattened tubes, connected to the uterus by oviducts, and surrounded by muscles. They lie in the ventral part of the opisthosoma and extend from the fourth to around the ninth segment. They have a narrow lumen (channel) and thick walls, in which the oocytes develop. As they become larger, the oocytes grow out of the ovary walls and into the surrounding haemolymph. They remain attached to the ovary by a stalk. Just before oviposition, the oocytes move back into the ovaries.

The uterus opens out into an elongate chamber called the genital atrium, which exits the body at the gonopore, covered by an operculum (see p.92 top). Attached behind this genital operculum are the gonopods – small projections that are used to pick up the sperm masses or packages from the spermatophore during mating. They may be two soft 'cushions', each with a thin, finger-like projection that can be erected by an increase in haemolymph pressure. The projections may be sucker-like, or hard and claw-like. Some species don't have protruding structures at all, instead, a sharp margin of the genital operculum attaches to hooks on the spermatophore. The gonopods of a female of a particular species can only pick up the sperm packages from the spermatophore of a male of the same species, thus ensuring that hybridization can't occur between closely related species.

The male reproductive system comprises a pair of testes and vasa deferentia, a pair of sperm reservoirs, a pair of ventral glands and lateral glands, a pair of secretory reservoirs, and the genital atrium with the spermatophore organ. The paired testes are two long tubes, which are connected to the sperm reservoirs by the vasa deferentia (slim tubes) and are surrounded by muscles. The testes lie in the dorsal part of the opisthosoma and extend from the third to around the eighth segment. They have a flattish lumen and thick walls, in which the spermatocytes develop. The sperm reservoirs are usually filled with spermatozoa, which have passed down the vasa deferentia from the testes. When active, a spermatozoan is long and corkscrew-like, with a flagellum. During transfer to the female in the spermatophore, a spermatozoan is immobile because it is coiled up like a spring in a ball with a covering to hold it together. Immobility is important at this stage, because the female stores the sperm in her genital atrium until she is ready to lay her eggs.

The ventral glands are, like the testes, long and tubular. They lie almost parallel to the testes, and lie in the ventral part of the opisthosoma. They open into the secretory reservoirs. Grainy secretions formed in the ventral glands are stored in these reservoirs. The lateral glands, not surprisingly, are found laterally in the anterior part of the opisthosoma. They produce a transparent secretion that is also stored in the reservoirs. Unlike that of the female, the genital atrium in the male is mainly filled with the spermatophore organ. This complex organ varies between species. It has a lobed structure enclosing an intricate set of internal cavities – the main one is called the inner genital atrium. This cavity acts as a mould to produce the spermatophore. The vasa deferentia, the sperm reservoirs and the secretory reservoirs open into the inner genital atrium at the front of the spermatophore organ. If the spermatophore organ is erect, then the inner genital atrium opens to the exterior of the body at the posterior. The spermatophore organ is erected by an increase in haemolymph pressure.

Distribution and habitats

Amblypygids are found worldwide in the tropics and subtropics, with a few genera in temperate zones. Tropical species have sensitivity to low temperatures, so cannot live in regions with cold winters. Most genera and families have a restricted distribution and only the genus *Charinus* has an almost global distribution. As a general rule, amblypygids are not found in Europe. The only exception is *Charinus ioanniticus*, which has been recorded on the Greek islands of Kos and Rhodes. It is also found further afield in Israel and Turkey. The genus *Phrynichus*, which contains 14 species, is a bit of a 'globe-trotter', with a distribution ranging across the Far East, Cambodia, Thailand, Malaysia, India, Sri Lanka, the Seychelles, Africa and Madagascar. In contrast, the genus *Xerophrynus*, containing just one species, is very restricted, being endemic to the Namib Desert in Namibia and Angola. The only amblypygids found in Australia are represented by *Charon* and the almost universal *Charinus*, both from the Northern Territories. The family Phrynidae is only found in the Neotropics, with the subfamily Phryninae found in Central America, and the other subfamily Heterophryninae found in South America.

Habitats

Amblypygids live in stable habitats, such as rainforests, where the majority of species are found. During the day, amblypygids are found in leaf litter, under stones, within hollow trees, under tree bark, in small mammal burrows, and in rock crevices. Amblypygids do not burrow, so have to make use of existing shelters.

BELOW At night, amblypygids are found in more exposed places such as on the bark of trees and rocky outcrops (Jatun Sacha, Ecuador).

TROGLOPHILES AND TROGLOBIONTS Many species can be found in caves. Interestingly, there are many rainforest species that inhabit caves, e.g. *Damon diadema* from Kenya. Such species are usually found close to the cave entrance. Caves can act as refuges for forest species after surrounding forest has been cut down. Troglobitic amblypygids (those that only inhabit caves), e.g. *Tricharinus caribensis* from Jamaica, are found deeper in cave systems, in total darkness, and have a reduced number of eyes. Another adaptation to life in caves is elongated antenniform and walking legs. The species *Stygophrynus longispina*, which is found in caves in Malaysia, has a massive leg span of 50 cm (20 in)!

DESERT DWELLERS A few amblypygid species are found in deserts. How are these species able to withstand such extreme conditions? The answer is that they are behaviourally, and also sometimes morphologically, adapted to such an environment. These tough arachnids hide away from the fierce sun in rock fissures and caves leading to groundwater, hence avoiding desiccation. Amblypygids obtain a certain amount of moisture from their food, but also drink water by sucking it in using their powerful pharyngeal pump. *Xerophrynus machadoi*, which is found in the Namib Desert, resourcefully imbibes the dew that may form during the chilly desert nights.

THE NESTING ENIGMA The totally blind amblypygid *Paracharon caecus* lives deep inside termite nests in Guinea Bissau in West Africa. This small beast, which measures a diminutive 7 mm (¼ in), is enigmatic, because nothing is known about its ecology, how it disperses, and so on.

AQUATIC AMBLYPYGIDS The spider *Argyroneta aquatica* is able to live under water – but can amblypygids? Mostly, the resounding answer is no; they are terrestrial animals which become affected within minutes and drown in less than an hour when immersed in water. However, individuals of *Phrynus marginemaculatus* are able to stay under water for up to 8 hours at one time, without being affected. *Phrynus marginemaculatus* doesn't actually live under water, but is found under stones on beaches in the Florida Keys and on the islands of the Antilles. This ability is therefore thought to be an adaptation to surviving periodic flooding in their seaside homes. Like *Argyroneta*, *P. marginemaculatus* survives by breathing through a plastron, which covers the book lung slits on the ventral side of the opisthosoma. This sheet of air clings to fine, branched and rigid cuticular structures.

General biology and behaviour

Feeding and digestion

Arthropods, especially insects, are the main prey for amblypygids, and small vertebrates, such as lizards and frogs, are also a good food source. For example, *Phrynus longipes* in Puerto Rico has been known to eat adult *Anolis stratulus* lizards. Potential food doesn't need to be on the move for amblypygids to be interested though, as they are known to accept dead insects and pieces of meat when in captivity. Surprisingly, these fearsome-looking creatures sometimes show non-carnivore tendencies, and have been known to eat rolled oats!

ABOVE Arthropods such as crickets, and sphingid or noctuid moths are favourite foods for amblypygids (Jatun Sacha, Ecuador).

AMBUSH PREDATORS Amblypygids usually lie in wait for their prey. They stay completely still until a potential prey item passes close to them, at which point they strike rapidly with their raptorial palps. The strike is so fast that amblypygids are able to capture a moth in flight! Smaller animals, such as the Charinidae, actively forage, and they might also scavenge. They will come out after dark and take up position on the bark of a tree, or a favourite rock often coming back to the same place each night.

ORIENTATION TOWARDS PREY Prey-capture behaviour in amblypygids involves three main steps – orientation towards the prey, approaching the prey while extending the palps, and the actual attack. The main sensory structures involved in the capture of moving prey are the trichobothria on the tibiae of the walking legs. So, even if an amblypygid has lost its antenniform legs, it can still capture prey easily. But, if it is unfortunate enough to lose its trichobothria on the walking legs as well, it can't capture prey at all. If an animal loses the trichobothria, but still retains the antenniform legs, then it won't go hungry; it is still able to locate dead or very slow-moving prey, because the antenniform legs have chemoreceptive hairs. These legs are also likely to be involved in estimating the distance towards the prey, when it is close enough to touch.

SPEED OF PREY CAPTURE Amblypygids can move extremely quickly. However, they can also readily adapt the speed of their prey-capture response. For example, a fast-moving moth is caught rapidly, whereas rolled oats (if you can call them 'prey') are handled cautiously and slowly. If the prey is large, both palps rapidly unfold and grab it, then pull it towards the chelicerae where it is held by the palps. If the prey is small, the amblypygid might use only one palp for capture, but shrewdly use the other to grab at other small prey at the same time!

WHO NEEDS A FISHING ROD? In 2003, amblypygids were discovered fishing for prawns in mountain streams! An individual of *Heterophrynus cheiracanthus* from Tobago was seen catching freshwater *Macrobrachium* prawns. It immersed its antenniform legs in the water and held its palps aloft. It then made a very speedy grab into the water with its palps and successfully caught a prawn. This is very interesting, because the walking legs usually detect the prey, while the antenniform legs are used secondarily in chemoreception. This indicates that physical contact with the antenniform legs is another way that amblypygids detect prey. It clearly wasn't a one off, because other individuals were seen to be consuming prawn catches.

A LIQUID LUNCH As with all arachnids, digestion takes place outside of the body. The chelicerae tear a hole in the prey's body, and digestive fluid is quickly disgorged into it from the amblypygid's mouth. The tissues are gradually digested until they become fluid, which is then drawn up through the grooves of the papal endites by the powerful sucking stomach. The grooves are covered in a dense fringe of hairs, which filter the liquid as it passes through, thus removing any particles. The prey's body is slowly macerated by the chelicerae, until all the tissue 'consommé' has been ingested, and the remaining husk is then discarded. It takes an amblypygid one to two hours to feed on a large cricket or moth.

Behaviour, aggression and fighting

The way an amblypygid reacts to a potentially hostile situation depends very much on whether it is in the company of another amblypygid (when they are more likely to act aggressively), or some other organism.

ON THE DEFENSIVE The first line of defence for an amblypygid is usually to run away. Unlike other arachnids that produce venom, like spiders and scorpions, or uropygids that produce an exoskeleton-dissolving liquid, amblypygids have no chemical means to repel, incapacitate or kill predators. They also don't have huge chelicerae like many tarantulas, or heavily armoured palps like many scorpions. Amblypygids therefore rely on looking and sounding tough, rather than actually *being* tough.

A few species also rub their chelicerae together, creating a hissing sound. This is produced by specialized club-shaped hairs on the inside surface of the chelicerae. Amblypygids also autotomize parts of their legs, or even whole legs, in order to escape a potential predator, which may include large lizards, shrews, hedgehogs and mongooses.

ON THE OFFENSIVE Amblypygids are not particularly social animals. Individuals of particularly aggressive species, such as *Heterophrynus alces* from Brazil, are nearly always found on their own. However, individuals of more tolerant or sociable species, such as *Phrynus ceylonicus* from Sri Lanka, have been seen resting almost within touching distance of one another. But the key point is that they have their own 'personal space' or territory, no matter how small, and they *don't touch* one another. So usually there is no aggression among amblypygids, because of this 'avoidance' strategy. Problems arise when amblypygids are kept in captivity and are overcrowded – fights, and possibly even cannibalism, may result, depending on the species. Inevitably, there is always an exception that proves the rule. The species *Heterophrynus longicornis* is exceptionally tolerant, and usually lives in small groups in a defended territory. Even in captivity, males and females can live together for months.

Aggressiveness varies depending on the developmental stage. Nymphs are not aggressive towards each other. Amblypygids become increasingly aggressive just before maturity. Adults are more tolerant, particularly mothers, which show no aggression towards their offspring.

RITUALISTIC FIGHTING Amblypygids are very ritualistic when it comes to fighting, and this is intended to communicate strength while avoiding injury. Fighting is mainly performed by males, but also by females of some species. Certain ritualistic fighting behaviours are species specific, but a basic fighting pattern has been observed in nearly all species that don't have elongated palps.

An amblypygid fight is very much like a fencing match. The two competitors face each other, unfold their palps and tap each other with their antenniform legs. The animals then stand at an angle to each other, while folding in the palp nearest their opponent, and

LEFT Two adult male *Phrynus parvulus* engaged in a territorial fight (Costa Rica).

extending the antenniform leg on that side. The other palp and antenniform leg are extended away from the opponent. The antenniform leg acts as the 'sword' and is used to stiffly tap the opponent for a few minutes. Instead of the bout ending in a handshake, however, it finishes with the two opponents suddenly unfolding their palps and rushing towards each other. They push hard, palps against palps. The loser then leaves the scene of defeat, or allows the winner to step over him while he adopts a submissive crouching position.

VARIATIONS ON A THEME For those species with slender and delicate palps, the fighting pattern described above would be potentially very injurious. Different fighting patterns have therefore developed in such species. The fighting pattern of *Euphrynichus bacillifer* misses out the 'palp-pushing'. Instead, the antagonists move apart until they can barely reach each other. Then one of them shoots a palpal hand at the other and tries to grab its palp, which usually decides the outcome. This is because during this type of fighting, males communicate to each other the length of their palps. The longer the palps within the same species, the larger and stronger the male; so the communication of larger palp size is enough to win the contest. The loser admits defeat in the same way as in the basic fighting pattern. Variations within this species include 'chelicerae-pushing' and vibrating palps held out to the sides. Other fighting patterns are less 'polite' and more like a wrestling match. For example, opponents of *Trichodamon froesi* (from Brazil) try to lift each other off the vertical surface where they live, which tends to result in the loser being physically jettisoned!

Grooming

Like many insects and arachnids, amblypygids groom themselves, and spend quite large amounts of time engaged in the process. Grooming is important to keep their sensilla clean (especially on the antenniform legs and walking leg tibiae), so that they remain effective at receiving stimuli. Amblypygids have special cleaning 'equipment' on the palps and the chelicerae especially for the purpose. A palp has both a cleaning 'organ' or brush, and long fringe setae. The cleaning organ usually consists of two rows of bent and serrated setae surrounded by long setae that are probably chemoreceptors. The chelicerae also have setae along their inner or ventral surface.

An antenniform leg is groomed by being pulled between the chelicerae, which move in alternate circular motions. The setae on the inside surface brush off the dirt, which may be helped by salivary fluid. The part of the leg nearest to the body is groomed by the cleaning organ on the palp. This organ will also be used to groom the length of the walking leg, whereas the chelicerae cleans the tibia and tarsus. The femur and tibia of one palp are groomed by the cleaning organ on the other palp. The cleaning organs themselves are kept clean by being pulled through the chelicerae, and the cleaning organs keep the chelicerae clean.

Moulting

The first rule of a good moult is to find a suitable site. Amblypygids need a vertical site or one that is overhanging, and which is rough enough to get a firm grip. It positions itself head downwards. The moult begins when the old exoskeleton ruptures around the front and margins of the carapace. This flaps open and allows the animal to escape, firstly by its body and palps, followed by the walking legs. The antenniform legs are the last to be freed and the animal uses its palps to pull the legs out of the old exoskeleton.

The animal emerges from its moult soft and white, with a greenish opisthosoma, and the body contents are visible through the semi-transparent exoskeleton. The amblypygid then takes itself off to a hiding place and stays there for up to three days, to allow its exoskeleton to harden. It takes young amblypygids approximately one year to reach maturity, and they moult between four and five, or six and eight times in this time period, depending on the species. Apart from tarantulas, amblypygids are the only arachnids that continue to grow even though they have reached maturity. However, the frequency of moulting is lower once they reach this stage, i.e. from once a year to every three years, and their size increase becomes less each time. Interestingly, in mature females, reproduction and growth are linked. Basically, after reproduction, a female moults. Because the cuticle of the reproductive organs is also shed, the female effectively becomes a virgin again!

Autotomy and regeneration

As with all arachnids, amblypygids are able to shed their legs through autotomy. Although a spider's leg separates between the coxa and trochanter, an amblypygid's leg usually separates between the patella and tibia. The legs most likely to be lost are the antenniform legs. Legs are usually regenerated at the next moult, as long as the leg has not been severed at the femur. The number of articles of the regenerated leg is far greater than that in a normal leg. The regenerated leg is only initially shorter than a normal leg, because each article is shorter. However, after subsequent moults, the articles increase in length, leading to a much longer leg overall.

Life history

Reproduction

Reproduction in amblypygids can essentially be divided into five phases – courtship, formation of the spermatophore, luring of the female to the spermatophore, transfer of the spermatophore, and behaviour after insemination.

TAPPING, TREMBLING AND TOUCHING: AMBLYPYGID COURTSHIP One of the most important things for a male, for successful reproduction, is for the female to be receptive and pick up his spermatophore. He therefore needs to calm her to get her in the mood, and to synchronize her movements with his – this is the function of courtship. However, it has also been suggested that males and females of *Heterophrynus longicornis* indulge in courtship not to reproduce, but in order to prevent aggression between them. Not surprisingly, the order Amblypygi, containing around 120 species, demonstrates much less variation within courtship than the order Araneae, which contains over 40,000 species! However, although amblypygid courtship follows certain common patterns, parts of it, such as the rhythm and pattern of tapping with antenniform legs by the male, are still varied enough to be species specific. Within amblypygid courtship, there are hardly any distinct courtship phases. This is in contrast to scorpion courtship, for example (see p.264). Amblypygid courtship is more like a series of patterns of movement through which the male and female communicate that intensify towards the climax of spermatophore formation.

Courtship in *Phrynichus scaber* was described in detail by Peter Weygoldt in 2000. The first encounter between a potential mating pair is normally initiated by the male. During courtship, the pair stands face to face. Movement starts with the male vibrating his antenniform legs. He partly unfolds his palps, and often jerks his body up and down. The pair then move closer

to each other. With his palpal hands, the male slowly and gently scratches the female's palps. This is frequently interrupted by the male's jerking movement and by the cleaning of his hands. In a gesture that humans might consider romantic, the male often grasps the female's palp and gently 'caresses' her hand with his chelicerae. The male slowly taps the female with his antenniform legs, the female might tap him back. The courtship continues with the male moving from side to side in front of the female, stroking her palps. The male moves his opisthosoma under the female's, where he jerks up and down for a couple of minutes, during which time he trembles his antenniform legs. He then returns to his previous courtship movements. The male alternates between these two types of movement, which build in intensity as the intervals between become shorter and the jerks more numerous. After about 5–7 alternations, the male puts his opisthosoma under the female's for up to eight minutes, jerks up and down, and then at long last produce the spermatophore. Variations are found in the length of courtship; it takes 1–2 hours in *Charinus* and an exhausting 5–8 hours in *P. scaber*.

FORMATION OF THE SPERMATOPHORE Amblypygid spermatophores are highly complex structures, which are species specific and so vary greatly. They may have extremely long and thin stalks, as in *P. scaber, Heterophrynus longicornis* and *Euphrynichus bacillifer*; reduced stalks, as in *Charinus seychellarum*; or no stalks at all, as in *Damon gracilis* (this is likely to be because this species lives in narrow crevices, and there just wouldn't be room for one!). Some spermatophores have large protuberances, as in *Charinus montanus, Phrynichodamon scullyi* and *Euphrynichus bacillifer*. The sperm masses also vary, as they are separate in some spermatophores, such as in *E. bacillifer*, and fused in others, such as *P. scullyi*. Spermatophore structures are highly specific.

The spermatophore can take from five to twenty minutes to produce. The male stands in front of the female, with both of them facing in the same direction. Secretory products flow from the secretory reservoirs along the ventral channels of his spermatophore organ and out of the opening on the ventral side of his body. The male touches the ground with the tip of his spermatophore organ, so that the secretions attach to the ground. He slowly lifts his body and the secretions pull out and then quickly harden into a stalk, depending on the species. He keeps still for several minutes while the secretory masses flow from the reservoirs into the inner genital atrium of his spermatophore organ, and spermatozoa are pressed out of the sperm reservoirs and into the areas where the sperm packages are formed. Imagine that the inner genital atrium is like a complex jelly mould, and that the spermatozoa and bodily secretions are the jelly. The 'jelly' solidifies inside the mould to form the spermatophore head, and is attached to the stalk by secretions. The dorsal arms of the spermatophore are moulded in the blind dorsal pockets. The male then pulls his body up and the spermatophore head is released from his spermatophore organ. The male maintains contact with the female throughout spermatophore formation, by touching her with his antenniform legs, which are extended behind him.

LURING THE FEMALE To lure the female to pick up the spermatophore, the male performs a variety of movements, depending on the species. He may scratch her hands and tap her with his antenniform legs, jerk his body up and down, or vibrate his antenniform legs in 'beckoning' gestures, and do a rocking dance with his palps opening and closing to the rhythm. The male then either quickly pulls the female by her palpal hands over the spermatophore, or the female moves towards the spermatophore herself.

TRANSFER OF THE SPERMATOPHORE The actual process of spermatophore transfer is variable. The female's gonopods may grasp or attach themselves to the protuberances of the sperm masses, and the spermatophore is taken into her genital chamber. Sometimes the protuberances are damaged or completely torn off by the female's gonopods during transfer, such as in *Phrynichodamon scullyi*. In some species, when the female presses down on the proximal end of the spermatophore head, the distal end rises, and the sperm masses move into her gonopore. After mating, the spermatozoa are stored there, and some species have receptacles to house them.

AFTER INSEMINATION The couple separate once the sperm packages have been picked up. The female might vaguely threaten the male in a half-hearted way, but there is no trouble. A male can produce two spermatophores in one night, but he then needs to 'recharge' again before he can produce any more. This might take a few days. Females may indulge in multiple matings.

Egg laying and maternal care

Up to two months after mating, the female lays her eggs. These pass out of the uterus, along with stored secretions from the ovaries, which later harden to form a covering for the eggs. As the eggs and egg sac fluid pass through the genital atrium, the spermatozoa that are stored there are washed out, so the eggs are fertilized in transit. Egg-laying always takes place at night. Ovipositing females are understandably very sensitive, and if disturbed will not continue, so oviposition is rarely observed. The female positions herself head upwards on a vertical

LEFT Species that live in dry habitats, such as *Damon variegatus*, have evolved a 'brood pouch' to further protect their eggs. This is formed from folds in the pleural membrane of the opisthosoma, which cover a large part of the egg sac.

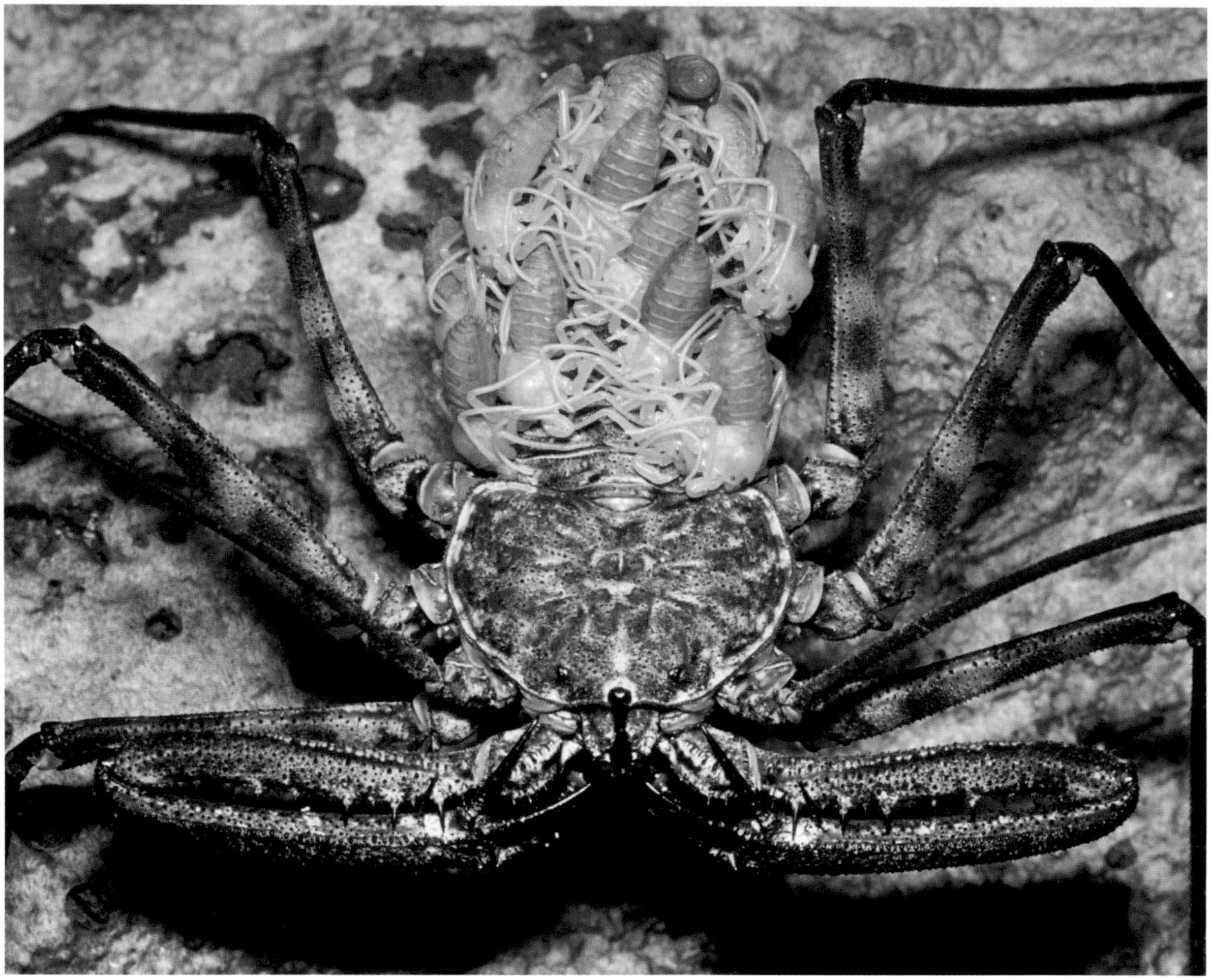

ABOVE A female carrying her young on her back. They are green as their exoskeleton hasn't yet sclerotized.

surface, and exudes her eggs within a filamentous, viscous fluid. The eggs and fluid ooze slowly down and cover the ventral side of her opisthosoma. At this point, she changes into a head-downwards position, and bends her opisthosoma away from the vertical surface, so that is almost horizontal, with the ventral side upper-most. The ventral side of the opisthosoma is held in a concave, dish-shaped position, with the eggs and fluid in the curved hollow.

The external surface of the fluid then hardens slowly to form a sac around the eggs. In less than two hours, the egg sac has hardened sufficiently so that the female can bend her opisthosoma back in line with her prosoma without the eggs slipping off, but it takes up to four days to fully harden. The egg sac now stays firmly glued on the ventral side of the opisthosoma, no matter what position the female adopts, so the female is able to run around without hindrance. She is also capable of catching food like this. However, she does not usually eat, because she has to maintain a concave opisthosoma. If she ate a large meal, her opisthosoma would swell and her precious cargo would drop off! The egg sacs of amblypygids are often quite large and may project out from either side beneath the opisthosoma. They can appear quite membranous and unprotected.

A female usually produces one clutch of eggs per year. Larger females produce more eggs than smaller ones. So, as females continue to grow after maturity, a long life means

good reproductive success. In addition, old females cumulatively produce more egg clutches. Smaller species tend to lay around 6–10 eggs, but larger species can lay up to 50. The eggs can range in size from 1 mm to 3 mm (1⁄32–3⁄32 in) in diameter, depending on the species. The eggs take around 13–15 weeks to develop, during which time the female has to go hungry. Finally, the female's long wait is rewarded by the emergence of many white and green praenymphae, which clamber onto both dorsal and ventral surfaces of her lean opisthosoma, and attach themselves by their pulvilli. They need to hang on tight though, because if they fall off this means almost certain death – their appendages have yet to develop, so they are unable to walk. This developmental stage takes around ten days, during which the long-suffering female has to stay completely still, or she and her micro-sized family may accidentally part company.

Post-embryonic development

BECOMING FREE-LIVING After the first 6–12 days, the tiny praenymphae moult into the first free-living stage – the protonymphae. To aid her miniature offspring in this process, the devoted mother attaches herself head uppermost on a vertical surface and remains motionless. One at a time, the praenymphae move to the tip of her opisthosoma, where they moult. As soon as their moult is complete, they climb off their mother and start a new life on their own. However, some *Charinus* protonymphae remain attached to the mother for a bit longer. At this stage, a protonymph looks like a typical amblypygid, but is obviously much smaller than a mature animal, ranging in size between 2 mm and 7 mm (1⁄16 and ¼ in), depending on the species. The palps are short compared to the body length, and in general they only have a small number of setae and spines. In the Phrynichidae however, the spines are already well developed at this stage. The genitalia have yet to develop, so the different sexes are indistinguishable. Interestingly, protonymphs develop a band of white on both antenniform legs, which remains through subsequent instars until maturity. As yet, the function of this banding is unknown.

INSTARS THREE TO EIGHT In the second free-living instar, the developing genitalia can be distinguished between the sexes, although they are rudimentary. In tritonymphae (third instar) amblypygids, the gonopods of the female have developed their erect structures and the male genitalia have become larger. In *Euphrynichus*, the palps have already lost all spines accept those that form the 'hand'. At the fourth instar, the genitalia are, externally at least, fully developed. However, males and females won't mate if put together. The fifth instar is the earliest developmental stage where reproduction might take place. Depending on the species, there may be between five and eight instars in total before maturity.

Longevity

It normally takes two years for amblypygids to reach maturity, at which stage they still continue to grow and moult. In adults of a large number of species, the palps develop into elongate structures that can measure up to four times the length of the carapace, due to positive allometric growth. Large species, such as *Heterophrynus longicornis*, have lived up to ten years in captivity. Smaller species have shorter lifespans, but these still run into several years.

4 Uropygi
Whip scorpions, vinegaroons

Uropygids are robust, formidable-looking nocturnal predators. Like amblypygids, they have a pair of elongated antenniform legs. They possess a pair of large, heavily armoured raptorial palps a little like those of scorpions, hence the common name 'whip scorpion'. Instead of having a 'tail' ending in a venomous sting, they have a very distinctive whip-like 'tail' or flagellum. This gave them their scientific name, as *oura* is Greek for tail. They spray a noxious-smelling chemical mixture from anal glands, as a form of defence. This chemical mixture contains acetic acid, the active component of vinegar, hence the quirky common name of 'vinegaroon'! The body length (not including the flagellum) of uropygids ranges in size from 2.5 cm (1 in) to just over 8 cm (3¼ in) in the aptly named giant whip scorpion, *Mastigoproctus giganteus*. This big beast is found in Florida, and the Chihuahuan and Sonoran deserts of southwestern USA and Mexico. *Mastigoproctus maximus* is also in this size range.

Classification and diversity

The systematic position of the Uropygi and Schizomida (short-tailed whip scorpions) is by no means firmly established. It seems that the sister-group relationship (those taxa descended from a common ancestor) between the two has been agreed, but their order status is still disputed. Some treat them as two separate orders (e.g. Harvey) while others place the Schizomida and Uropygi together as suborders within a single order (e.g. Weygoldt; Scharff and Enghoff). Yet others class them as two separate orders within the superorder Uropygi, and use the term Holopeltida for the vinegaroons (e.g. van der Hammen). Several arachnologists refer to the Uropygi as Thelyphonida, and reserve the name Uropygi for the taxon that includes the Thelyphonida and Schizomida. Some consider the Uropygi to be in a monophyletic clade (a group that includes all the descendants of the most recent common ancestor) called the Pedipalpi, which includes the Uropygi, Schizomida and the Amblypygi. Whatever the current thinking about their relationship, in this book the Uropygi (Thelyphonida/Holopeltida) and the Schizomida are considered in two separate, but consecutive, chapters. There are currently 108 valid uropygid species in 18 genera, all placed in one family, the Thelyphonidae, and four subfamilies: Hypoctoninae, Mastigoproctinae, Typopeltinae and Thelyphoninae. Nearly half of all the species are placed in just two genera: *Thelyphonus* and *Hypoctonus*.

As with amblypygids, uropygids are morphologically far less varied than spiders and are quite homogeneous as an order. They are rather primitive creatures that are sometimes

OPPOSITE The distinctive segmented opisthoma can be seen clearly on this female *Typopeltis crucifer*.

thought of as 'living fossils' because their body plan really hasn't changed all that much over 300 million years. Until very recently, fossil uropygids found in Carboniferous coal swamps were placed in the same family as living species. However, they have now been excluded from the modern family based on slight differences in the palps. Although uropygids aren't a colourful order, they may vary between reds and browns depending on their maturity. Their appendages are often lighter in colour than their body.

Anatomy

External anatomy

The exoskeleton can be very granular all over, granular in specific parts such as the palps, or relatively smooth. It also has spines, setae, trichobothria and many slit organs, e.g. there are around 3,600 slit organs covering more than 50% of the body of *Thelyphonus sepiaris*! These are located on all of the segments (especially on the ventral surface) and on the pleural membranes. They are also found in large numbers at all joints on the palps especially on the tarsi.

BELOW Dorsal view (left) and ventral view (right) of *Mastigoproctus giganteus*.

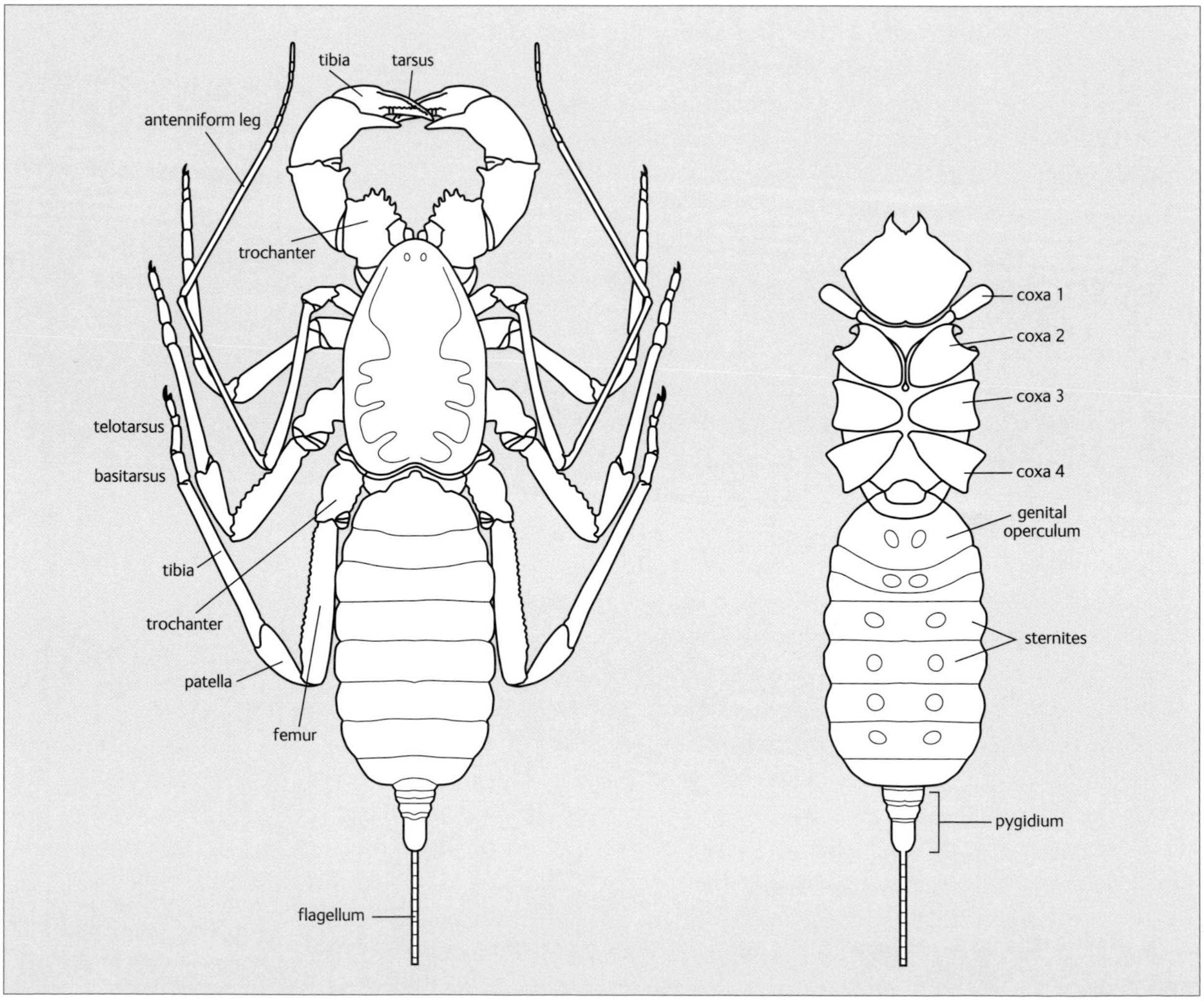

ABOVE The almost rectangular prosoma of a uropygid (Danum Valley, Sabah)

PROSOMA Dorsally, the prosoma is covered by a one-piece carapace, which is almost rectangular in shape. On the carapace, a keel may run from each group of lateral eyes to near the median eyes. In some species, the keel may be absent. Ventrally, there are six segments and each carries a pair of appendages – the chelicerae on the first, palps on the second, and a pair of legs on the other four. There is a small triangular sternum as the coxae of the legs take up most of the space.

EYES There are eight or 12 simple eyes arranged in three groups on the dorsal surface of the whip scorpion. There are two groups of three or five eyes situated either side of the prosoma near the first pair of legs, and the third group consists of two eyes in the centre of the carapace near the anterior edge. In some species of *Mastigoproctus* these median eyes may be located on a chitinous ridge. Where this ridge is well developed, the eyes are directed to the front and sides. The median eyes are able to produce a basic image, although the lateral eyes are only used in light detection.

CHELICERAE Between the palps are the chelicerae, each consists of two parts: a large base attached to the first ventral segment of the prosoma, and a movable finger or fang. The fang closes against two 'teeth' on the base to form a grasping mechanism. The fang moves in the vertical plane, as with the fangs of mygalomorph spiders and amblypygids.

RIGHT The palps of both sexes tend to be held curled in toward the mouth when the animal is walking around, but as soon as a potential prey item is sensed, the palps are extended outwards. The palps are also used in fighting, courtship, and even to dig burrows.

PALPS Like amblypygid palps, uropygid palps are armed with spines because they are also used to catch and hold prey. However, uropygid palps are shorter and more robust than those of amblypygids. A closer look at a palp reveals what a wonderful food-catching implement it is. The tarsus and tibia together form a movable pincer, which can grab prey (see p.112 left). The patellar finger opposes the tibia, and so forms a second pincer. Once the food is caught by the prehensile parts of the palp, it is crushed and masticated by the strong teeth on the trochanters.

ANTENNIFORM LEGS Two of the most important structures in the uropygid's 'sensory armoury' are the elongate antenniform legs, used in a similar way as in amblypygids. These legs have a covering of sensitive hairs, especially on the tarsi. In addition, there is a pair of trichobothria, arranged at a 90° angle to each other on each tibia. As the animal moves around, the antenniform legs gently touch and feel the surroundings, like a blindfolded person extending his or her arms when walking. These legs are also moved towards a stimulus in the same way as the flagellum. Secretory cells have been found at the tips of the antenniform legs in females. It is thought that these may produce pheromones, used during courtship. Although this hasn't yet been confirmed, it is quite probable, as males of certain genera, e.g. *Typopeltis*, chew the female's antenniform legs during courtship.

WALKING LEGS The walking legs are each divided into the coxa, trochanter, femur, patella, tibia, basitarsus and the telotarsus, which is divided into three segments. At the tip is a claw, called the apotele. When walking, three legs are moved at a time, with the first and third walking legs on one side and the second walking leg of the opposite side moving in synchrony. Although uropygids might seem rather slow moving and tentative, they are actually cursorial hunters (hunters adapted for running), so can really scoot along when necessary. The walking legs also play a part in sensing the environment, because they have a trichobothrium on the dorsal surface of each tibia, located near the joint between the tibia and tarsus. These

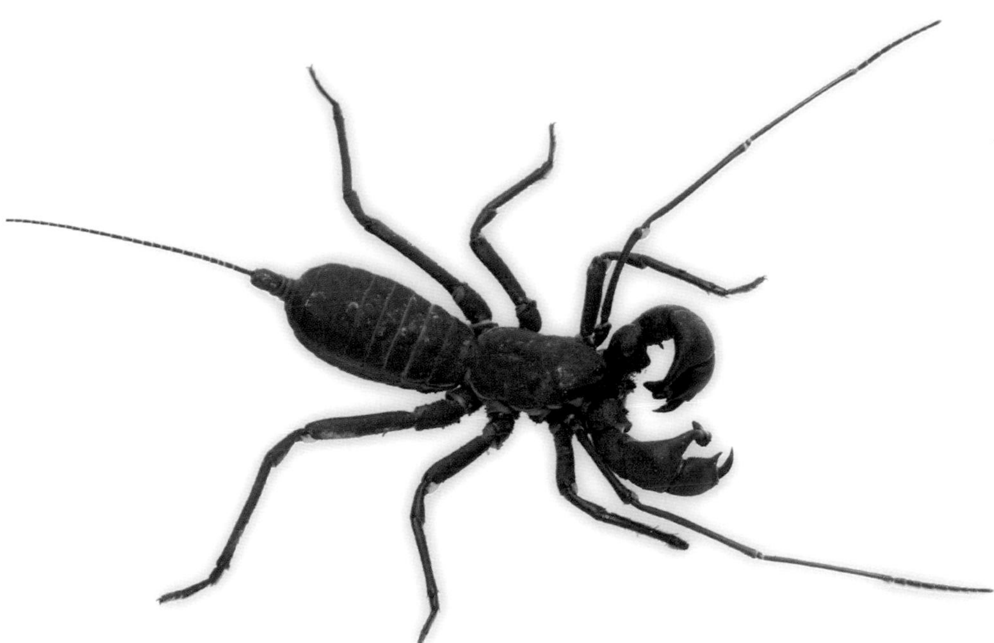

LEFT With its flagellum and antenniform legs, it is easy to see the sensory nature of a uropygid.

trichobothria on the walking legs, plus those on the antenniform legs, enable the animal to have a 360° angle of sensitivity. Additionally, the legs have a lyriform organ on each metatarsus.

OPISTHOSOMA The opisthosoma is distinctly segmented, with tergites and sternites. All the tergites of the pre-opisthosoma have corresponding sternites. In adults, the tergites and sternites almost meet laterally, whereas in immatures the plates are somewhat shorter, so that the pleural membrane between them shows, giving the appearance of a rather bloated opisthosoma. The opisthosoma is divided into a large pre-opisthosoma consisting of nine segments and the much smaller post-opisthosoma (pygidium) of three segments (see p.112 right).

EXTERNAL GENITALIA Ventrally, both adult males and females have a genital operculum covering the genital chamber, which is poorly sclerotized. Sternite 2 is known as the genital sternite because the genitalia are located centrally to the rear of this plate (see p.112 right). In adult specimens of *Mastigoproctus* and *Thelyphonellus*, the operculum has developed into a more obvious structure than in immatures, because it extends outwards and becomes enlarged.

RESPIRATORY STRUCTURES Uropygids breathe via two pairs of book lungs. The openings of the first pair of book lungs are located to the rear of sternite 2 of the opisthosoma, while the openings to the second pair are at the rear of sternite 3.

PYGIDIUM The plates on the three segments of the pygidium are fused, and so form a cylinder (see p.112 right). The last segment (segment 12) is known as the anal segment because it bears the small slit-like anus. At either side of the anus are the two inconspicuous pygidial glands, which produce defensive chemicals (see *Defensive behaviour*, p.120). In addition, the sides of the anal segment often have one or two pairs of pale dorsal spots, which are often superficially lens-like and covered by thin cuticle. These are known as ommatidia, and supposedly have a photoreceptive function, though this has not yet been proven.

SEXUAL DIMORPHISM In females, the palps are short and stout, but in males they are longer. This is because positive allometric growth of the male femora and patella occurs in the subadult stage. In Mastigoproctus, Typopeltis and Thelyphonellus, the palps have 'accessory tips' on the male's movable tarsal 'fingers', which are used to manipulate the sperm carriers (see *Reproduction*, p.123). Such differences are not present in nymphs, so the sexes are very difficult to tell apart when immature. It is easy to see how adult males have evolved large palps compared to those of females. Larger palps are an advantage when fighting other males and males with stronger palps may be able to displace smaller males that have started to mate. There is therefore a selection pressure for stronger and longer palps.

Flagellum

The flagellum is a long, multi-segmented, whip-like structure protruding from the anal segment on the pygidium (see p.112 left), and considered to be homologous (originating from the same evolutionary origins) as the sting, or telson, of scorpions. The flagellum looks a bit like a radio aerial and, indeed, it does act like a 'biological receiver', picking up chemical stimuli and the movement of air currents, via its sensory hairs.

It is thought it may be light sensitive too. Unfortunately, flagella are frequently lost in dead, preserved specimens, and it is only by observing live specimens that the importance of this structure can be truly understood.

Uropygids are very sensitive to both airborne and substrate-borne disturbances. A small puff of air in the direction of a uropygid is enough to evoke an immediate reaction. The animal rapidly extends its antenniform legs, and raises its body off the ground. It then orientates its flagellum and antenniform legs towards the source and, if the stimulus continues, it turns around to face the source with its palps stretched out wide and its flagellum pointing towards it.

BELOW The flagellum is very mobile, and so can be directed towards a stimulus from any direction. It can also be bent over the uropygid's back, so that it is directed towards the animal's head region.

Internal anatomy

Uropygid internal anatomy has not been studied in nearly as much depth as that of other arachnids. There isn't much known about the central nervous system. However, there is a supraoesophageal 'syncerebrum' (basic brain), a suboesophageal ganglion and an abdominal mass. There are also giant fibres in the peripheral nerves of the ventral nerve cord, which enable rapid responses to certain stimuli. A uropygid's nervous system is certainly sophisticated enough for it to have the ability to learn. For example, specimens of *Mastigoproctus giganteus* were placed in complex mazes and demonstrated spatial awareness, as well as learning quickly how to avoid a mild electric shock when placed in a specially adapted box. They also demonstrated habituation to a harmless regular stimulus, so that they stopped responding in an aggressive way. All these results show that uropygids are well adapted to their environment. They are able to learn more efficient foraging patterns as they associate an increase in quantity of potential prey with a specific location. This means that less energy is used in searching for food, and less time is spent exposed to potential predators.

The circulatory system is well developed. The segmented heart has nine pairs of ostia, and is located in the opisthosoma, but it can also extend into the prosoma. The two pairs of book lungs operate in the same way as spider book lungs. The two pairs of coxal glands (opening through pores on the coxae of the first and third pairs of legs), nephrocytes and Malpighian tubules are responsible for excretion. The coxal gland extends from segment 4 to segment 6 (from the anterior of the body).

INTERNAL GENITALIA Behind the genital operculum of an adult female, there is a large chamber. Attached just inside are two specialized finger-like projections, which are basically just strongly sclerotized plates. These are used to grasp the male's sperm carriers during mating (see *Reproduction*, p.123). The genitalia comprise paired ovaries and oviducts, a uterus and a seminal receptacle. Behind the genital operculum of a male is a smaller chamber with two strongly sclerotized bars running along its sides, which turn up towards the centre of the opening. These bars can help to sex the animal in the nymphal stages, before the palps have enlarged. The genitalia comprise a pair of testes and paired vasa deferentia, which secrete sperm cells and secretory fluids. They also have spermatophore organs, which produce the spermatophore.

BELOW In tropical lowland forests, uropygids can be found at night on the trunks of trees. They are usually only active above ground when it is dark (Danum Valley, Sabah).

Distribution and habitats

All four uropygid subfamilies are found in Asia. The greatest diversity is in Southeast Asia. In addition, the Hypoctoninae are also found in West Africa, the Mastigoproctinae in the Americas, and the Thelyphoninae in the southwest Pacific. All 21 species of *Hypoctonus* are found in Southeast Asia, especially in Myanmar. The 18 species of *Mastigoproctus* range across the southern USA down to Brazil. Considering the size and sheer attractiveness of uropygids, there are surprisingly few studies of their ecology. Most uropygids inhabit rainforest ecosystems, but some, such as *Mastigoproctus giganteus*, are found in deserts and other arid areas. During the day, they are found under logs, rocks and other such objects.

General biology and behaviour

Feeding and digestion

Apart from a wide variety of arthropod prey, uropygids have also been reported to eat small vertebrates such as frogs and toads. Tropical uropygids also feed on slugs and worms. They are known to cannibalize each other when starved, although they can survive for up to four months without food. Uropygids are therefore opportunistic predators that can consume nearly any type of prey, if they can capture and subdue it! The phrase 'pick on someone your own size' doesn't apply to the choice of prey; uropygids attack large and active prey, even if it means marshalling their forces and attacking several times. If, however, the prey is too troublesome, the uropygid eventually gives up. Stronger uropygids are known to snatch the prey of weaker individuals, although sharing of prey is also possible, with up to six animals feasting on the same prey item at the same time. Uropygids go off their food when they are preparing for moulting and reproducing females don't feed either. However, once freshly moulted animals emerge at the onset of rain, they make up for lost time by feeding voraciously.

Immature uropygids appear to avoid prey with hard exoskeletons, they prefer softer-bodied organisms such as termites, fruit flies, bush cricket nymphs and collembolans. Apparently, uropygids are not always exclusively carnivorous; it has been reported that hungry animals accept plant-based foods such as cooked rice and ripe banana! Some species, for example *Thelyphonus schimkewitschi*, are known to consume dead insects, demonstrating that some uropygids are also scavengers.

PREY CAPTURE Uropygids rely on the combined input from various sensilla on their legs and bodies to locate their prey very accurately. The prey animal produces both airborne and substrate-borne vibrations. Airborne vibrations are detected by the trichobothria on the tactile leg tibiae. Prey catching is a complex behavioural phenomenon, and some movements are thought to be characteristic to individual species. Firstly, the antenniform legs and the flagellum are directed towards the prey. The uropygid then turns towards the prey, extends and opens its palps, and approaches very slowly and cautiously. When the uropygid is close, it touches the prey gently, several times, with the tarsi of the antenniform legs, as if to get a measure of its size and shape. As the tarsi have chemoreceptive hairs, the uropygid is probably 'tasting' the potential food to see whether it is edible. It then extends the palps widely, and with a quick forward thrust of the body, seizes the prey. As with scorpions, uropygids tend to hold their prey by the head, bring it close to the mouth and then dig the chelicerae in. It crushes the prey and masticates it between the teeth on the trochanters. Some animals, e.g. *Typopeltis crucifer*, have been seen to 'tickle' the prey so that it moves towards the palps of its own accord.

DIGESTION The digestive system comprises the mouth, pharynx, oesophagus, midgut, caeca (up to four pairs in the prosoma and eight pairs in the opisthosoma), hindgut and anus. The excretory system works in the same way as in other arachnids. The labrum, labium and coxae of the palps (which are fused ventrally) produce a 'pre-oral' cavity (a cavity in front of the mouth). As with amblypygids, uropygids tear out the flesh of their prey with their chelicerae. The flesh is delivered to the trough in the 'floor' of the pre-oral cavity, where it is digested by disgorged digestive fluids. The trough then funnels the liquefied food to the

OPPOSITE Uropygids are very efficient and powerful predators, feeding on a wide variety of arthropod prey, such as grasshoppers, beetles, bugs, cockroaches, spiders and caterpillars. *Mastigopractus giganteus* eating a grasshopper.

mouth. Dense setae on the palpal coxae and chelicerae filter the food, and only allow very small particles to reach the gut. The remains of a uropygid's meal therefore consists only of hard parts such as exoskeleton and wings, which cannot be liquefied.

Defensive behaviour and aggression

Uropygids are often known as 'vinegaroons', as well as 'stinking scorpions' and 'vinagrillos' (Spanish). These wonderful names have come about because uropygids produce noxious-smelling chemical secretions that are sprayed in a mist from the paired pygidial glands as an extraordinarily effective defence. The pygidial glands are located either side of the anal slit. Interestingly, they are not quite symmetrical, the right gland is usually further under the body than the one on the left. They are thin-walled sacs, which can become quite large because they have concertinaed striate tissue. They have an outer layer of muscles that contract and discharge the spray. The glands taper towards the rear into two narrow ducts, opening through thin slits that can be opened by a pair of muscles.

The components of the defensive secretions vary greatly between species and even between individuals of the same species. It has been suggested that individuals of the same species produce a different concentration of chemicals under different feeding conditions. In nearly all species studied, the dominant component is acetic acid, from which the order derives its common name of 'vinegaroon'. The defensive secretion of *Mastigoproctus giganteus* consists of 83% acetic acid, 15% octanoic acid, plus some other acids in minute quantities. This highly pungent cocktail is diluted with water. Other individuals studied yielded secretions of 84% acetic acid, 5% caprylic acid and 11% water. The mix from *Typopeltis crucifer* is very similar to that of *M. giganteus*, even though it is from a different genus. In *Typopeltis stimpsoni*, the defensive cocktail is made up from 78.3% acetic acid, 5.8% caprylic acid and 15% water.

To aim the spray, a uropygid directs its pygidium towards the unwanted stimulus, and then very accurately blasts out the 'liquid deterrent' in fine droplets. The spray is spread over a broad area, covering the assailant. If the stimulus is towards the front of the uropygid, then it lifts its opisthosoma up at a sharp angle, so the pygidium can still be aimed accurately. The furthest that a uropygid's spray has reached is 80 cm (31½ in), although force rather than distance is important, because predators are usually close to the animal when the spray is released. Adult individuals have been known to spray up to 19 times before the defensive chemical runs out; it takes about 24 hours to recharge before they can spray again.

EFFECTS ON PREDATORS Invertebrate predators of uropygids include ants and solifugids. It is thought that the caprylic acid in the defensive chemical mix acts as a 'wetting agent', helping to spread the spray over a greater surface area of the exoskeleton of the predator. It is also possible that it disrupts the lipids in the exoskeleton, thereby increasing its permeability, and possibly causing the predator future problems because of dehydration. It also enables a much quicker penetration of the acetic acid. Its role is therefore to maximize the effectiveness of the spray on impact. Residual spray that gets onto the uropygid's body, also acts as a deterrent. It is very interesting to note that a uropygid never seems to be affected by its own spray, although it is not certain whether it affects other uropygids. In studies, as a result of being sprayed by uropygid defensive chemicals, invertebrate predators showed aberrant behaviour, such as dragging their bodies and mouthparts on the substrate and persistently cleaning for some time after the attack. Their antennae, cerci (as in cockroaches) and tarsi are particularly sensitive to acetic acid.

Vertebrate predators of uropygids include birds, amphibians, reptiles such as *Anolis* lizards, and mammals such as mice. The defensive chemical mix affects the eyes and respiratory tracts of these potential predators. Although caprylic acid doesn't have a role as a wetting agent in vertebrates (as there is no exoskeleton to permeate), it still helps the residue of the spray on the body of the uropygid to persist, and may ward off further attacks. Allegedly, these secretions are able to actually blister the skin of humans, although this has not been confirmed. The vapour of the spray is also repellent.

AGGRESSION Captive uropygids show aggression as a result of overcrowding and starvation. An aggressive uropygid grabs another by its palps and has a kind of 'arm wrestle', while flicking its flagellum rapidly. Equally matched beasts quickly take the measure of each other and withdraw after a short fight, whereas a stronger uropygid eventually grabs the weaker individual's body in its palps. However, it has been observed that adults and immatures may live under the same shelter without aggression if there is plenty of food available.

Adaptations to water loss

Surprisingly for such heavily armoured invertebrates, uropygids appear to be rather susceptible to desiccation. Because of this, some researchers believe that uropygids don't have a waterproofing wax layer in their exoskeleton, although the literature is divided on this. Many species of arachnids (and insects) do have a waxy layer that limits permeability at low temperatures, but permeability changes with increased temperature. Gregory Ahearn in 1970 concluded that uropygids do have some kind of a waterproofing layer in their cuticle, because his work on water balance in *Mastigoproctus giganteus* showed that the cuticle became much more permeable above 37.5°C (99.5°F).

Although it is not certain whether uropygids have a waterproofing layer, they clearly exhibit other ways of avoiding desiccation. They are nocturnal, thereby avoiding the heat of the day, and are seldom found in the open except after rain. They live in deep burrows, which helps to avoid desiccation, and they appear to have a preference for humidity. They also obtain moisture from their prey, and drink frequently if open water is available; a dehydrated uropygid in captivity can drink continuously for up to eight minutes. It scoops the water into its mouth using its palps and alternate cheliceral movements. When open water is not available, dehydrated uropygids undertake a form of water 'divining' – they are able to detect moist substrate with their antenniform legs, indicating that these legs possess hygroreceptors (water detectors).

Uropygids are able to remove moisture from the substrate by placing the mouth directly onto the substrate, and sucking up the water, or by scooping some of the substrate towards the mouth using the palps. Such activity has been called 'substrate drinking'. Dehydrated creatures may regain up to 10% of their original body weight in this way.

ABOVE Uropygids are efficient burrowers due in no small part to their well-developed palps, which they use to scoop up substrate. They deposit the substrate a short distance away from the burrow, before returning to the exact spot to dig some more.

LIFE UNDERGROUND Uropygids are rarely seen out in the open except in wet conditions and inhabit deep burrows, often constructed under rocks or other surface structures, to avoid dehydration. Burrows are also used for reproduction and brooding, as moulting chambers, and for prey capture. An animal resting in its burrow often extends its antenniform legs outside the entrance, and when potential prey passes close by, the uropygid extends its palps rapidly and seizes the hapless creature, which is pulled into the burrow and consumed.

Tunnels of 2 cm (¾ in) diameter and 15 cm (6 in) long have been observed in the laboratory, and young uropygids take around 6–8 hours to construct burrows as long as this. Burrows often

end in a small chamber, which houses a female's brooding nest. Sometimes, the chambers of two burrows may connect. Uropygids are not above saving themselves some hard work, because they are known to take over abandoned rodent burrows. Adults are also known to assert their 'authority' over youngsters. Older uropygids often take over occupancy of younger uropygids' freshly made burrows, the youngsters then have to go off and start all over again!

Studies on *M. giganteus* have shown that their abundance at any given locality is strongly influenced by soil hardness. Soils with very little cohesion cave in easily when a hole is dug. *Mastigoproctus giganteus* was shown to have a preference for sand loam soils, which provide the best burrowing conditions for that species. Other species have a preference for different soil hardnesses. Uropygids also make use of natural hideaways, as they often seek shelter within crevices and under rocks, vegetation and other surface debris.

Grooming

As with all arachnids, uropygids groom themselves and for this they use mainly their palps and chelicerae. It is important to keep sensory hairs and slit sensilla clean, so that they remain effective at receiving stimuli. Grooming is usually undertaken in a sequence, starting at the front end and working towards the back end. The palps are cleaned first, by being brushed against each other and the spines are rubbed against the chelicerae. The tarsi of the antenniform legs are frequently groomed as they are particularly important sensory structures for the uropygid and so need regular maintenance. Each tarsus is inserted between the chelicerae, which have a thick brush of hairs, and is swept from base to tip.

Grooming of the walking legs is less common, and they tend to be groomed at random, so that not all legs are groomed in one session. They are either cleaned in the same way as the antenniform legs, or they are grasped and pulled between the spines of the palp. A uropygid only very occasionally grooms its flagellum. The uropygid tilts up its opisthosoma so that the flagellum lies almost flat along it, and is directed by the palps towards the chelicerae. If the flagellum can't quite reach the chelicerae, then it is brushed against the hairs on the spines of the palps. The body is groomed by the fourth pair of walking legs.

Moulting

Moulting is likely to occur once a year, in early summer. It can however, be postponed for a while if the uropygid cannot construct a moulting chamber. It can take several weeks or even months to prepare for moulting in the chamber, during which time the uropygid does not feed, and the opisthosoma increases in size, inflating like a balloon.

Moulting consists of two stages and is hormonally controlled (*see also* chapter 2, Spiders). During the first stage, apolysis, the old cuticle separates from the new cuticle, which is developing beneath. This separation is in preparation for the second stage, ecdysis, which consists of the actual shedding of the old cuticle. Interestingly, the uropygid is still a sensory animal even during this most vulnerable time.

Ecdysis starts with the exoskeleton rupturing at the front of the prosoma, and along the sides just beneath the carapace. Once the animal has moulted, it is ghostly white in colour because of the lack of pigment. However, after about three days, it starts to darken. Complete pigmentation and sclerotization takes around 3–4 weeks in older creatures, but less time in youngsters. Unlike tarantulas, adult uropygids don't moult when they reach maturity, which normally takes four years.

Life history

Reproduction

The general mating behaviour of uropygids is quite similar in various genera. There are a few minor differences, which are discussed where appropriate. Peter Weygoldt described reproduction in *M. giganteus* in depth in 1970. He roughly divided the pattern of reproduction into four phases – two phases of courtship, which involve the initial approach of the male and the second (main) approach of the male, the production of a spermatophore, and the insemination of the female.

COURTSHIP In the initial approach, the male rushes forward to the female, and makes contact using his palps. He then grasps her antenniform legs with his palpal claws. She retreats backwards, with the male following. They then move in the opposite direction, with the male walking backwards while holding the tips of her antenniform legs with his chelicerae. This is followed by some stroking of the female's antenniform legs and palps, before the female retreats. After this comes the second phase. As the male approaches for the second time, he steps onto her prosoma, while his antenniform legs tremble, rapidly touching the ground. Although the female steps backwards, the male also steps backwards, pulling her with him, and turn to face the same direction. Once the female touches the male's opisthosoma with her palps, he begins to search for a suitable place to deposit his spermatophore.

BELOW The male rushes to the female and uses his palps to make contact in the initial approach of courtship.

RIGHT During the second phase of courtship, the male, whilst holding the female's antenniform legs in his chelicerae, turns until they are both facing in the same direction.

PRODUCTION OF THE SPERMATOPHORE Once the male has found a suitable patch of ground, he produces the spermatophore. He rubs his gonopore against the ground, and the base of the spermatophore is secreted. During much leg trembling, the male then secrets the sperm carriers onto the stalk.

In uropygids, there are two different types of spermatophore structure and these are transferred in two different ways. In *Thelyphonus linganus*, which exemplifies the first type of transfer, the spermatophore has a small base with a short stalk. This carries a pair of curved sperm carriers, each with a slim pointed tip near the base, a hooked tip at the other end and a thin joint-like section in the centre. There is also a small cushion-like structure attached to the end of the stalk, between the two sperm packages. A reservoir filled with corkscrew-like spermatozoa is in the pointed tip. In this design of spermatophore, there is no ejaculatory duct (unlike in the spermatophore of *Mastigoproctus*). The spermatozoa do not have to be forced out of the reservoir, because the sperm packages break, allowing a more 'passive' exit. During transfer the female pulls the sperm packages from the spermatophore, and because there is no ejaculatory opening in the sperm package, the long and narrow tips containing the reservoir have to break in order for spermatozoa to be transferred to the female's seminal receptacles. Then the female's seminal receptacles become filled not just with spermatozoa, but with parts of the sperm packages, such as the long tips.

In *Mastigoproctus giganteus*, which demonstrates the second type of spermatophore structure and transfer (as do *Typopeltis* and *Thelyphonellus*), the spermatophore is composed of a small base carrying two hard sperm packages, about 7 mm (¼ in) long, with blunt posterior ends and pointed anterior ends. The sperm packages in the second type of structure are larger than in the first type. Each sperm package consists of a reservoir filled with spermatozoa and an ejaculatory duct, surrounded by protective packaging and a conductor. The ejaculatory duct leads from the reservoir to an opening in the narrow anterior tip. It is thought that the sperm package undergoes compression to force the spermatozoa out of the reservoir. During transfer the female pulls the sperm packages from the spermatophore, and the male pushes the sperm packages into the female's seminal receptacles. The male continues to press on the sperm packages and the spermatozoa are ejaculated into the seminal receptacle – as well as transferring the sperm properly, this pumping by the male probably removes sperm from previous matings. Only spermatozoa are transferred into the seminal receptacles, no parts of the sperm packages are transferred.

This second system is considered to be more effective, because the spermatozoa are transferred properly, without parts of the sperm packages being left behind in the genital orifice. With both types of spermatophore, the protection that the sperm package gives to the sperm reservoir is considered to be an adaptation to life in dry environments, preventing the sperm from drying out.

INSEMINATION

The last phase involves the insemination of the female. As described above, *M. giganteus* has the second type of spermatophore structure and transfer. The male pulls the female slowly towards the spermatophore until she is standing above it. The female lowers her opisthosoma and presses down on the blunt posterior ends of the sperm carriers. The base of the spermatophore acts as a pivot, so the pointed anterior ends see-saw upwards. At the same time, the blunt ends move apart and the pointed ends move together. As the female opens her gonopore, specialized finger-like projections inside grasp the tips of the pointed sperm carriers, thereby preventing them from falling out of her gonopore. The couple then release their grip on each other, swapping position, so that the male can grasp the female's opisthosoma from above with his palps.

With the movable finger of his palpal claw, he then pushes the sperm carriers (with ejaculatory ducts innermost) firmly into the female's gonopore, until only the blunt tips protrude. The male then manipulates the sperm carriers for over two hours by pulling the ends apart and then pushing them together again, over and over. This action must help to transfer the spermatozoa from the reservoir into the female's gonopore. The female's insemination is the pinnacle of the 'dance' – the male has finally achieved his goal, and releases the female. By flexing her opisthosoma, she pushes the sperm carriers out of her gonopore and they are discarded. The couple go their separate ways, although they may mate again a few days later.

The males of some species do not assist the female in taking up the sperm at all. For example, males of *Thelyphonus* don't push the sperm packages into the female's gonopore, because they have the first type of spermatophore structure and transfer. The only role that the male has in sperm transfer is to ensure that the female pulls the sperm packages off the spermatophore. He does this by leading her over the spermatophore, and then pulling her sharply forward. The female stands over the spermatophore, lowers her body and presses

down on the hooked ends of the sperm packages. The pressure on the cushion between the sperm packages causes the tips of the sperm packages to move into the female's gonopore. She holds onto them by closing her gonopore, and then moves forward so that the curved sperm packages are pulled straight, bending at the joint in the centre. The sperm packages either break at this point, or retain their curved shape and the ends poke out of the genital opening. The female has to rub her ventral surface against the ground to rid herself of these extraneous structures. Unlike spiders and scorpions, uropygids don't seem to indulge very often in sexual cannibalism, although this phenomenon hasn't been studied thoroughly.

COURTSHIP VARIATION WITHIN A SPECIES In 2001, Fred Punzo and Carla Reeves studied the courtship behaviour of *M. giganteus* from a dry habitat in southwestern Texas, and from a humid, subtropical habitat in southern Florida. They discovered that there is geographical variation in male courtship – the Texan males were far speedier than the slower Florida males. This time difference might be due in part to the dry conditions in Texas – the longer the uropygids are exposed to dry air on the surface, the greater the chance of dehydration. Therefore, to speed up courtship is to increase the chance of survival. Punzo and Reeves surmize that if such differences in courtship displays between populations become more marked, they could result in reproductive isolation and ultimately to speciation.

Egg laying and maternal care

The female is able to store sperm until it is required for fertilization. When a female is ready to lay her eggs, she goes into her chambered burrow. When the young hatch from the eggs, they are known as larvae or 'pre-nymphs'. These miniature creatures climb onto their mother's back, forming a ring around her opisthosoma. While carrying her young, the mother becomes aggressive to any mechanical stimulation. She stiffens her body, arches her opisthosoma, elevates her prosoma off the ground and extends her palps. This is in contrast to the rather timid response that she would normally give. If, however, the young are removed, the female loses her aggressive tendencies. But, she becomes cannibalistic and gobbles down her young as soon as they are separated from her! The presence of the brood therefore changes her behaviour, making her much more aggressive, while at the same time reducing feeding. Such behaviour would increase the survival chances of her offspring. The mother remains still for many days without feeding, while she is carrying her precious load. In fact, she carries them around for a few weeks, in the same way as some spiders, scorpions and amblypygids do. After their first moult into protonymphs, the young leave their mother and the safety of their burrow. The starved mother dies soon afterwards. In captivity though, she may live for several years.

Post-embryonic development

In uropygids, there are four instars with associated moults before maturity – the protonymph, the deutonymph, the tritonymph and the tetranymph. *M. giganteus* goes through around one instar per year, so it takes around 3–4 years to get to maturity.

Nymphs are rather similar to adults but, not surprisingly, they lack secondary sexual characters. Additionally, their pygidial glands are incompletely developed. The positive allometric growth of the male palp starts in the tetranymph, so before this stage the palp of a male is identical to that of a female. Other minor differences include the shape of the prosomal

ABOVE A female *Typopeltis crucifer* lays between seven and 35 eggs in a membranous sac, which hangs from her genital aperture.

sternites, which change through the four stages. The most anterior sternite develops from a short and wide structure to an elongate and narrow structure. Interestingly, and perhaps a little surprisingly, the number and distribution of trichobothria remains the same throughout all stages.

Longevity

There is no real data on the longevity of uropygids in their natural environment. However, there is good evidence that they can survive in captivity for at least three years after they've reached maturity, making them more than seven years old at death.

5 Schizomida
Short-tailed whip scorpions

Small, but perfectly formed, the Schizomida are one of the lesser known of the Arachnida. What sets them apart from the other arachnid orders is the unique structure of the dorsal surface of the prosoma, which is divided into platelets. Schizomids are commonly known as 'short-tailed' whip scorpions, because they have a short, stubby flagellum. 'Schizo' comes from the Greek word meaning split, and here refers to the carapace. They are closely related to uropygids, some scientists consider them to be in the same order, while others consider them to be in their own order. They use chemical defences, and also have antenniform legs, as in uropygids. Few species exceed 1 cm (½ in) in size; the majority are under 5 mm (3⁄16 in) long.

OPPOSITE *Schizomus portoricensis* from a lava tube cave (La Palma, Canary Islands).

BELOW *Paradraculoides bythius* is adapted to cave life, with long legs and palps, a complete lack of eye spots, a large body and a reduction in pigment.

Classification and diversity

The Schizomida and Uropygi have a very close relationship, but it has not yet been totally resolved. Here the Uropygi and the Schizomida have their own chapters and the Schizomida are placed directly after the Uropygi chapter to try to reflect their close relationship.

There are currently 258 described Schizomida species in 46 genera that are unequally divided into two families – the family Hubbardiidae and the Protoschizomidae. However, there are probably many undescribed species that live in tropical leaf litter and cave systems. Schizomids are a homogeneous bunch. Indeed, they show very little diversity even between continents! The only real differences are found in schizomids that have morphological specialization for life in caves, and the degree to which they are adapted varies between different species, and tends to be on a sliding scale. The following Australian species illustrate these variations: *Bamazomus subsolanus* and *B. vespertinus* are partially adapted to cave life, with enlarged bodies and no eye spots while *Apozomus howarthi* has no adaptations to cave life, even though it is found in caves, so indicating that it is a troglophilic species (organisms that are found in caves, but are also found in other habitats).

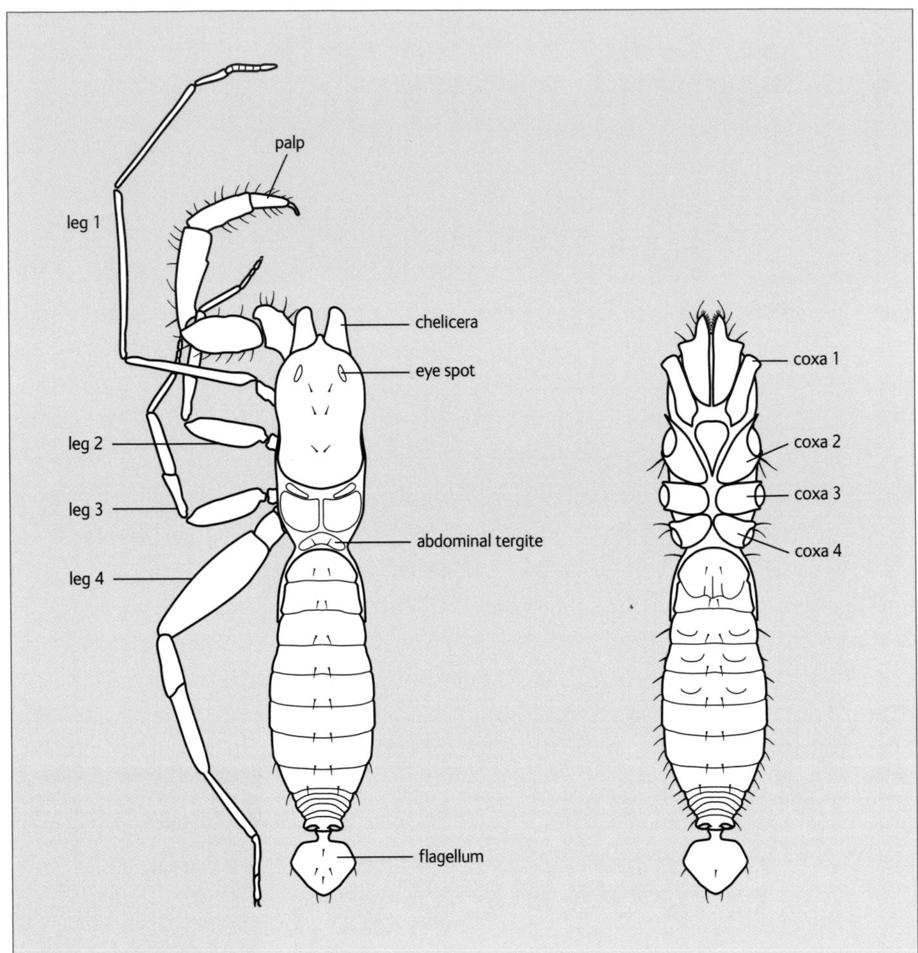

RIGHT Dorsal view (left) of general schizomid, and ventral view (right) of male *Trithyreus borregoensis*.

Anatomy

External anatomy

The structure of a schizomid is not typical of an arachnid structure, as there is no distinct prosomal region - the opisthosoma seems to be a mere extension, and the flagellum too is strange. The sclerotization of the exoskeleton is rather weak compared to that of uropygids. This, together with their small size, means that they have a high rate of water loss and are unable to resist desiccation, so they are constrained to life in humid habitats.

PROSOMA The dorsal surface of the schizomid, equivalent to a prosoma, is divided into platelets and is unique to this order. It consists of the carapace-like propeltidium, the mesopeltidium and the metapeltidium (front, middle and back dorsal shields). The mesopeltidium consists of two wedge-shaped plates, and the metapeltidium consists of a long rectangular plate that is sometimes divided in the centre. The front of the propeltidium comes to a point called the anterior process, which has up to three setae, depending on the family. The ventral surface of the prosoma is mainly covered by the leg coxae, although there is a triangular sternum between the coxae of legs 1 and 2.

EYES SPOTS Most schizomids don't have eyes. Occasionally, in the area where the eyes would be there are 'eye-spots', which are pale areas that don't seem to have any function, although the fact that troglobionts have lost their eye spots, indicates that they may be simple light receptors. There are only around five species that have a pair of eyes – these have rounded lenses and are slightly raised on the lateral edge of the prosoma.

CHELICERAE The chelicerae have two segments – one movable and one fixed. There are several 'teeth' on the fixed digit, and these vary in number. For example, there are two large teeth on the fixed digit in the Protoschizomidae, whereas in the Hubbardiidae, there are three or more. There can also be an 'accessory' tooth or up to five teeth on the movable digit and these are used to determine different genera. In some genera, such as *Pacal* for example, there is a keel.

PREY-CATCHING PALPS The palps are leg-like and raptorial, and used to capture prey. They have six segments: the coxa, trochanter, femur, patella, tibia and tarsus. The tarsus has a pair of small spurs and a distinctive claw. In some species, there are prominent spurs on the femur and tibia, and the patella is distinctly curved. In the Protoschizomidae, there are spiny setae. Some species have dimorphic males (of two different forms) – with simple palps like those of females, or with modified palps, which may be elongated or enlarged. In some species, the trochanter has a pointed projection, which may have setae. Troglobitic schizomids have extremely elongated palps. Such an adaptation is found in many arachnids, e.g. amblypygids.

ANTENNIFORM LEGS Like the uropygids and amblypygids, schizomids have a particularly long pair of highly flexible first legs, which are antenniform. They have seven segments: the coxa, trochanter, femur, patella, tibia, basitarsus and telotarsus, which is divided into seven articles. These legs have numerous trichobothria and are the major sense organs. As a schizomid walks

BELOW The distinctive antenniform legs, clawed palps, rotund opisthosoma and stubbey flagellum can be clearly seen on this female *Hansenochrus* sp. (Saba, Netherlands Antilles).

along, the antenniform legs constantly touch the substrate, gently probing every crevice and crack thoroughly for potential food sources. Schizomids move in a jerky fashion, because they move forward a short distance, then stop and search, then move on again.

WALKING LEGS The second, third and fourth pairs of legs are the walking legs. Each leg has seven segments: the coxa, trochanter, femur, patella, tibia, basitarsus and a telotarsus, which is divided into three segments, with an apotele ('claw') with three prongs. Troglobionts often have extremely elongated legs. Schizomids can move quickly both forwards and backwards and are capable of running at great speed, albeit for short distances. In fact, they have been seen to run backwards for several inches! *Schizomus crassicaudatus* is able to jump backwards if touched from the front, in order to avoid capture. Most schizomids have enlarged hind femora so it is thought that all species have the potential to jump short distances.

OPISTHOSOMA The opisthosoma has 12 segments in total. Segment 1 is very narrowed and shaped like a boomerang. Segments 2 to 9 have tergites and corresponding sternites, which are connected at the sides by pleural membranes. Segments 10 to 12 comprise the metasoma, and are markedly narrow and tubular. The flagellum is classed as a 'telson'. The anus opens at the base of the flagellum, on segment 12. On segment 12, males of some genera have a dorsal structure that extends above the flagellum. The structure ranges in size from small and inconspicuous to enormous and pointed. The opisthosoma may be quite elongated in males, depending on the species. There is also a pair of defensive anal glands, either side of the flagellum, as in uropygids.

FLAGELLUM Like their sister-group the Uropygi, schizomids have a flagellum, though it is certainly not whip-like! The female flagellum is short but slender. In adults, it is segmented (3–6 segments), with sensory hairs. The exoskeleton may have thin areas that allow the flagellum to bend. The male flagellum is usually unsegmented, and is quite peculiar. For example, it is enlarged and bulbous in *Luisarmasius*, and elongated in *Clavizomus*. It tends to be elongate in species with an elongate opisthosoma. The shape of the male flagellum is very important in determining different species, but it is generally similar in each genus. Bizarrely, there are holes that pass right through the flagellum in males of *Tayos ashmolei*. In some species, the flagellum additionally has long, thick, protruding setae (as in *Protoschizomus pachypalpus*) that make the animal resemble a medieval weapon of war.

But why is the male's flagellum so different to that of the female? The shape makes great sense when you know its function. The male's flagellum is clasped by the female during courtship, and its spatulate or bulbous shape would be much easier to grip than one that is slender.

EXTERNAL GENITALIA The genital atrium opens between sternites 2 and 3. In males, sternite 2 overlaps sternite 3 in the centre, whereas in females, sternite 2 turns up slightly.

RESPIRATORY STRUCTURES Schizomids have one pair of book lungs. The pair of stigmata is located at the lateral junction of sternites 2 and 3.

SEXUAL DIMORPHISM This is quite distinct in schizomids. The main difference is that the flagellum is enlarged and quite bulbous in males, whereas it is segmented and rather slender in females. There is also considerable sexual dimorphism in some species regarding the palps.

Internal anatomy

The internal anatomy of schizomids is very similar to that of uropygids. There isn't much known about the central nervous system. However, there is a supraoesophageal ganglion (the 'brain'), a suboesophageal ganglion and an abdominal ganglion. The brain is linked to the chelicerae, while the suboesophageal ganglion supplies all of the rest of the prosomal region, and the abdominal ganglion obviously supplies the opisthosoma. The circulatory system is well developed. The segmented heart has five pairs of ostia, and is located in the opisthosoma, but it can also extend into the prosoma. The pair of book lungs operates in the same way as spider book lungs. The digestive system is comprised of the mouth, pharynx, oesophagus, midgut, caeca, hindgut and anus.

A SPECIAL EXCRETORY SYSTEM The excretory system has the usual Malpighian tubules, coxal glands and nephrocytes. However, the excretory product of schizomids appears to differ from that of most other arachnids. Most arachnids excrete guanine and uric acid in chalky white spots; in contrast, *Draculoides vinei*, a cave-dwelling schizomid from Australia, excretes watery-looking clear droplets of fluid. It is possible that schizomids are 'ammonotelic' i.e. they excrete excess nitrogen as ammonia instead of uric acid. This is so far unknown in arachnids, but is found in terrestrial arthropods that have access to surplus water. The advantages of excreting ammonia rather than uric acid are that carbon is not lost in the process, and it is a low-energy process. Certainly, schizomids have access to surplus water in their humid habitats.

INTERNAL GENITALIA In schizomids, there is a certain level of variation in the female genitalia. The genital atrium opens into the spermathecae. These are very interesting structures, and unlike those of other arachnids. Instead of being paired, they may have many branches or lobes, looking like chilli peppers or small growths of mushrooms! They may have distinct stalks and bulbs. There are secretory glands that surround the spermathecae. These probably discharge secretions into the spermathecae via small ducts.

In *Orientzomus sawadai*, the system has a single central ovary (lying beneath the gut) that is connected to a pair of lateral narrow oviducts, and a single common oviduct with small seminal receptacles in bunches on the ventral wall. Oocytes of different stages of development are scattered in the ovary wall, with oocytes in advanced developmental stages protruding on short stalks. When eggs are produced in the oocytes, they pass through the lumen of the ovary, through the lateral oviducts and into the common oviduct. The eggs are fertilized in the common oviduct by spermatozoa that were stored in the seminal receptacles. The fertilized eggs then pass out through the genital pore.

The male genitalia comprise a genital atrium, a pair of testes and paired vasa deferentia, which secrete sperm cells and secretions. They also have spermatophore organs, which produce the spermatophore. Spermatogenesis occurs in cysts in paired testes located along the sides of the opisthosoma.

ABOVE A species of *Hansenochrus* (family Hubbardiidae) moving around in damp mossy foliage (Saba, Netherlands Antilles).

Distribution and habitats

Tropical and temperate

Schizomids are found in most tropical and subtropical zones of the world, such as Southeast Asia, India, Australia, Africa, Central and South America, although a few species extend into neighbouring temperate regions. Although it's been suggested that the distribution of schizomids is limited by low temperature, species in the genus *Hubbardia* have been collected from under snow-covered rocks in California, and *Hubbardia briggsi* is often found in snow-covered habitats during the winter.

The family Hubbardiidae is distributed worldwide. The family Protoschizomidae has a much more restricted distribution, and is found only in Texas and Mexico. Schizomids are not naturally found in Europe, but have been discovered in botanic gardens in France and Britain. Such unexpected distributions are due to introductions through the transportation of nursery stock from abroad. In this way, the Sri Lankan *Schizomus crassicaudatus* has found itself in France and the South American *Stenochrus portoricensis* has been found in England! The presence of schizomids on various islands in the Indian and Pacific Oceans, including those of Galapagos and Hawaii, may be because of introduction through human activity, or by rafting on floating debris. Interestingly, although schizomids are found almost worldwide, nearly all species have very limited distributions indeed and many are only known from one locality.

From caves and crevices, to litter and logs

Relative humidity is an important factor in determining the habitats of schizomids. They are generally found in humid conditions under stones and rocks, in rotten logs and in rainforest leaf litter, particularly in the organic soil layer. Although schizomids are generally restricted to rainforest habitats, they may also be found at rainforest margins, such as in eucalypt woodlands. The aptly named *Surazomus arboreus* lives in areas in South America

that are seasonally inundated with water, and escapes drowning by moving up into the trees. Troglobitic fauna probably originated from wet forest litter fauna. As their original wet forest habitats dramatically decreased in size, these creatures would have taken refuge underground. For example, *Draculoides vinei* (from Australia) is confined to interconnected caves below ground that have a high humidity. Some species, e.g. *Afrozomus machadoi*, are found in more obscure places such as termite mounds, where they probably feast on the inhabitants. One species has been discovered in the nest of an African mouse! There are also schizomids that are myrmecophiles (species found in association with ants), such as *Stenochrus portoricensis*. Some species are found in columns of ants, or in the nests of ants. Schizomids have also been recorded in artificial environments, such as greenhouses.

General biology and behaviour

Prey capture and feeding

In their natural environment, schizomids feed on a wide range of prey such as collembolans, isopods, millipedes, cockroaches, other schizomids, termites, psocids, zorapterans and also worms. Observations of *Draculoides vinei* reveal that schizomids go for prey of a range of sizes, from 10% all the way up to 100% of their own body size! Schizomids are clearly a hardy bunch – studies have shown that *Hubbardia pentapeltis* is able to survive a surprisingly long five months without a scrap of food.

Schizomids are actively predaceous, constantly probing the substratum for prey with their antenniform legs. When prey is encountered, the schizomid uses its antenniform legs to stroke it, paying particular attention to its extremities – possibly to assess the size of the prey animal. If the schizomid decides not to run away, it will suddenly lunge forward and grab the prey in its raptorial palps. As soon as the prey has been subdued, feeding starts. The prey might be carried off to a nearby crevice to be eaten. The prey is dismembered using the chelicerae, and then the tissues are macerated, and the liquefied tissues sucked through the mouth.

Predators and parasites

There is very little data on schizomid predators and parasites. However, it *is* known that amblypygids predate schizomids, e.g. *Phrynus marginemaculatus* was observed feeding on the schizomid *Stenochrus portoricensis*. Even though schizomids are so small, there are parasitic nematodes that are even smaller. A specimen of *Stenochrus goodnightorum* was found to have almost its entire opisthosoma filled by a nematode!

Defensive behaviour and aggression

Many papers have been written about uropygid defensive secretions, but for schizomids the amount of research in this area is in proportion to their size – very little. *Megaschizomus mossambicus* does however emit a smell of acetone when disturbed, and *Schizomus crassicaudatus* emits a smell of acetic acid. So it seems likely that they are very similar to uropygids in that they use chemical deterrents against predators. Schizomids don't just defend themselves; *Stenochrus portoricensis* has actually been seen defending its burrows and crevices against intruders, by displays. These displays are made up of short advances and retreats, with the antenniform legs and palps held up in a threatening way. However, some schizomids appear to be quite gregarious, e.g. *Stenochrus portoricensis*, which lives in large groups under logs.

Life history

Reproduction

STRUCTURE OF THE SPERMATOPHORE Sperm transfer in schizomids is indirect, by a weakly sclerotized spermatophore that is deposited by the male and then picked up by the female. The spermatophore basically consists of a basal plate, which is attached to the ground, and a stalk that is round in cross-section with two pairs of lateral, wing-like distal structures. One of these is larger and has an opening with a complex shape, while the other is narrower with a tip (which is not always easy to see) close to the spermatophore packets. The surfaces of both wings have a fine slat-like or thorn-like structure. At the apex, the spermatophore has two symmetrical vesicles, each bearing an apical thorn.

COURTSHIP AND SPERMATOPHORE TRANSFER There is courtship behaviour in Schizomida. The first phase is a courtship 'dance' by the male, consisting of a sequence of rhythmic leg vibrations, leg shaking, and pulling backwards of the first pair of legs. The male does not touch the female at all during this phase. The stimulation of the female is achieved through air vibrations (induced by rhythmic leg movement).

The second phase is called a courtship 'march', which consists of a jerky forward movement. The female's chelicerae grasp the male's enlarged and bulbous flagellum, and he pulls her forward step by step above the spermatophore, which he has just deposited. The female may resist from time to time, pulling the male back a few millimetres, before he pulls her forward again. The female then moves over the spermatophore and inserts it into her genital atrium. She stores the male's sperm in her spermathecae until it is required for fertilization.

PARTHENOGENESIS Some species are parthenogenic, e.g. *Afrozomus machadoi* and *Stenochrus portoricensis* (as no males have ever been found). Others are facultatively parthenogenic (it may or may not occur).

Egg laying and maternal care

The brooding habits of schizomids are, not surprisingly, rather understudied. The following observations (made in a laboratory) were of brooding in the species *Hubbardia pentapeltis*. Fifty-five days after mating, the female schizomid started excavating a brooding chamber, a process that took about five days even though it was only about 3 cm (1¼ in) deep. Her opisthosoma then began to distend noticeably, so much so that the small white eggs inside could be seen through the pleural membranes. Around two weeks after her frantic nest building, the rotund female laid her eggs at the bottom of the chamber (as with uropygids). She produced approximately 30 eggs (as an order, schizomids lay between six and 30 eggs), glued together in a pile on the ventral side of her opisthosoma, which she arched over her prosoma. This is in contrast to uropygids, which lay their eggs into a membranous sac hanging from the gonopore (see p.127). The egg laying took less than 24 hours.

Around 35 days later, the eggs hatched and their appendages became visible, although they were rather undeveloped. The mother positioned the larvae so that they surrounded her arched opisthosoma. They held on to each other by linking legs and were sticky on the ventral surfaces. The nymphs continued to develop *in situ* for another 30 days or so, until they moulted and dropped off the mother. Just over a week later the female, freed from her

motherly duties, broke out of the chamber. She was found dead 10 days later, totally devoid of any food in her now malnourished body.

Variations in brooding are found in other species. For example, *Stenochrus portoricensis* doesn't hold the opisthosoma erect while carrying eggs. Also, it doesn't construct a brood chamber, but wanders freely instead. This has been seen in other species too. Male schizomids have been observed to cannibalize their newly moulted offspring, just emerged from the brood nest, but this might be an artefact of laboratory life.

Certain subtropical species, such as *Surazomus mirim*, appear to produce several broods in a season, because in certain environments (e.g. the organic soil layer in a secondary upland forest in Manaus, Brazil) adults and juveniles are found throughout the year. Additionally, there does not appear to be a distinct reproductive period. With a temperate species such as *Stenochrus portoricensis*, however, there does seem to be a definite breeding season, as in late summer there was an exceptionally large number of second instars found in forest litter in Florida.

Post-embryonic development

Most species probably have five post-embryonic moults, and it may take two or three years to reach maturity. Hatching takes place around 35 days after the eggs are laid. At this stage, the young schizomids can be described as 'eggs with legs' and are around 1.8 mm (7/100 in) long. As with all arachnids, nymphs look like adults, but are not highly sclerotized – in fact, the first instar nymph is ghostly white because of the lack of sclerotization – and they also lack spines and setae. The larval tarsi are very different to those of other instars, because each tarsus has an adhesive disc and a pair of spines, which helps the larva to stay attached to the mother (along with its sticky ventral surface).

The telson, chelicerae and abdominal segmentation become defined after approximately 70 days after hatching. Around this time, a moult occurs and the nymphs leave their mother. They are rather sluggish at this stage. Setae are noticeable on the legs and the flagellum, and particularly on the underside of the opisthosoma. Just over a week later, the nymphs become much more active and their chelicerae start to darken. Subadult males have a flagellum that still looks like that of a female, but it hasn't yet thickened up. At the last moult, sexual dimorphism becomes apparent, as the male's flagellum becomes markedly different to that of the female. Additionally, adults obtain their final livery – straw yellow to brownish green colouring.

Longevity

Some workers record that schizomids are relatively long-lived in culture. *Hubbardia pentapeltis* has been kept in the lab for many months but a specific time frame was not been mentioned.

6 Palpigradi
Micro whip scorpions

OPPOSITE The multi-segmented flagellum with long setae is very distinctive on this specimen of *Eukoenenia austriaca*.

Palpigrades are small, enigmatic creatures that have often been considered the most primitive of arachnids. They are commonly known as 'micro whip scorpions' because of their size and because they resemble immatures of their much larger relatives, the 'whip scorpions'. This resemblance is also reflected in their original scientific name, Microthelyphonida ('small thelyphonids'), Thelyphonida being a name that is often applied to Uropygi.

As palpigrades walk, they touch the substrate repeatedly with their first pair of legs. This behaviour is reflected in their current scientific name, since 'palpo' is Latin for touch, stroke or feel, and 'gradatus' means step-by-step. Unlike any other arachnid, they use their palps for walking, and not for any other function. They are colourless and translucent, and possess a very distinctive flagellum. They were the last order of arachnids to be discovered, the first having been described in 1889. These little beasts range in size from the microscopic 0.65 mm (3/100 in) *Eukoenenia grassii*, to the largest species *E. draco* (from Spain) at a 'staggering' 2.8 mm (3/32 in)!

Classification and diversity

The systematic placement of palpigrades is still rather in doubt, with several hypotheses in competition. There are currently 82 described extant species, with 75 of these contained in four genera within the family Eukoeneniidae: *Allokoenenia*, *Eukoenenia*, *Koeneniodes* and *Leptokoenenia*. The other seven species are contained in two genera within the family Prokoeneniidae: *Prokoenenia* and *Triadokoenenia*. Palpigrades are, in general, a very homogeneous order. The only real variation is found between endogean (soil-dwelling) species and troglobitic species (species adapted to living in caves). Troglobitic species are larger in size, have elongated appendages and an increased number of sensory receptors on the prosoma. For example, these differences are found in *Eukoenenia spelaea* (from the Alps of Provence, France), *E. hispanica* (from the Spanish Pyrenees), and the largest species *E. draco*, from caves in Mallorca, Spain.

Anatomy

External anatomy

The exoskeleton is thin and flexible, almost colourless and translucent, with small cone-shaped projections that almost entirely cover the external surface (apart from the pleural membranes of the prosoma and parts of the chelicerae). There are no lyriform organs.

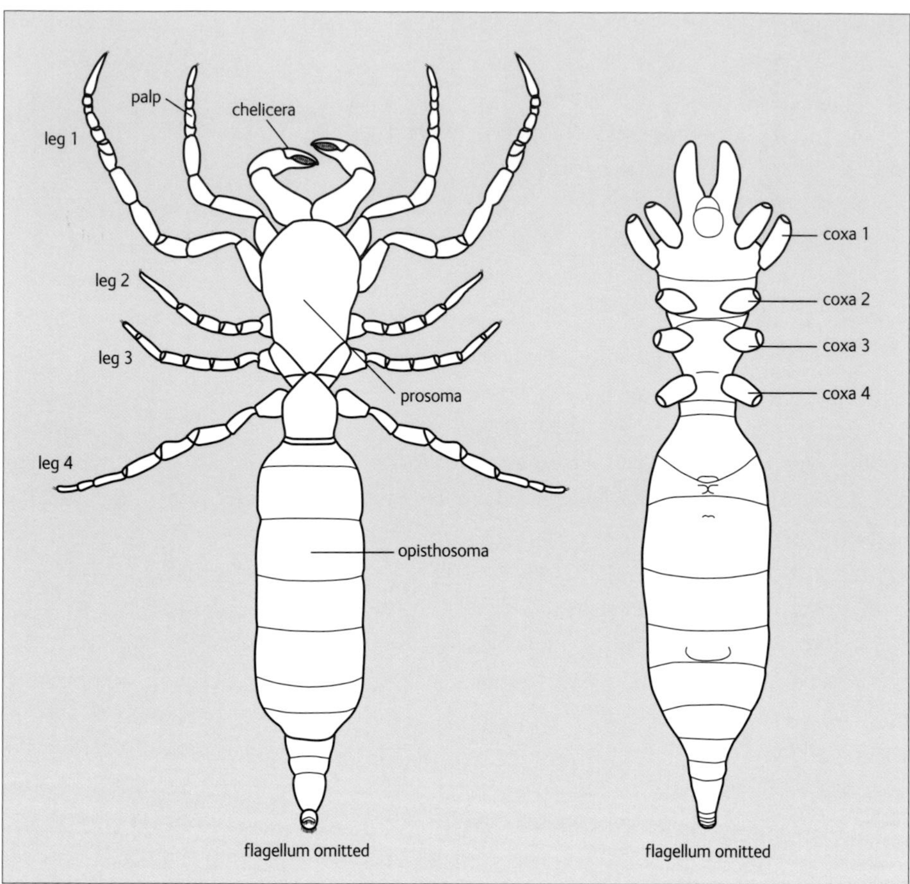

RIGHT Dorsal view (left) and ventral view (right) of female *Eukoenenia mirabilis*.

PROSOMA Unlike that of most arachnids, the dorsal surface of the prosoma in a palpigrade is composed of four platelets. Because these platelets, or sclerites, are soft with an indistinct edge, they are called pseudosclerites. There is a propeltidium, two small mesopeltidia and a metapeltidium, which are separated by soft tissue, as in schizomids. The propeltidium, which covers most of the prosoma, extends from above the chelicerae to the level of the third pair of legs. It is basically a rigid shield. The mesopeltidia are on the lateral edge of the posterior margin. Behind these is the metapeltidium, which is undivided. The soft pleural membrane around the pseudosclerites allows the palpigrade to move not just in the horizontal plane, but vertically too. The ventral surface of the prosoma has five sternal platelets.

The front edge of the prosoma bears a single pair of modified, blade-like setae. There are no eyes, although there are modified setae positioned either side of the carapace, which are considered to be sensory organs. These setae are important in identifying palpigrades. Depending on the species, there are 1–13 blades. The number of blades is usually consistent in a species, although there may be variations within a population or even between individuals.

CHELICERAE The rather large chelicerae extend in front of the carapace, and are divided into three parts. They are strongly developed and have great mobility, and articulate almost at right angles to the plane of the body and the tips rest just near the mouth. The inner surface of the fixed digit of the second segment and the inner surface of the movable digit have between seven and ten serrated little teeth. The other parts have many setae.

PALPS The palps are located immediately behind the chelicerae and are composed of trochanter, femur, patella, tibia and tarsus (with five segments). There are well-developed claws on the tip of the tarsus. Although palps in all other arachnids are used in feeding, or modified for sperm transfer, this is not their function in palpigrades. So what *is* their function? Interestingly, they are only used for walking.

LEGS Instead of all legs having seven segments (as with most other arachnids), palpigrades have a varying number. The first pair of legs has 12 segments, the second and third pairs have the usual seven segments, and the fourth pair of legs has eight segments. Each leg consists of the trochanter, femur 1, femur 2, patella, tibia and the tarsus. However, what makes them variable is the number of tarsal segments. So, leg 1 has seven tarsal segments, legs 2 and 3 have two segments, and leg 4 has three tarsal segments. All the legs have three claws – two large and one small. Palpigrades can move quickly on their differing lengths of legs.

The first pair of legs are long, thin appendages, with many sensory receptors, such as trichobothria and sensilla. These legs function mainly as antenniform tactile structures, as with those of Uropygi and Amblypygi. As the animal walks along, the first pair of legs is held outwards and forwards, touching the substrate repeatedly in exploration.

OPISTHOSOMA The tergites and sternites of the opisthosoma are not clearly defined, which is a similar situation to the pseudosclerites on the prosoma. They are more like a thicker version of the surrounding pleural membrane. The segments on the ventral surface are unequal in length – segment 3 appears to be shaped like a bow tie, because it is overlaid by the genital valve of sternite 2. Segment 4 is broader than all the other segments. The genital organs open to the exterior on sternite 2. The last three segments (9–11) have a lot of movement

LEFT The opisthosoma of a palpigrad is long, with eleven segments, and broadly joined to the prosoma. The segments are separated by small and shallow furrows, which are more obvious on the dorsal surface.

and can telescope together. They are similar to a uropygid's pygidium. They also possess a very distinctive flagellum that looks like the stem of a horsetail plant. The anus is on the last segment of the opisthosoma (segment 17 of the entire body).

SPECIAL STRUCTURES – VERRUCAE AND LUNG SACS In palpigrades, the segments of the opisthosoma have various structures such as setae, lobes and in Prokoeneniidae, sacs or verrucae. There are often lobes on both abdominal segments 4 and 6, which have setae containing ducts of glands. From a side-on view, these can be seen clearly. Not all palpigrades have lobes on the same sternites and some do not seem to have them at all. They are really just membranous areas that can be inflated slightly. Preserved specimens usually have the membranous area retracted, hence no tubercle is apparent. In the family Prokoeneniidae, there are three pairs of sacs (one pair per sternite) on sternites 4 to 6. These are sometimes called 'lung' sacs and thought by some to relate to respiration, although this is not yet proven. They appear to be eversible (they can turn inside out), like those of amblypygids. There are no ventral sacs in species of the family Eukoeneniidae.

FLAGELLUM The long, narrow and rather fragile flagellum has 14 or 15 segments. Each segment has 6–8 long setae, which curve upwards towards the tip of the flagellum, making the structure resemble a plant stem. The flagellum is easily broken off. The opisthosoma can be lifted up and bent forward, so the flagellum can be directed to the front of the animal. The flagellum is moveable in the same way as that of a uropygid and can be actively lifted or lowered through muscle control. While the animal is still, the flagellum is kept horizontal, but it is moved around when the animal is walking.

EXTERNAL GENITALIA The genital opening is found between abdominal segments 2 and 3 in both sexes. It is covered by two valves in the female and three in the male (the first valve is subdivided). In the male, the first valve sometimes has two or three pairs of thread-like structures on the rear edge that are attached to large glands. It's not currently known what the function of these thread-like structures is, but it's thought that they are ducts for a secretion from the gland and that they may play a role in reproduction. It has been suggested that they might be involved in silk production, as the histological structure of the glands is similar to that of the silk glands of spiders.

SEXUAL DIMORPHISM As is often the case in arachnids, female palpigrades are usually larger than males. Females and males often differ in the number and arrangement of glandular setae on the ventral surface of their opisthosoma.

Internal anatomy

CENTRAL NERVOUS SYSTEM A palpigrade has both a large supraoesophageal ganglion, and a smaller suboesophageal ganglion from which a nerve cord passes to the opisthosoma. The nerves leading to the chelicerae run from the supraoesophageal ganglion, whereas the nerves to the palps lead from the suboesophageal ganglion.

CIRCULATORY SYSTEM There is not much known about the circulatory system, although researchers agree that it is very simple. The heart is found in the opisthosoma, which runs

within the second to the fifth segments. There are between one and four ostia. There is a dorsal vessel, the aorta, in the prosoma.

RESPIRATORY SYSTEM Palpigrades lack both book lungs and tracheae, so how does gas exchange occur? They actually exchange gases through their cuticle by diffusion, as with many mites. Their small body size and thin cuticle allows this process to take place. It is necessary for the cuticle to be moist. Although diffusion of oxygen occurs through the cuticle, it is also considered to be waterproof. As mentioned above, members of the Prokoeneniidae have three pairs of sacs that might be involved in respiration.

DIGESTIVE AND EXCRETORY SYSTEMS The mouth has a very anterior position (between the bases of the chelicerae) unlike the more posterior position in other arachnid orders, and the opening is shaped like a crescent moon. The mouthparts consists of the labrum (upper lip), the rounded labium (lower lip), the mouth, pharynx and the chelicerae. The labrum has many small hair-like glandular cells. It overhangs the labium and forms the front part of the preoral cavity. The labium has many glandular (mucous) cells. The buccal cavity, which is found above the mouth, acts as the equivalent of baleen in whales, because it has little rows of protuberances that sieve the externally digested food. As the liquefied food passes from the buccal cavity, it enters the muscular pharynx, then moves though the oesophagus and into the midgut. The midgut tube has an epithelium composed of secretory and digestive cells. Leading off the midgut are diverticula. There is one pair that extend into the prosoma, and six pairs in the opisthosoma. Waste matter in the midgut empties into the hindgut or rectum and the anus opens on segment 17 (segment 11 of the opisthosoma).

The excretory system is rather basic, consisting of just coxal glands and absorption cells in the gut. The coxal gland tubules start in the third abdominal segment, and lead forward into the prosoma, where they exit at the coxae of the first pair of legs. There are no Malpighian tubules.

REPRODUCTIVE SYSTEM There is little information available regarding the reproductive system of palpigrades. However, in males, there is an elongate pair of testes located in the third to eighth abdominal segments. The female has a pair of ovaries with two oviducts that join together into the uterus.

Distribution and habitats

Palpigrades are found around the world, with the greatest numbers in tropical and subtropical regions. They are found in the southern USA (as far north as Oregon, but only in a cave), the Mediterranean, Asia and Africa. Only species of the Eukoeneniidae occur in Europe. In the north of Europe, they are found in caves; some of these look more like tropical species than local endogean species (those found in soil or plant litter), though in the south they are also endogean. The genus *Eukoenenia* is spread over quite a few parts of the world, with *E. florenciae* and *E. mirabilis* being widely distributed. For example, *E. mirabilis* is found in Madagascar, northern and southern Africa, Western Australia, Chile, and southern Europe. How did this species become so widely distributed? It's thought that this is due to accidental introduction through human activity. For example, sea-faring trade in the Mediterranean may

have been responsible for the wide distribution of the species in that region. Palpigrades are quite rigid in their requirements; they need a humid environment within a certain temperature range. So they are found in habitats that provide this, such as soil up to 1 m (40 in) deep, beneath stones and in cracks in the ground. In forests, they are found in the mineral subsoil, because they prefer the deeper layers. However, they are also found in the organic layer, particularly when it has rained heavily. Soil-dwelling palpigrades move down into the deeper layers when those above become too dry. Many species are restricted to cave systems. Although most palpigrades are terrestrial, *Leptokoenenia scurra* has been discovered living on the seashore and is apparently capable of swimming. The most unusual habitat may the 'holy' home of *Eukoenenia austriaca*, which is found in the catacombs of St. Stephen's Cathedral in the centre of Vienna, Austria.

General biology and behaviour

Details of prey capture and diet in palpigrades are still almost completely unknown, although it has been established from observations in the laboratory that they are predatory beasts that feed on very small arthropods (such as collembolans), which they catch with their chelicerae.

Life history

Reproduction

Unfortunately, as with prey capture and diet, very little indeed is known about a palpigrade's reproductive habits. However, this is of no real surprise, given that these beasts are not easy to keep for long in the laboratory and are very difficult to study in their natural habitat. Although it is agreed that males produce spermatophores, it is not known how they are transferred. In some species, the spermatophores may fill a large part of the male's opisthosoma. In the tropical species *Eukoenenia janetscheki*, from Central Amazonia, Brazil, there is a lack of a distinct reproductive period. Therefore, adults and juveniles can be found throughout the year.

Palpigrade eggs are large, and consequently only a few are laid at a time; troglobitic species lay 1–2, while endogean forms lay 3–4. Thus the eggs are larger and clutch size is smaller in troglobitic species. It is likely that the female does not lay all of her eggs at the same time, because it has been discovered that there are ovules at different developmental stages within the female's opisthosoma.

Some species, such as *Eukoenenia mirabilis*, are considered to consist of mainly parthenogenic populations, because the sex ratios are extremely skewed in some localities. For example, in Pouilles, Italy up to 500 females of *E. mirabilis* were collected, but only 2 males were found! However, in other species, such as *Prokoenenia wheeleri*, males appear to be more numerous than females.

Post-embryonic development

Palpigrades are thought to follow these developmental stages: prelarva, larva, protonymph, deutonymph, tritonymph and adult. However, the prelarva and larva have not been observed and in Eukoeneniidae only two nymphal stages are known. The different instars vary in terms of size, genital development, the structure of the flagellum, and the arrangement of setae.

In some species, the different developmental stages can be distinguished by the number of setae on the basitarsus of leg 4, which increases with increasing age. The number of sacs on the opisthosoma of Prokoeneniidae also varies depending on the developmental stage. The first instar has no sacs, the second instar has two pairs, while the third instar and adult have three pairs. In immature *Eukoenenia* species, there may be only one blade in the lateral organ of the prosoma, but in the adults, there are multiple blades.

7 Ricinulei

Hooded tick spiders, tick beetles

Ricinuleids used to be one of the least known and rarest of the arachnids, because up until fairly recently there were very few collected. However, modern explorations into Mexican caves have shown that they are more widespread than previously thought. These little beasts are commonly known as 'hooded tick spiders' because of the hood-like structure or cucullus that covers the chelicerae and mouth – a unique feature to the order. Another common name is 'tick beetle'. Although these creatures don't look that much like ticks, they *are* currently thought to be closely related to ticks and mites, and their scientific name Ricinulei is possibly derived from the tick species *ricinus*.

Apart from the cucullus, ricinuleids have other unique features including copulatory organs on the third pair of legs of the adult male, a telescopic pygidium and a locking device that couples the prosoma and opisthosoma. Interestingly, the Ricinulei is the only arachnid order that was first known from a fossil, and this was originally thought to be a beetle! They range in size from the 4 mm (¼ in) long *Cryptocellus emarginatus* from Costa Rica, to the 10 mm (½ in) long *Ricinoides afzelii* from tropical West Africa.

OPPOSITE *Cryptocellus* is found as far north as Honduras and as far south as Brazil. This specimen of *C. goodnighti* is seen here walking around in soil.

Classification and diversity

There are currently only around 55 described extant species worldwide, contained in three genera *Cryptocellus, Pseudocellus* and *Ricinoides* within a single family, the Ricinoididae. It is not a surprise, since there are only a small number of species, that they are very similar in appearance. However, they do have some variation in their exoskeleton. They are very often a rich, reddish brown in colour. They are thought to be most closely related to the Acari.

BELOW This male *Cryptocellus narino* from the rainforest of Colombia has a very granular look.

Anatomy

External anatomy

Ricinuleids have a leathery, tough and thick exoskeleton covered in a variety of raised bumps, or tubercles. Tubercles are found on all sclerotized parts of the body, including the legs, and vary in structure – they may be cup-shaped, broadly flattened and button-shaped, low conical or tall conical. They have fine grooves on their surface that twist slightly towards a smooth apex. The intersegmental membranes may also be

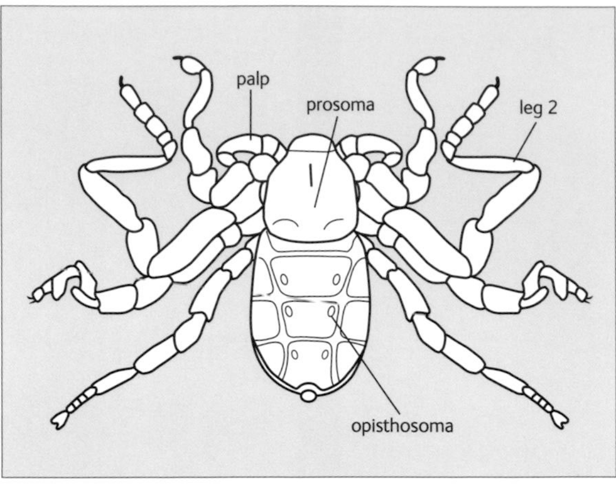

ABOVE Dorsal view of a generalized ricinuleid.

covered with tubercles. It is possible that some types of tubercle may have connecting ducts through the different layers of the exoskeleton. In some species, the prosoma, opisthosoma and legs are covered with large, white, comma-shaped setae.

PROSOMA The prosoma has six segments. The carapace is slightly broader than it is long. In contrast, the posterior edge has a ridge, which forms part of the locking apparatus between prosoma and opisthosoma. In adults, there is a concentration of tubercles along the posterior border. The sternum is mostly obscured by the leg coxae. The only part of the sternum that is still visible is the tritosternum, which is located between the coxa of legs 1 and 2. The tritosternum has the longest two or three setae possessed by ricinuleids. Between coxae 1 and 2 is an extension of the cuticle, which has a bulbous enlargement with an opening; this is the exit of the coxal gland.

EYES Ricinulei usually do not have eyes. However, some species have a pair of rudimentary organs on either side of the carapace that are thought to be very basic eyes, because these ricinulei have been observed to be highly sensitive to light. Fossil species have two pairs of lateral eyes.

CHELICERAE The chelicerae can't be seen from above when they are covered by the cucullus. They can be withdrawn or extended. They are two-segmented, with a large fixed joint, a small fixed finger and movable finger that is dorsal rather than ventral. The fixed and movable fingers possess teeth with rounded, rather than sharp tips. The chelicerae are very variable in size and in the arrangement of the teeth, even within a species, and between right and left chelicerae in one individual. The chelicerae are used in prey capture and feeding, as well as possibly in digging.

BELOW Attached to the chelicerae, are two long thick brushes of setae, which are interspersed with setae that have pointed tips. These setae are thought to trap bodily fluids from prey items.

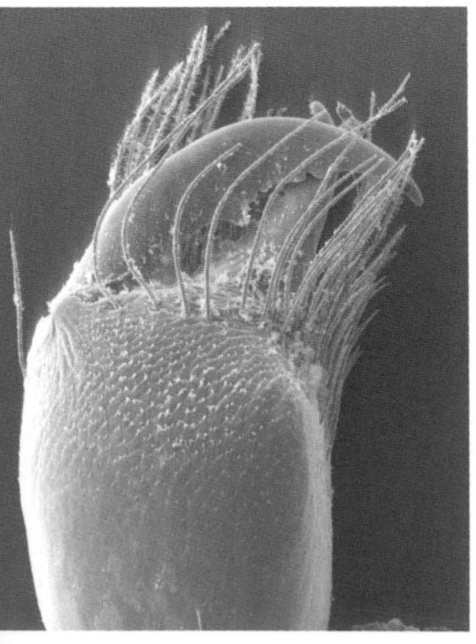

The cucullus: a unique structure

ABOVE The distinctive cucullus can be tightly shut over the chelicerae and mouth or raised up like the hood on a pram.

The distinctive cucullus is formed from a separate 'platelet' that is broader than it is long, with a convex anterior surface, and a concave posterior surface with two depressions into which the chelicerae fit. The surface may be covered in tubercles, setae and pits. The cucullus is articulated to the anterior margin of the carapace, with special musculature that allows it to be raised or lowered. It has several functions: it aids the capture of prey, and it helps to hold the prey while it is being eaten, and this function may be assisted by a row of large tubercles on the ventral edge. Additionally, it is used with the chelicerae and palps to carry eggs.

MOUTH The coxae of the palps are fused and form part of the preoral cavity. The front is covered by the anterior half of the labrum (lip), which can be raised or lowered. The labrum and the borders of the fused coxae are covered with setae. Additionally, the 'floor' of the preoral cavity is carpeted in extremely fine setae in regular, overlapping rows, which point forward.

ROTATIONAL PALPS Ricinulei possess slender palps with small chelae that consist of one small fixed finger and a large moveable finger – the tarsus. This is unique among arachnids for being dorsal rather than ventral. The inner surfaces of the fingers have rows of flattened teeth. The divisions of the palp are coxa, trochanter 1 and 2, femur, tibia and tarsus. The palps are also unique among arachnids in having two trochanters, and a femur that can rotate in a 180° arc, giving it great mobility.

The palps of a cavernicolous Mexican species, *Pseudocellus pearsei* (see p.150) were recently studied, and the different characteristic surface structures discovered include sensilla (which also occur on the other appendages), slender setae, a single, short, clubbed seta, chemosensory terminal sensilla in the fingers of the palpal chela, a single pore 'organ' containing one olfactory sensillum, and conspicuous long, mechanoreceptive slit sensilla. These sensory structures enable the palps to be highly effective short-range sensory organs, used in conjunction with the second legs (see below). They touch the substratum with the tips of their palps when they walk. Ricinuleids also use their palps in prey capture (as with scorpions and uropygids, for example), and manipulation in copulation (males) and in egg carrying (females).

LEGS: FROM LOCOMOTION TO INSEMINATION Ricinuleid legs possess tubercles, pits, spines, and setae of different shapes (such as smooth, barbed with one or two projections, tree-like, spiked, hooked and spiralled). The legs also have many slit organs – sometimes instead of the usual single organ there may be two or three, plus extensive sensilla on the distal parts of the first and second pairs of legs. All the legs have the same divisions but differing numbers of divisions on the tarsus.

LEFT Most adult males have an enlarged femur on leg 2, and in many individuals it is at least twice as large in diameter as that of females.

RIGHT It is easy to see how much longer a ricinuleid's second pair of legs is in this male *Pseudocellus pearsei*, a cavernous species from Mexico.

The second pair of legs are used for locomotion, but also have a long-range sensory function, and so are often used as 'functional antennae'. Then they are carried horizontally, and are vibrated with up-and-down movements. Every so often, they touch the substrate around the body, as do the palps. Because they are used in walking, they can't be considered as antenniform legs, like those of amblypygids and uropygids. In adult males, the metatarsi and tarsi of the third pair of legs are modified into special structures which are used to hold and transfer a spermatophore to the female's genital opening. These structures are so variable that they are used to identify species.

RESPIRATORY STRUCTURES Ricinuleids have tracheae for respiration, but no book lungs. Their one pair of spiracles open on the prosoma in the region of coxae 3 and 4, and cannot be seen when the prosoma and opisthosoma are locked together. Such a location for spiracles is unusual among the arachnids, as they usually open on the opisthosoma. The only other arachnids that possess spiracles opening on the prosoma are some mites, e.g. in the Prostigmata and solifugids, although solifugids additionally have spiracles opening on the opisthosoma.

Ricinuleid spiracles are crescent shaped, with the opening lined by rows of short setae. The spiracles are surrounded by rigid sclerotization, embedded in the membrane that makes up part of the pedicel. Some ricinuleids use a plastron to help maintain an oxygen supply in waterlogged conditions, e.g. *Cryptocellus adisi*. This species is found in subsoils of primary and secondary upland forest of Central Amazonia. When the subsoil in which it lives is temporarily saturated, it is able to retain an air bubble on the comma-shaped, cup-like setae on its body and legs. It is also bedecked in honeycombed depressions covered in minute hairs.

LEFT The distinctive dorsal surface of a ricinuleid opisthosoma, reminiscent of scutes on a turtle.

OPISTHOSOMA The bulbous opisthosoma is attached to the prosoma by a narrow pedicel, which can't be seen from above. The opisthosoma is considered by many to have ten segments, although not all authorities agree, as some think it is 11 (if the pygidium is considered to have four segments). The dorsal surface of the opisthosoma is very distinctive, because the four main tergites are subdivided into three smaller platelets surrounded by intersegmental membrane.

The ninth tergite is very important for the coupling process, as it forms the anterior ridge that fits into the carapace. The tenth tergite is actually the first visible tergite when a ricinuleid is coupled. On the membrane of the midline of the opisthosoma, are four long and narrow pleurites, or side sclerites. Apart from a few pseudoscorpions, ricinuleids are unusual in possessing these lateral plates. The entire opisthosoma is curved upward, because the sternites are convex in shape.

PEDICEL AND EXTERNAL GENITALIA It is only in the ricinuleids that the genital opening in both sexes is located on the pedicel, and it is normally concealed when the prosoma and opisthosoma are hooked together. It is therefore necessary to be 'unhooked' during mating and the laying of eggs. The female's genital opening is much larger than that of the male and has two sclerotized lips that are shaped liked a crescent moon, whereas in the male the genital opening is long and tubular. Some authorities class it as a 'penis', but it is not used as such. In mites, the term 'spermatopositor' is used, which might be more appropriate here.

COUPLING MECHANISM The prosoma and opisthosoma have the unique ability to lock together. The coupling mechanism that enables them to do this involves several structures. Coupling of the prosoma and opisthosoma occurs on both the dorsal and ventral surfaces, so ricinuleids are really able to 'batten down the hatches'. There is a deep groove that runs

widthways across, on the dorsal surface of the anterior edge of the opisthosoma. In front of this is a ridge that is covered by tubercles. The carapace has a corresponding ridge at the ventral edge that fits neatly into the abdominal ridge and the two ridges hold gently together.

There is another coupling mechanism on the ventral surface. Keels on the borders of coxae 4 fit into two sockets on sternite 10. This enables the prosoma and opisthosoma to be tightly coupled. Interestingly, larvae are able to keep their prosoma and opisthosoma coupled, even though they don't yet possess a fourth pair of legs.

TELESCOPIC PYGIDIUM The last three segments (14–16) at the end of the opisthosoma constrict tightly and the tergites and sternites are fused to form a short, conical, tail-like structure called a pygidium. There are a few other arachnids that possess a pygidium – the Uropygida, Schizomida and Palpigradi. Adults tend to have a partially retracted pygidium, whereas larvae and nymphs wander around with their pygidium fully extended. The reason for this is not clear.

SEXUAL DIMORPHISM In adults, the main sexually dimorphic character is the male copulatory apparatus on the third pair of legs. Other differences include the following: females may have a broader carapace than males; the cucullus of the male is usually wider, e.g. as in *Cryptocellus foedus* and slightly longer than that of the female; the femur of the second pair of legs is often sexually dimorphic – in adult males, it is usually enlarged and can be more than twice the diameter of that in females; the patella in adult males is longer and more curved.

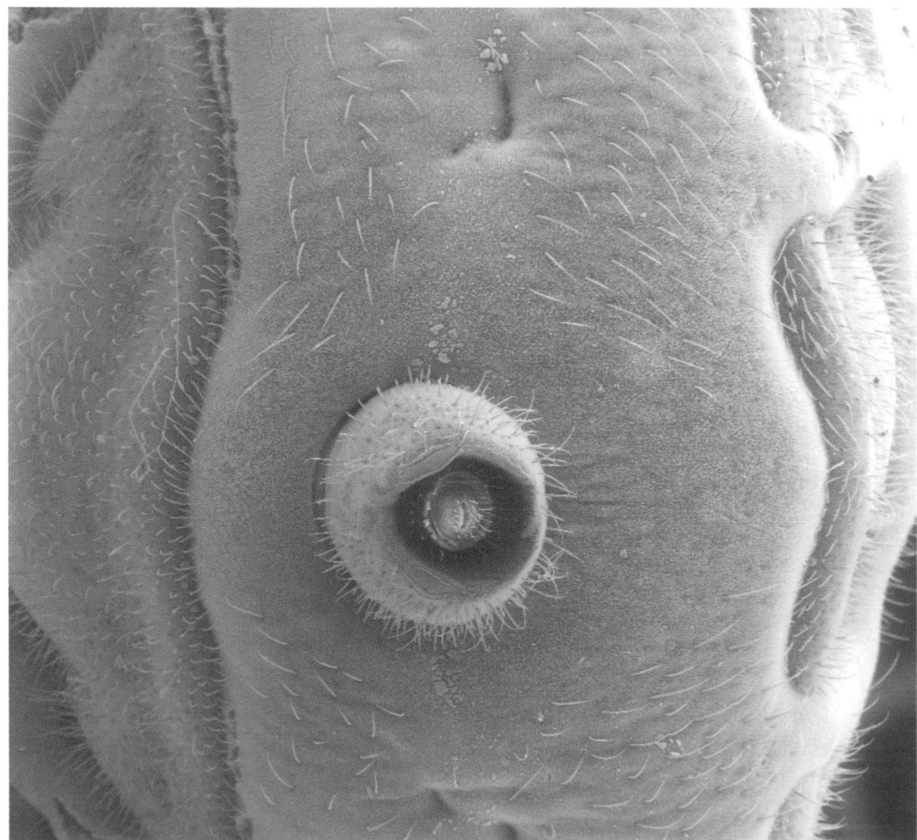

RIGHT What sets the Ricinulei apart is that their pygidium is telescopic. Segment 16, known as the anal or terminal segment, can be withdrawn into the preceding segment, and so on – and the whole pygidium can be drawn into the opisthosoma.

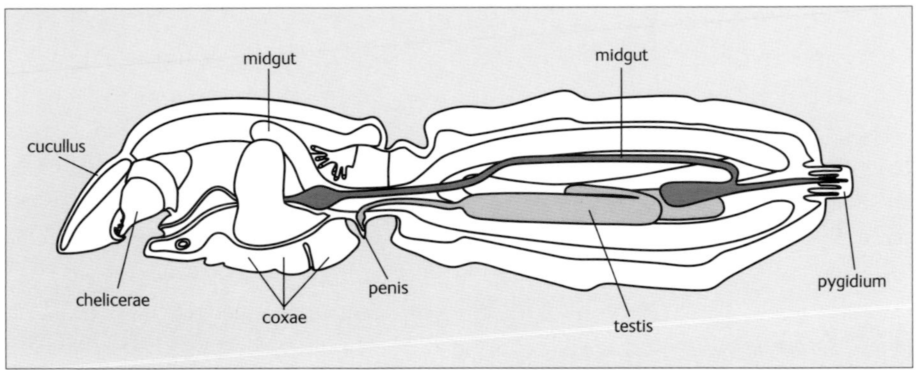

LEFT Cross-section through a generalized male ricinuleid.

Internal anatomy

Not surprisingly, given their rarity up until recently, little is known about the internal anatomy of ricinuleids. However, the excretory system does consist of Malpighian tubules and a pair of coxal glands, and the circulatory system is not highly developed. Recently, the midgut of *Cryptocellus boneti* was studied. It was discovered that the epithelial cells of the anterior part of the midgut and the diverticula have both secretory and digestive roles. However, in contrast, the epithelial cells of the posterior part of the midgut, as well as the stercoral pocket, only have one type of cell. The cells of the midgut don't seem to be involved in food absorption, which is unusual, as that's where it happens in other arachnids.

There is more known about the genitalia. Females have one of three types of spermathecae – those with two sacs, those with four and those with multiple sacs. Males have tubular, paired testes, which are located in the opisthosoma and are ventral to the midgut. They have paired ducts that merge into the seminal vesicle. There are spermatozoa at different developmental stages in the testes at one time. Although the complexity of the male's copulatory structure often correlates exactly with the female's internal genitalia (i.e. the tip of the tarsal process correlates exactly with the number and shape of the spermathecal sacks) though there are species, e.g. *Cryptocellus narino* where this is not the case.

Distribution and habitats

Ricinuleids are found in Central and South America, and western and central Africa. *Pseudocellus* is found from Texas in the north to Panama in the south, and *Ricinoides* is found only in Africa. Many species have very small distributions and are highly localized. Ricinuleids are normally found in dryland forest and tropical forests, where they live in dark, humid habitats. The humidity is important as they appear to have a great sensitivity to desiccation. They mainly live in leaf litter (especially in the thinner parts of the humus layer) and soil, but they can also be found beneath rotten wood and other plant matter, and under the thin outer covering of roots. Occasionally they are found in sandy soils as well as in cave ecosystems.

General biology and behaviour

Food

Ricinuleids predate on other small arthropods, such as collembolans, *Drosophila* (Diptera) larvae and other invertebrates, such as nematode worms, spider larvae and spiderlings. They

have a tendency to go for slow-moving prey, but a couple of species, *Cryptocellus lampeli* and *Ricinoides afzelii*, have been recorded as capturing more speedy prey, such as small termites. Some species, such as *C. lampeli*, appear to be rather fussy as to what they like to consume. The more bizarre diet choices belong to cave-dwelling ricinuleids. For example, *C. osorioi* usually goes for the larvae of dipteran bat parasites. *Pseudocellus pelaezi* prefers dead bodies of other cave dwellers, such as crickets, amblypygids and millipedes, and also feeds on the faeces of bats! So, they are also detritivores and scavengers. Ricinuleids in Central Amazonia have been seen to become prey themselves to *Peripatus* (Onychophora).

The extensive sensilla on the distal parts of the first and second pair of legs are important in prey detection, the second pair of legs act as functional antennae. The prey does not have to be touched for feeding to start. Prey is captured by the chelae on the palps, pushed into the preoral cavity and then held by the chelicerae. The lowered cucullus helps the chelicerae to hold onto the prey. In the species *Cryptocellus lampeli*, the cucullus is actually used in prey capture, to hold down the head of the prey (a termite) while the palps find a good grip elsewhere on the body.

The prey is squashed to a pulp in the preoral cavity by the alternate movements of the chelicerae, along with movements of the cucullus, which has to be raised in order for feeding to take place. If the prey is particularly large, it will be partly dismembered to make it a more manageable size. The discarded parts won't be wasted, however, because the ricinuleid will retrieve them later on. Digestive juices are then discharged onto the pulp, and the soup is sucked into the mouth, a process repeated several times until the pulp has been totally ingested. Total feeding time to consume a *Drosophila* larva is about 40 minutes. The undigested remains are then dropped.

Moulting

Moulting normally takes places deep within burrows. An adult that has just moulted is a distinctive reddish colouring, which is interesting, because other arachnids that have just moulted are pale until full sclerotization has taken place.

Life history

Reproduction

Not a great deal is known about the mating habits of ricinuleids. When a male meets a receptive female, he immediately makes his presence felt by tapping and stroking her with his long second pair of legs. This courtship doesn't last long though, a maximum of five minutes. The male then mounts the back of the female, with his fourth legs grasping her opisthosoma.

To get into position, the male wedges the edge of his cucullus between the female's opisthosoma and carapace, and then they both turn to face in the same direction. The male continues to stroke and tap the female with his legs for another five minutes. At this point, the male tilts his opisthosoma upwards in order to expose his genital aperture and moves one of his third pair of legs forwards. By twisting and manipulating the metatarsal and tarsal processes together, he is able to form a ring-like structure.

He then moves the manipulated third leg forward under his body and collects a spermatophore. The spermatophore is a small white ball of around 0.5 mm (1/50 in) diameter, with a fairly hard surface. It is collected into the ring on the third leg – much like an egg in a

spoon in an egg-and-spoon race! Once the male has his spermatophore in position, he pushes it against the female's exposed genital opening. He then appears to rub the spermatophore inside, by moving his leg backwards and forwards for about 15 minutes. The couple stay together for around a further 60 minutes before going their separate ways. The female then stores the sperm in spermathecae until fertilization.

Some species, e.g. *Ricinoides hanseni,* transfer a spermatophore to their copulatory structures before they look for a female (as with male spiders). *Cryptocellus becki* from Brazil has been recorded as having no distinct reproductive period i.e. larvae can be found all through the year, so this may hold true for all ricinuleids

EGG LAYING AND MATERNAL CARE Ricinuleids normally lay 1–2 eggs. The analogy of the ricinuleid hood being like the hood on a pram is quite appropriate it seems, because in most species, the female carries her eggs under it until they hatch. In some species, the female carries her eggs using just her chelicerae and palps, or the palps on their own.

Post-embryonic development

The different developmental stages of ricinuleids are larva, protonymph, deutonymph, tritonymph and adult. In *Ricinoides afzelii* it takes about 1–2 years to reach maturity. There

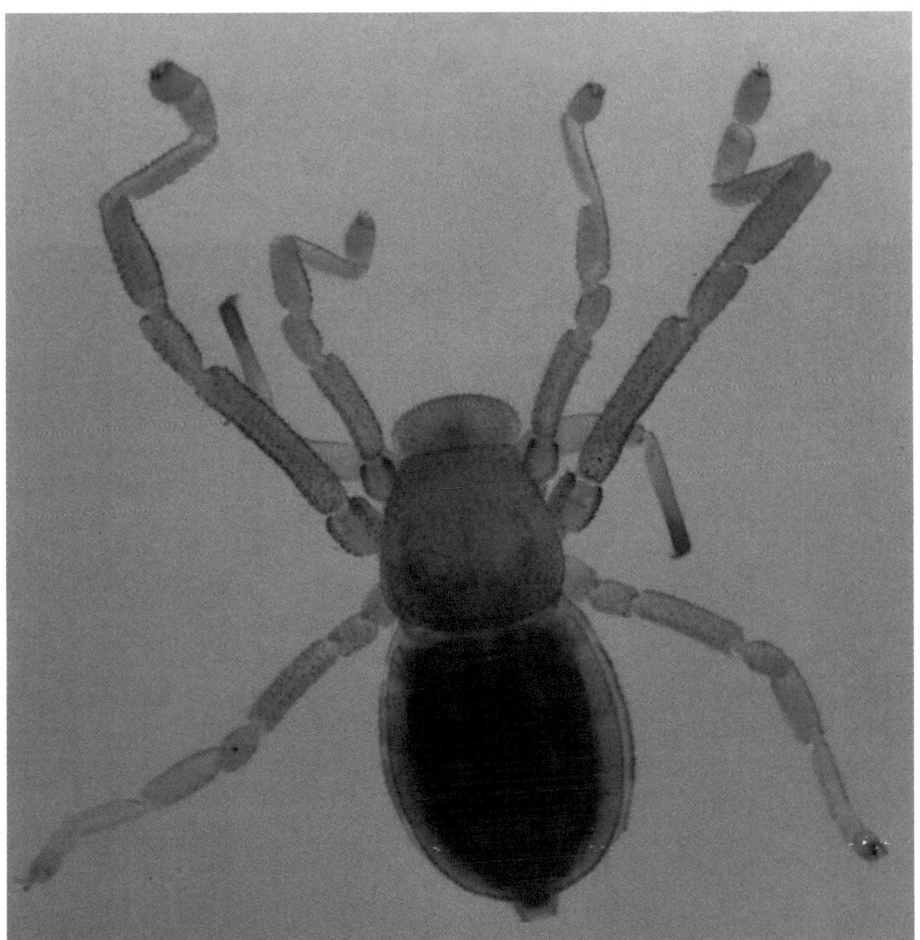

LEFT Ricinuleid larvae (here *Pseudocellus pearsei* from Mexico), don't have a fourth pair of legs; instead they have a stubby pair of small limb buds, which lack claws. When they moult into the next instar, their fourth legs appear. The Acari, which Ricinulei are closely related to, also have six-legged larvae.

ABOVE Two nymphs of *Ricinoides atewa* (Atewa forest in southeast Ghana).

are quite a few differences between the different developmental stages. In immatures, the tergites and sternites are more widely spaced, and there are fewer of them than in the adult. In the adult, they almost touch. They are also lighter in colour, being a pale brownish-yellow. Tritonymphs and adults have differences in the degree of the fusion of the abdominal sternites.

The exoskeleton of immatures has many more tubercles than in the adult. Immatures have a uniform covering of tall conical tubercles, even on the cucullus and legs, and are found on practically all of the body's sclerites. These tubercles are much more prominent than those in the adult, which are more button-like. After each successive moult, tubercles 'migrate', until they are concentrated on the posterior edge of the carapace. Pre-anal setae, which are flattened broad setae on the last segment before the anal segment, are present in larvae and nymphs, but not in adults. Immatures are creamish-brown, whereas adults are dark reddish brown, even when they have just moulted. In the larva, the carapace is almost square, but with each successive moult, it becomes wider just past the halfway point. The carapace length stays fairly constant, whereas the body length varies at each developmental stage.

Immatures have only a few teeth on their chelicerae, e.g. in *Cryptocellus lampeli* the larvae have only one large and one small tooth on the fixed finger, and five teeth on the moveable finger, whereas adults have ten largish teeth on their chelicerae. Larvae have only six legs when they emerge, but once they moult into the protonymph stage, they develop eight legs.

This is the same in Acari. The adult male's impressive copulatory organs on the third pair of legs first appear in the deutonymph, as tarsomeres 1 and 2 and the metatarsus become noticeably different to those of the female. By the tritonymph, the metatarsus has widened out and formed a cavity on the dorsal surface.

Longevity

Quite amazingly, these little beasts are very long lived given their diminutive size. Studies on *Ricinoides afzelii* have established that their lifespan is between five and ten years! Certainly, a specimen of *Cryptocellus lampeli* was kept alive in captivity for three-and-a-half years.

8 Acari
Mites and ticks

Mites form the most diverse group of all the Arachnida, found throughout the world, in decomposing organic matter and soil, on plants, in fresh water, in deep marine habitats, on and in animals. The microhabitats on animals in which they are found are astounding – inside bat anuses, inside the nasal passages of seals, in porcupine ears, in the lungs of owls, within the skin of the legs of chickens and in the digestive systems of sea urchins!

Mites are the only arachnids to have a large range of feeding habits. As well as predators, the order includes internal and external parasites of vertebrates, invertebrates and plants, scavengers, detritivores and fungivores. Many are omnivorous, while others are specialist feeders. Ticks are ectoparasitic mites that only feed on blood. Because of their feeding habits, some acarines are of medical and veterinary importance. Ticks transmit a greater variety and number of disease organisms than any other arthropod worldwide. They are almost as important as mosquitoes as vectors of disease in humans and are certainly *the* most important vectors of disease in domestic animals. Mites can cause conditions such as scrub typhus, dermatitis, asthma and scabies in humans, as well as mange and respiratory problems in some animals. They are also of commercial and agricultural importance, being parasites of crop and ornamental plants, parasites of honey bees, and infesters of foodstuffs. However, mites are extremely useful as bio control agents, especially on other mites that are plant parasites.

Mites are the smallest of all arachnids. The smallest mite is the worldwide microscopic *Eriophyes ribis* that is smaller than a full stop at a miniscule 0.1 mm (1⁄250 in)! The largest acarine is the tick *Amblyomma clypeolatum* (from India), which can reach a length of 30 mm (1¼ in) after feeding and is the size of a large grape!

Classification

The higher classification of the Acari is by no means firmly established. Most people consider the Acari to be a subclass, but many classify it as a superorder. Some classifications list the superorder rank as order, and the order rank as suborder. For example, Hillyard's *Ticks of North-West Europe* classes Ixodida (ticks) as a suborder, whereas Baker's *Mites and Ticks of Domestic Animals* classes ticks as an order. Be that as it may, most classifications divide the Acari into seven orders (suborders) within three superorders. The order (suborder) of ticks, Ixodida, is divided into three families – Ixodidae (hard ticks), Argasidae (soft ticks) and Nuttalliellidae. There are 889 species of ticks in total – 702 species in the Ixodidae, 186 species in the Argasidae, and just one species (*Nuttalliella namaqua*) in the Nuttalliellidae.

OPPOSITE The setae of the peacock mite are used to defend them against predators, as well as possibly aiding in dispersal by the wind.

RIGHT Summary of the acarine superorders and orders, giving the numbers of described families, genera and species (after Walter and Proctor, 1999).

Acarine superorders and orders (with old names in brackets)	Families	Genera	Species
Opilioacariformes			
Opilioacarida	1	9	17
Parasitiformes			
Ixodida – ticks (Metastigmata)	3	12	880
Holothyrida (Tetrastigmata)	3	9	32
Mesostigmata (Gamasida)	73	637	11,632
Acariformes			
Oribatida (Cryptostigmata)	150	1,100	11,000
Astigmata (Sarcoptiformes)	70	627	4,500
Prostigmata (Trombidiformes, Actinedida)	131	1,348	17,170
Total	**431**	**3,742**	**45,231**

Diversity

The diversity of form of acarines is staggering, which is not surprising, given that there are seven orders in the group with over 45,000 species. A.D. Michael was absolutely right when he said in 1883 that 'only those who are not acquainted with the Acarina can suppose that none is beautiful'. There are three superorders, the Opiloacariformes, the Parasitiformes and the Acariformes. The Opilioacariformes are relatively large (greater than 1 mm, or 1/32 in), and are thought to be the most primitive living mites. They got their name because they look superficially like opilionid species in the suborder Cyphophthalmi. They have a leathery exoskeleton (like argasid ticks), long legs with white banding, and a brown and purplish striped body. The name Parasitiformes: comes from the parasitic lifestyle many of them have. There is a particularly large size range in the Parasitiformes from exceptionally large ticks (up to 30 mm, or 1¼ in long after feeding) to tiny mesostigmatids at just 200 µm (1/125 in). The Acariformes are the most diverse form. Acariform mites are some of the smallest terrestrial arthropods at less than 250 µm (1/100 in) in length, and can be as small as 149 µm (3/500 in). The following are some examples of the diversity of shape and form within some of the main orders.

BELOW Ixodidiae have a hard, plate-like shield called a scutum and this and prominent mouthparts can be seen from above, as in this male *Amblyomma hebraeum*.

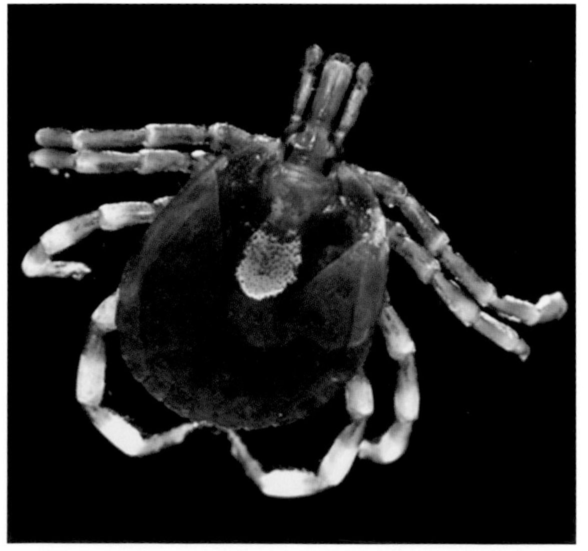

Ixodida

ARGASIDAE (SOFT TICKS) Argasids are considered to be soft because they don't have a scutum – instead they have a leathery integument across the whole surface. The mouthparts are not usually visible from above. *Argas reflexus* is elongate with a surface covered in various sized discs. *Argas vespertilionis* has a very rounded shape with an interesting pattern of radiating lines on its dorsal surface. *Ornithodorus savignyi* has an almost shield-shaped body, which is covered in tubes.

Order/family	Characteristics
Parasitiformes	
Ixodida (ticks)	• quite variable – the major variations are found between the 'hard' and 'soft' ticks • up to 3 mm (1¼ in) in size
Holothyrida	• a small order of beetle-like mites • relatively large – up to 7 mm (¼ in) in length • heavily sclerotized body
Mesostigmata	• relatively large, ranging from 4 mm, or 1/8 in, (in the genus Megisthanus) to 200 µm, or 1/125 in • body usually dorso-ventrally flattened, may be vaguely rectangular or oval • adults characterized by a series of dorsal and ventral sclerotized plates • found in a wide range of habitats, including leaf litter, vertebrate nests • most are predatory
Acariformes	
Oribatida	• strongly sclerotized • some species can withdraw their legs into the gap around their head region • can fold the anterior part of their body against the posterior • 0.2–1.5 mm (1/125–6/100 in) in size • usually found in leaf litter and soil, occasionally on foliage and tree trunks • intermediate hosts of tapeworms
Astigmata	• includes several families that are ectoparasites or endoparasites of vertebrates • includes free-living fungivorous mites found in stored foods and animal bedding, which cause allergies such as asthma and dermatitis
Prostigmata	• has the greatest number of species of all the orders, and the most diverse morphology • parasitic and free living

LEFT Characteristics of the six main orders.

Mesostigmata

PARASITIDAE Mites in the genus *Poecilochirus* are classic examples of mesostigmatids, with a very distinctive dorsal shield. *P. carabi* is phoretic on the beetle *Necrophorus vespiloides*. The varroa mite, *Varroa destructor*, is an ectoparasite of honey bees. It is distinctively shaped like a crab. Adult females are brown, and between 1.5 and 1.99 mm (6/100 and 1/12 in) wide. Adult males are yellowish in colour, 0.7 to 0.88 mm (3/100–1/50 in) wide and with a spherical body.

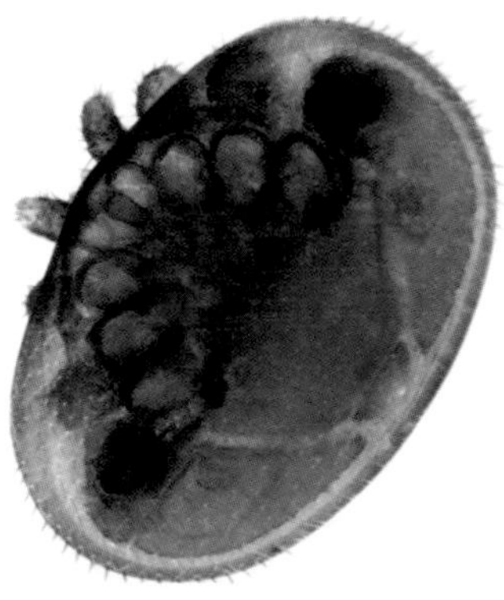

FAR LEFT Mites in the genus *Dendrolaelaspis* of the Parasitidae are small and rather rectangular in shape, with long setae all over the dorsal surface. They are usually predators on small invertebrates such as bark beetle larvae and nematodes.

LEFT The distinct flat and elliptical body of the varroa mite, which fits snugly into the abdominal folds of the adult bee, protected from the bee cleaning them off.

RIGHT Feather mites in the Astigmata usually have amazingly shaped setae that enable them to lodge themselves between the fine striate structures of a feather.

ABOVE The heavily sclerotized nature of this oribatid from the Damaeidae family is clear; it creates an armour against predators.

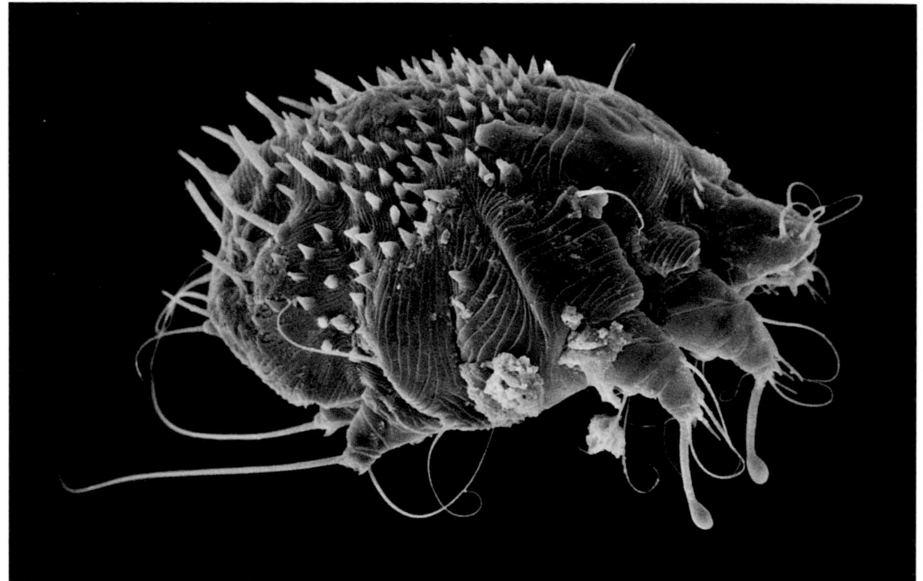

RIGHT The notorious mite *Sarcoptes scabiei* in the Astigmata causes scabies in both humans and animals. It has a globular body, short stubby legs, striated exoskeleton and long setae.

Oribatida

The oribatids are known as the 'armoured box mites' or beetle mites, as they are strongly sclerotized. They can close their globular body by folding the anterior part of their body against the posterior. Some can withdraw their legs into the gap in their armour around their head region.

Astigmata

Specialized feather mites are found in various families including Pterolichidae and Analgidae. *Ctenoglyphus plumiger* in the Glycyphagidae family is found in stored food and can cause allergies in humans and animals. It is well named, because it has the most ornate plumed setae around the margin of the opisthosoma.

Prostigmata (Trombidiformes)

TUCKERELLIDAE The stunning mites in the genus *Tuckerella* are particularly attractive beasts (see p.158). They have ornate leaf-shaped setae, which cover their protective dorsal shields, as well as 5 to 7 pairs of long trailing setae from the posterior, which give them the common name of peacock mites (also known as ornate false spider mites). These mites are important agriculturally in the tropics, as they feed on stems, fruits and grass roots. The fruit feeders can be pests of citrus.

TROMBIDIIDAE Mites in the family Trombidiidae are called velvet mites, because they possess a plush carpet of fine setae on their dorsal surface. They are very conspicuous as they wander around in search of prey, because of their colourful livery. Adults of *Dinothrombium tinctorum* (the African giant red velvet mite) can reach up to 16 mm (¾ in) in length, whereas species in the British genus *Eutrombidium* are only 4 mm (⅛ in) long.

DEMODICIDAE The follicle mites are in the family Demodicidae. Females of *D. folliculorum* are shorter than the males and have more rounded bodies.

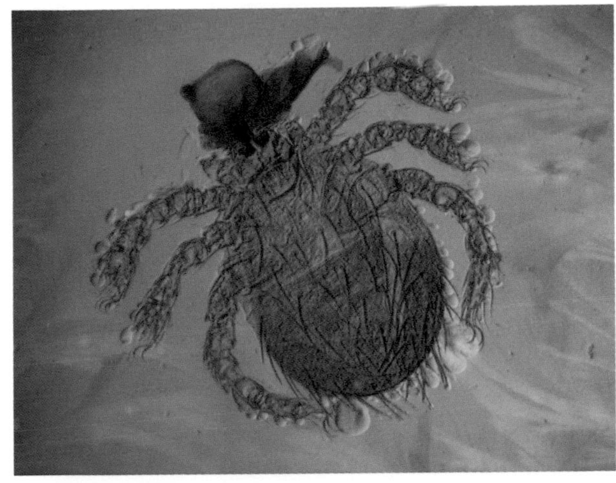

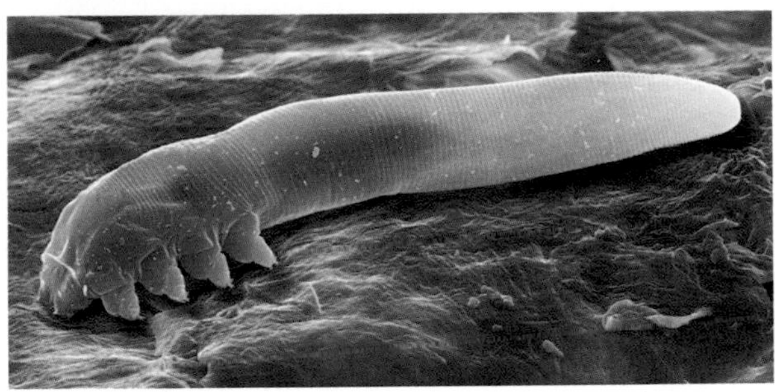

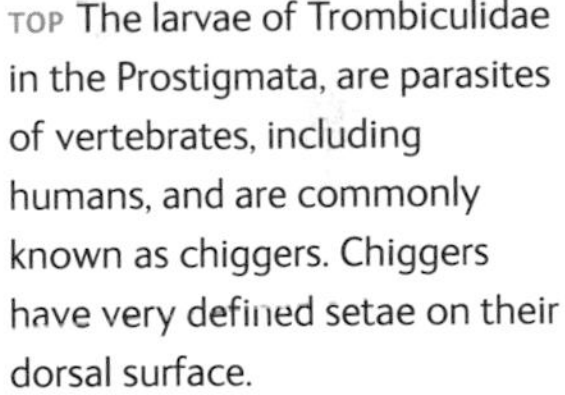

TOP The larvae of Trombiculidae in the Prostigmata, are parasites of vertebrates, including humans, and are commonly known as chiggers. Chiggers have very defined setae on their dorsal surface.

ABOVE The males of the species *Demodex folliculorum* and *D. brevis* in the Demodicidae family are peculiar, extremely small, bullet-shaped creatures with very elongate bodies that are found in eyebrow, eyelash and facial pores of humans! Their long bodies allow them to sit in the pores like a knife in its sheath.

LEFT Some trombidiids are a gorgeous ruby-red colour, and are therefore known as red velvet mites.

Colouration

Mites are a colourful bunch – some are brown, yellow, greenish, blue, and even ruby red. Some oribatids produce bright orange or red eggs. It is thought that redness evolved in terrestrial mites (such as predatory prostigmatids on leaves) for protection against UV radiation from the sun. Because bright colouration made them more vulnerable to predation, these red mites evolved the production of distasteful secretions. Other organisms are also deterred by their noxious taste e.g. hydras, notonectid bugs and dytiscid beetles. As terrestrial mites evolved to inhabit water, the red colour was retained in e.g. *Eylais discreta*, along with the ability to be distasteful and put off aquatic predators. This perhaps explains why water mites tend to avoid predation by fish, even though they may be very numerous in their aquatic habitats. Ticks don't display nearly the same range of colours as other mites, most are a shade of brown, but they can be attractively patterned.

RIGHT A male African cattle tick, *Amblyomma hebraeum* has beautiful patterns on its dorsal surface.

Anatomy

External anatomy

The morphological terminology of acarines can be problematic, because different workers use different terminology, sometimes even for the same group of mites. The two body parts are certainly not well defined. Instead of a distinct prosoma and opisthosoma, an acarine body is somewhat different. The mouthparts form a discrete region called the gnathosoma or capitulum at the anterior end of the body, which is called the idiosoma. The legs are attached to the anterior part of the idiosoma, in a region called the podosoma. The posterior part of the body is called the opisthosoma.

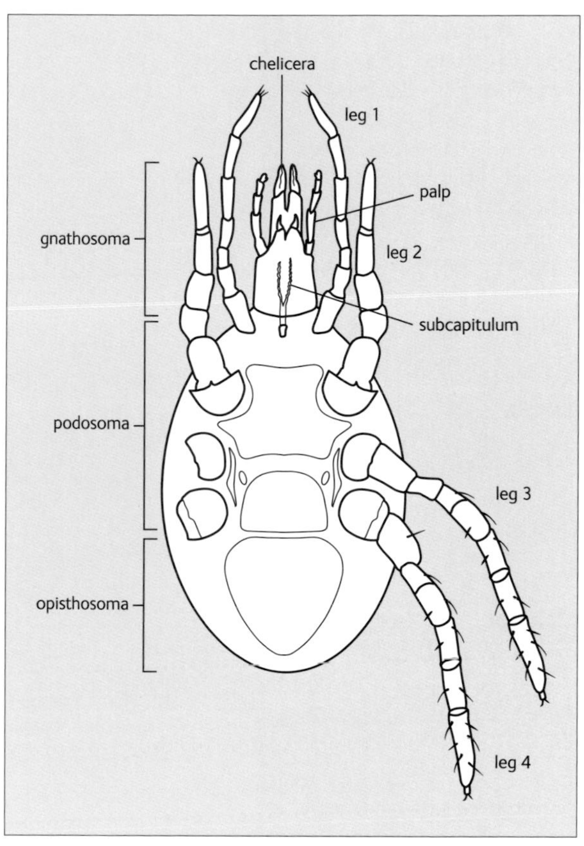

ABOVE Ventral view of *Macrocheles* sp. (Mesostigmata).

GNATHOSOMA The central part of the gnathosoma, which is vaguely triangular in shape, is called the subcapitulum and it is where the mouth is located. The hypostome is the anterior part and the basis capituli is the posterior part. Very large structures with teeth, called rutella, are found in most oribatids. A tritosternum is found on the ventral surface of many Mesostigmata. In ticks, the hypostome is an elongate, slim structure that has several rows of curved teeth on its ventral surface. This structure anchors the mouthparts to the host.

EYES Most mites have no eyes – the only mites that do are certain families in the Prostigmata, and they have a maximum of four and are found at the widest part of the body. The eyes are not easy to see, and only appear as more-or-less circular areas on the exoskeleton. A few genera of ticks have eyes just above coxae 2 and 3 in the Argasidae, and lateral to the scutum in the Ixodidae. Apparently, ticks with eyes use them for host seeking, but this has not yet been proven. Those ticks without eyes have photoreceptor cells, which detect light.

CHELICERAE In general, acarine chelicerae are adapted to chewing, sucking or piercing. A basic chelicera comprises a basal and middle segment, with a curved fixed digit and an opposing movable digit at the tip. The moveable digit is ventral and it moves in the vertical plane. There is a lot of variety in this basic form, e.g. parasitic species often have one chelicera modified to cut or pierce, like an old-fashioned can opener, while the other is much reduced. In the Oribatida, Prostigmata and Astigmata, the middle and basal segments are fused. Males of certain families (such as in the Mesostigmata) have structures for sperm transfer. There may be a spermadactyl (or spermatodactyl), which is a complicated finger-like structure on the base of the moveable digit, or there may be a spermatotreme (an opening) on the moveable digit of the chelicerae into which the spermatophore is extruded – this is found in males in the family Parasitidae. The chelicera of a tick is somewhat different to that of a mite, because it is not divided into segments. It also has two small, toothed digits at the tip. Only the teeth are exposed, as there is a membranous covering over the rest of the chelicera. Tick chelicerae are modified for tearing and cutting into vertebrate skin. In some species, males have modified chelicerae in order to transfer sperm to the female's genital opening. as with other mites.

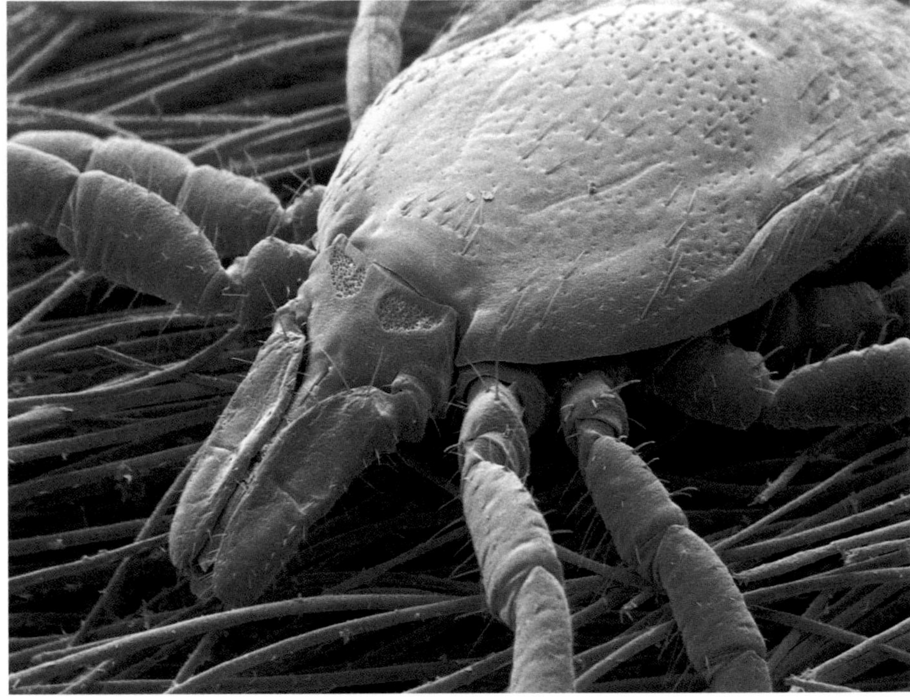

RIGHT Tick palps have only four segments. The palps of argasids are slim and all the segments are approximately the same, whereas in ixodids, the segments are variable in size – the segment at the base is short compared to that of segments 2 and 3.

PALPS Acarine palps are located on either side of the subcapitulum. They have a maximum of six segments – the trochanter, femur, genu, tibia, tarsus and apotele which has prominent claws. The tarsus often has a group of sensory setae at the tip. This basic blueprint is modified between different orders, with variation in shape and size, and fusion of segments. Only the Holothyrida and Mesostigmata have the full number of segments. The apotele on their palps is modified to form a membranous two- or three-pronged structure. Oribatids have palps that are inconspicuous and short, with five segments. The palps of the Astigmata are similar in structure, but only have two segments, while those of the Prostigmata are the most variable, with between one and five segments. Histiostomatid astigmatids are filter feeders of fine particulate matter and yeasts, and have modified palps with brushes to sweep particles towards the narrow mouth.

IDIOSOMA – DORSAL SURFACE The idiosoma can be quite variable. Although it is often oval in shape and flattened, it may be circular or extremely elongated. Mites possess a fairly typical arachnid exoskeleton. There are large pore canals, which are widespread in the procuticle, and most extend up to the epicuticle. Mites in the Parasitiformes (Holothyrida, Ixodida and Mesostigmata) have little variation in the exoskeleton. This is in contrast to the Acariformes (Endeostigmata, Oribatida, Astigmata and Prostigmata), which shows greater variation within the orders – especially in the Prostigmata.

Shields are basically localized areas of sclerotization. Sometimes they are very distinctive, sometimes just about detectable from the surrounding exoskeleton. The majority of the Prostigmata and Astigmata have an anterior shield (sometimes called a scutum), which is more lightly sclerotized than other shields. The dorsal surface of certain adult mesostigmatids, holothyrids and oribatids is covered by a single shield (the holodorsal shield). As with other arachnids, the exoskeleton in mites possesses other structures too. There are sensory setae that

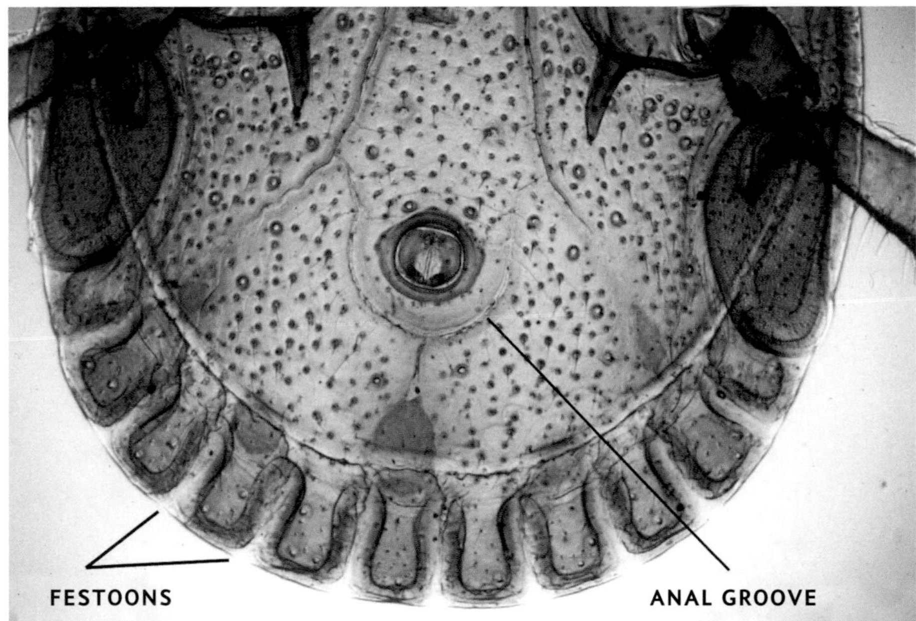

LEFT The rear edge of the idiosoma of most ixodid ticks has a number of folds called festoons, e.g. *Haemaphysalis inermis* has 11 festoons on the posterior edge; these are not found in argasids.

are chemosensory, sensitive to vibrations or sensitive to touch. Some mite orders have paired trichobothria – there is one pair in oribatids, along the anterior lateral edge and one or two pairs in some prostigmatids. Water mites have dorsal skin glands that exude foul-tasting secretions. Bending of the long sensory setae near the gland openings triggers secretion.

In ticks, wax canals have been detected only in the epicuticle of the genus *Ixodes*, in contrast to other mites, where wax canals are found only in the exocuticle and endocuticle. In immature ticks and adult females, the scutum covers only part of the dorsal surface, whereas in adult males, it covers the entire surface, and may have patterning or be plain brown. It doesn't stretch when the tick engorges on blood. As adult males don't engorge, the scutum doesn't restrict a possible increase in body size.

When ixodids have a blood meal, they have two feeding phases. In the first, there is a gradual uptake of blood, while in the second phase, the tick rapidly balloons out. The procuticle is quickly synthesized in the first phase, so it is able to stretch, while the epicuticle unfolds like a concertina. The exoskeleton is already very flexible in argasids, and there is no slow feeding phase that allows cuticle synthesis. Ticks in the genera *Haemaphysalis, Hyalomma, Dermacentor* and *Rhipicephalus* have pheromone glands, which open through a pair of foveal pores near the rear margin of the scutum in females, and in the centre of the dorsum in males.

LEGS STRUCTURE AND FUNCTION In acarines, each leg consists of seven segments – the coxa, trochanter, femur, genu (patella), tibia and tarsus. The terminal segment on the tip of the tarsus is called the ambulacrum; this structure is very variable among the different orders. The segment is formed from a pre-tarsus, which has a membranous structure called a pulvillus that often curves like the petals on a flower, and has up to three claws. Ixodida have two claws, Astigmata have one, Holothyrida and Mesostigmata have two, Prostigmata usually have two claws but a few families have only one, and there are between one and three claws in Oribatida. The pulvillus is similarly variable as it may be sucker-like in parasitic Astigmata. In Prostigmata, the pulvillus is replaced by an empodium, which may be claw-like but is often

membranous. The coxae are fixed in the Oribatida, Prostigmata and Astigmata, but free in Ixodida, Holothyrida and Mesostigmata.

Apart from locomotion, the legs have a sensory function, possessing a variety of sensory setae that detect potential food and changes in environmental conditions. These setae are mainly located on the tibiae and tarsi of the first two pairs of legs. The legs of some mites are modified for food capture. For example, species in the aquatic genus *Agauopsis* (Halacaroidea) have modified spines on their first pair of legs, used to capture prey. Other halacaroids, e.g. *A. filirostris,* have a much longer first pair of legs, which helps to capture large copepods.

Mites that live among hairs, i.e. ectoparasites of mammals, have hook-like structures on the tips of the legs which enable them to cling on. Other adaptations include suckers on the fourth pair of legs (in some Astigmata) and spurs on the second pair of legs (some Mesostigmata males). The fur mites – the families Atopomelidae and Chirodiscidae (Astigmata) – attach themselves to their hosts by their modified first and second pairs of legs. Some atopomelids have a ventral groove running from anterior to posterior, enabling them to clasp the hair shaft. The legs can also be modified for other reasons, such as for spermatophore transfer. Water mites may have any one pair of their legs modified for that purpose. Many mesostigmatids have modified legs in order to attack courting rivals, or to hold the female during mating.

Haller's organ is found only in ticks. It consists of two small patches of olfactory and chemosensory setae, located on the dorsal surface of the tarsus of leg 1. It is extremely sensitive to temperature, carbon dioxide, humidity, ammonia, pheromones *and* airborne vibrations.

RESPIRATORY STRUCTURES No acarines possess book lungs, as found in other orders of arachnids such as spiders and scorpions. They usually have a highly branched tracheal system with spiracles. These differ in position and number, depending on the species of mite or tick. For example, the Opilioacarida have four pairs of dorsal spiracles, whereas the Mesostigmata have only one pair, located laterally between coxae 2 and 4. The name Mesostigmata is derived from the mid-body position of the spiracles.

From each spiracle, a narrow channel of variable length, called the peritreme, extends forward. In ixodid ticks, a pair of spiracles is located below coxa 4. In argasid ticks, the spiracles are located near to coxae 3 and 4. Depending on the species, there may be a sclerotized plate (stigmatid plate) surrounding the spiracle. In ixodid ticks, there are muscles attached to the 'stigmatid plate' that dilate the spiracles.

EXTERNAL GENITALIA The external genitalia of acarines vary considerably between the groups. The genital opening is found between coxae 2, 3 or 4. It might be covered by one or two flaps or have associated shields. The females of many species have spermathecae or seminal receptacles near the external genitalia. These are temporary storage structures, which enable the female to store sperm until she is ready to fertilize her eggs. In many species, females have a secondary genital opening into which males deposit their sperm via an aedeagus. This is known as a bursa copulatrix and is usually located below the actual genital opening, on the posterior idiosomal margin. Although it is found in both astigmatids and prostigmatids, the term is typically associated with astigmatids. Males of these species may have associated copulatory structures.

SEXUAL DIMORPHISM Because there are so many species of Acari, there are obviously many different variations of sexual dimorphism.

Order	Evidence of sexual dimorphism	Examples of variations
Mesostigmata	common	• general arrangement of sclerites • leg modifications for fighting in males of some species • modifications of chelicerae for sperm transfer
Astigmata	extreme in feather mites	• many of the elaborate structures in males, the functions of which are little understood
Prostigmata family Demodecidae	common	• body shape and size • dorsal podosomal tubercles • slits and pores in exoskeleton • palps and claws • *Soricidex dimorphus* shows extreme sexual dimorphism
Ixodida family Ixodidae	well marked in adults	• scutum covers the entire dorsal surface in males, but is confined to the anterior surface only in females • females of *Ixodes* and *Haemaphysalis* have larger bodies than males, but males of *Hyalomma, Rhipicephalus* and *Dermacentor* often have larger bodies than those of females
Ixodida family Argasidae	not really evident	• genital aperture

LEFT Examples of sexual dimorphism across several orders.

Internal anatomy

CENTRAL NERVOUS SYSTEM Acarines have a well-developed central nervous system, comprising supraesophageal and subesophageal ganglia encircling the oesophagus. Nerves from these ganglia innervate the mouthparts, eyes (if they are present), reproductive organs, alimentary system and legs. In ticks, the supraesophageal region contains paired optic ganglia, and a branch from the palpal nerve innervates the salivary glands. There are different sensory structures on the body surface of acarines. Depending on the species, these include tactile setae, chemosensory setae, trichobothria and ocelli. Those species without such sensory structures may have areas of photosensitivity on the dorsal surface, or on the first pair of legs, e.g. *Ophionyssus natricis*.

CIRCULATORY SYSTEM In general, the acarine circulatory system is similar to that found in other arachnids. Some mite species in the Antarctic secrete an anti-freezing agent into their haemolymph, thereby preventing death by freezing (as do a few spider and scorpion species.) An interesting difference to other arachnids is that, although the quantity of haemolymph in a tick's body usually remains relatively constant in relation to the body weight, it increases greatly during feeding.

RESPIRATORY SYSTEM Some mites, such as mites in the superfamily Eriophyoidea (Prostigmata), don't have any respiratory structures; they exchange respiratory gases with the environment through

their exoskeleton. The folds of the thin, transparent exoskeleton of these mites possibly enlarge the area for gas exchange by diffusion. Mites living in polar soils that are often waterlogged have regularly spaced granulations on their exoskeleton, e.g. *Nanorchestes* spp. (Prostigmata), which help to maintain a layer of air over the body, aiding gas exchange across the exoskeleton.

The structure of tracheal systems in mites appears to particularly benefit those that live in aquatic or semi-aquatic environments. Because mites have small spiracles (due to their small body size), water is much less likely to enter into the tracheal tubes. Mites that inhabit the seashore often have no special adaptations to living in a watery environment, but can survive being submerged for long periods of time, because they hold air in their extensive tracheal systems. Some uropodid mites that live in the intertidal zone may capture air bubbles on their leg bases, or on their roughly structured integument, which act in the same way as the plastron used by the aquatic spider *Argyroneta aquatica*. In other aquatic mites, such as *Hydrozetes*, there are small projections near the spiracles, which help to retain a thin layer of air. Some species, e.g. *Limnesia maculata* alter their oxygen consumption to suit the levels in the water.

There is an atrium below each spiracle, which is connected to a maximum of eight large tracheal trunks; this leads to a system of branching tracheae. In general, larval ticks don't have tracheal systems, except in certain argasids, which have a simple system.

DIGESTIVE SYSTEM The structure of the digestive system varies between groups within the acarines. Most parasitiform and prostigmatid mites have similar digestive systems to other arachnids, as they ingest fluids that have been broken down externally. The mouth is connected to a muscular pharynx, which connects to a tubular oesophagus and then the midgut. The midgut may have up to seven caeca and can take up a lot of space in the body. For example, in *Myobia murismusculi* (Prostigmata), the entire body cavity is filled by the midgut and its caeca. From the midgut, a short intestine leads to the hindgut, then to the rectum, and opens at the anus. The anus is usually located on the ventral surface of the opisthosoma, although location varies with species. In Ixodida, Holothyrida, Mesostigmata and Oribatida it is covered by sclerotized shields, whereas in most Prostigmata and Astigmata, it is covered by a pair of cuticular flaps. There is a pair of anal suckers in some astigmatid males.

Opilioacariform and sarcoptiform mites are totally different, however, because they ingest food fragments. The chelicerae cut off pieces of food, aided by the rutella, which are then moved into the mouth. The fragments form a ball of food in the oesophagus, and then get compacted into a food bolus. This passes slowly from the midgut to the anus and is partly digested en route. The gut caeca are simple lobes, and accumulate some of the digestive products from the bolus. However, the bulk of the bolus remains intact, and is finally excreted as a faecal pellet. To aid expulsion, the anus has evolved into a rather large opening, covered by a pair of valves, like miniature trapdoors.

A tick's digestive system shares the same structural layout as that of other mites, although it is specialized for blood feeding. Unlike other mites, ticks possess a salivarium, which is a fusion of paired salivary ducts. A salivary duct extends back into the lateral regions of the body cavity, where it joins a large pair of salivary glands. The mouth is more of a small aperture than a cavity. In ixodids, the mouth leads straight to the pharyngeal valve, whereas in argasids it leads directly to the pharynx. The pharyngeal valve closes when the pharynx contracts. This minimizes or completely stops regurgitation. In argasid ticks, however, the valve can be much more basic or even non-existent; here movements of the labrum stop regurgitation when the pharynx contracts.

EXCRETORY SYSTEM The types of excretory organs present vary between species. Midgut cells serve as excretory organs in most acarine species, by absorbing waste products during digestion and later discharging them into the midgut. The waste products are then expelled from the body. In addition to the midgut cells, there may be one or two pairs of Malpighian tubules. Ticks have only one pair, while in some mite species they are absent altogether. Malpighian tubules are usually located at the join between the midgut and hindgut. Nitrogenous waste products accumulate in these tubules, and are then consolidated into guanine crystals, which are voided along with other waste products. Tick faeces are black, because they are iron-rich from the blood diet. The telltale signs of a tick feeding site are small black pellets accumulated in the host's fur. Coxal glands open on or near the coxae. In mites, the location of the coxal gland openings varies between species. Coxal glands are absent in all ixodid ticks, whereas in argasid ticks, there is a pair that open above the coxae of legs 1 and 2.

A well-developed excretory system is lacking in certain mites, such as those in the superfamily Eriophyoidea. However, the posterior midgut does have a role in water and ion regulation. Metabolic wastes are stored in the midgut cells or possibly in the connective tissue surrounding these cells.

INTERNAL GENITALIA The reproductive systems of both ticks and other mites are generally rather similar. In females, the ovaries may be single, paired or clustered. The ovaries have one or two oviducts, which may be long and complex, that join the single uterus. There may be specialized regions in the oviducts that produce the outer layers of the eggs. Usually, the uterus opens directly through the genital pore on the ventral surface, between the first and second pairs of legs. Sometimes there is a cuticle-lined vagina instead of the uterus. Seminal receptacles are connected to the uterus.

Males have single or paired testes, with paired or fused vasa deferentia. The spermatozoa travel from the testes, down the vasa deferentia and to the ejaculatory duct. Many mites, e.g. in the Mesostigmata and Prostigmata have indirect sperm transfer. Many acariformes produce spermatophores and the ejaculatory duct is often modified into a variety of internal structures that play a role in shaping the spermatophore. There are mite species that have a penis (called an aeoleagus in this group), and this is used to transfer the sperm directly into the female's genital pore.

Distribution and habitats

Mites are distributed around the world. For example, the bee mite *Varroa destructor* (Mesostigmata) is cosmopolitan, as are the ticks *Ornithodorus* and *Ixodes ricinus*. On the opposite end of the scale is *Aceria clianthi* (Prostigmata), which is a mite endemic to New Zealand.

Habitats

Mites live in every sort of habitat, both terrestrial and aquatic, apart from the open ocean. Within these habitats, the potential microhabitats are almost infinite. Ixodid ticks are particularly common in grasslands, woodlands and heath lands, especially where there are suitable hosts i.e. cattle and sheep. They occupy more types of habitat than argasid ticks, which are found mainly in dry places, and tend to live with their host, in its living place. Of the other mites, only free-living ones are discussed.

LIVING ON PLANTS There are many mite species that are found on plants. Arboreal acarines are found on leaves, branches and tree trunks, and in the soils that form in tree holes, and at the bases of branches and epiphytes. Arboreal mite families include Cymbaeremaeidae and Camisiidae (Oribatida). There are many mites that live on the surfaces of leaves. The different structures on leaf surfaces, e.g. density of hairs on leaf veins, affect the abundance and species of mites that live there. Hairy leaves often have five times more mites than smooth leaves at the same site. Predatory and fungivorous mites are the predominant dwellers within leaf domatia, which are tiny 'house-like' structures in vein axils on the undersides of leaves. They may be deep tufts of hairs, flap-like pockets or pouches, or even dome-like structures.

LIVING IN SOIL The soil-litter horizon is the most densely populated mite habitat – over 250,000 mites per square metre may be found in the top 10 cm (4 in) of soil alone! Dry and sandy soils, cold soils, and soils with well-developed organic layers are all home to soil mites, and dozens of species may be present. Mites, along with the rest of the soil fauna, are usually concentrated in the rhizosphere – the horizon made up of the surface litter layers and the area around roots and rhizomes – which provides most of the resources they require. Mites can also be found much deeper in the soil, at 2 m (6 ft 6 in) depths or more. The main soil predators are in the Mesostigmata (suborders Parasitina and Dermanyssina), and in the Prostigmata.

RIGHT This mite *Pergamasus* is a minute predator in forest and woodland leaf litter and soil and can number in the thousands per square metre of soil.

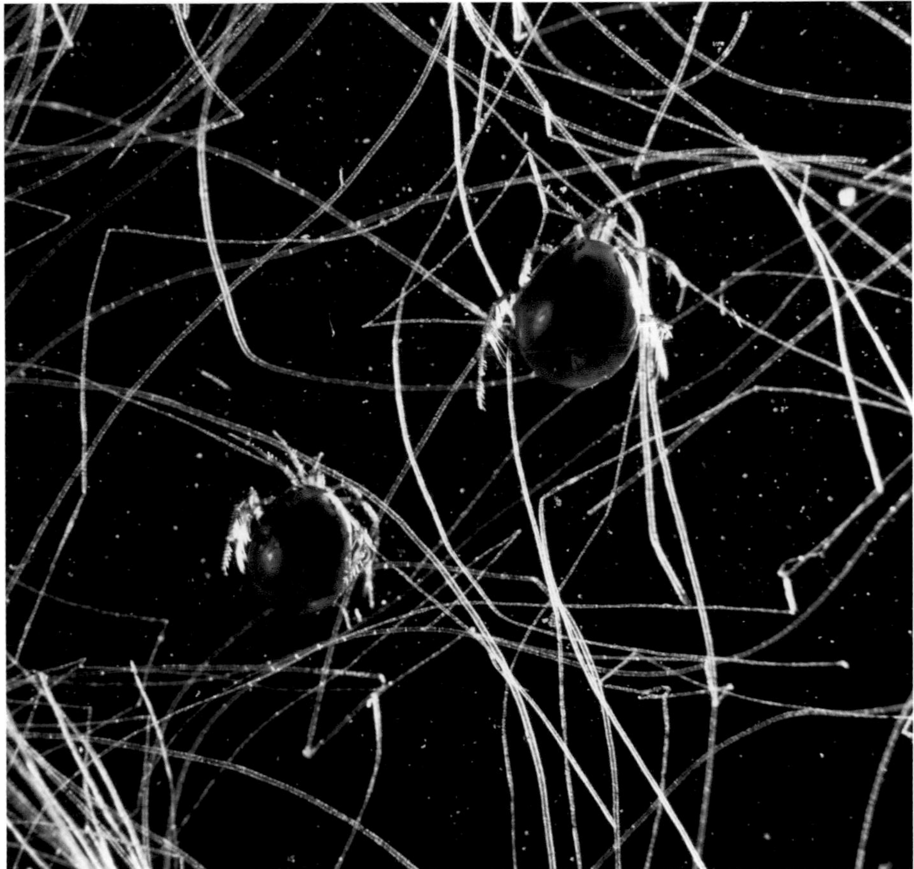

LEFT Water mites are found in almost any watery habitat, from a miniscule drop of water in a leaf axil, to the dark 4 km (2 ½ mile) deep marine abyss. They are also found in deep lakes, gushing waterfalls, rivers, ponds, bogs, hot springs, deep ground water, glacial meltwater, tidepools, and between the high and low tide marks on beaches.

LIVING IN WATER In contrast to many arachnid orders that have no truly aquatic representatives at all, mites include around 5,000 named aquatic species, usually grouped in Hydracarina (also known as the Hydrachnidia or Hydrachnellae). This contains families with some appropriately aquatic names as Hydrachnidae, Hydrodromidae and Hydryphantidae. It is thought that there are many undescribed species in this under studied group. Ameronothrids are a fascinating group that remain dormant in ephemeral potholes until these are filled with seasonal rains.

Water mites have a very wide range of thermal tolerance. *Thermacarus nevadensis* lives in thermal springs with a scorching temperature of nearly 50°C (122°F), while other species inhabit chilly ice-filled lakes. Water mites decrease in diversity and number with increasing depth. They are also rare when there is a lack of substrate, partly because they need plants as oviposition sites. Water mites may also be found on aquatic animals such as crabs, seals, otters and sea snakes.

Surprisingly few aquatic mites can swim. Many get around by crawling on the substrate e.g. aquatic plants. Those that *can* swim have swimming setae, which give the mite lift. Some swimming setae are spoon shaped, some are uniform in width with no special elaborations, while others are fringed with hairs – hairs upon a hair! Some aquatic mites have taken things to extremes – mites in the genus *Creutzeria* (Astigmata) possess extremely long whip-like swimming setae at the tips of their highly modified legs. Not surprisingly, those mites that are good swimmers, e.g. *Eylais*, are found in open water. Some aquatic mites have an even more

novel way of getting around. Mites in the family Hydrozetidae levitate up to the water surface – they collect air bubbles in their midgut, so that when they let go of the substrate, they float upwards quickly. They levitate to the surface when light conditions drop too low at their level in the water column. They may also use levitation as a rapid escape response.

ANTARCTIC MITES There are around 30 species of mites that live in the coldest place on Earth, Antarctica. They are found on rocks covered with lichen (upon which they feed), in bird nests, on penguin guano and among detritus on the seashore. Species in the genus *Alaskozetes* survive temperatures down to –30°C (–22°F) because they secrete a type of anti-freeze in their haemolymph.

Dispersal

Certain mites are able to migrate to new microhabitats. Spider mites can walk to new leaves, but they also disperse on the wind. Some species achieve this aerial dispersal by hanging down on a thread of silk – when the wind breaks the thread, they are blown off to a new place. This is similar to ballooning in spiders. Mites may be passively blown on the wind, while many eriophyoid mites actively seek dispersal by aligning themselves to air currents and springing up into the air using their caudal lobe (at the posterior of the body).

'Hitchhiking' or phoresy

One of the main ways that mites disperse is through phoresy. In the Mesostigmata, for example, many families have phoretic associations with insects. In the Prostigmata, *Microdispus lambi* (Microdispidae) from eastern Australia is phoretic on various species of flies associated with cultivated mushrooms. Its preferred host is the fly *Megaselia halterata* (Phoridae), but where necessary it attaches to almost anything that moves. The phorid flies disperse to newly established crops of mushrooms, thereby dispersing the mites as well. Perhaps the most unusual ride that a mite can take on another animal is that of mites in the genus *Rhinoseius* – these little creatures climb into the nostrils of hummingbirds to hitch a ride between the flowers on which they feed!

Phoresy is not only common in mites, but in pseudoscorpions too, as well as other non-arachnids such as insects, nematodes and molluscs. Unlike these other organisms though, mites have stages in their development in which they are specifically morphologically and behaviourally adapted to phoresy, e.g. deutonymphs of Astigmata have special attachment structures such as claspers and suckers, and have dorso-ventral flattening.

It is thought that phoresy has lead to parasitism in several groups of mites. For example, the species *Macrocheles glaber* (Macrochelidae, Mesostigmata) lives in rotting vegetation and manure, and individuals are often found attached to beetles and flies that visit the manure. Not only do the mites use these insects phoretically, but they may also suck their blood. In fact, mesostigmatids are often associated with incidental parasitism of butterflies on which they are phoretic.

Unlike other mites, ticks are not phoretic. However, as a consequence of their parasitic lifestyle, they may be carried around by their hosts. For example, when ticks are attached to birds, they can be very widely distributed and therefore able to transmit diseases to totally new areas.

General biology and behaviour

Food by parasitism

Acarines have many different ways to obtain food. Mites are parasites, predators, scavengers, detritivores and fungivores. Most groups of mites are adapted to only one of these methods, although in the superfamily Halacaroidea there are predators, parasites and fungivores. Ticks are all external parasites. Here, we look at the ways in which ticks and other mites obtain food through parasitism.

PARASITISM OF VERTEBRATES A parasite is an organism that lives in association with, and at the expense of, another organism, the host, from which it obtains organic nutrition. (The host is defined as an organism that has another organism living in or on it.) Parasites have a range of effects on their hosts; at one end of the scale, the host may continue to live and reproduce normally, while at the other extreme, the host dies. Mites have a very large number of hosts - vertebrates, invertebrates and plants. Within a parasitic mite or tick species, the different developmental stages in the life-cycle may have different hosts.

Mites live on most species of mammal. They live among hairs, in hair follicles and in skin glands, which provide wonderful microclimates. Various parts of the skin and its secretions are eaten and imbibed by parasitic mites. *Demodex brevis* (see p.163 middle) and *D. folliculorum* are found in the glands and follicles of human faces where they feed on fluid between skin tissues and secretions beneath the skin. As a genus, *Demodex* mites are not specific to humans, though. Around 150 species have been found on placental mammals and marsupials, but not on monotremes. Fur mites in the family Listrophoridae attach to the bases of hairs and eat fatty substances produced by the hair follicles. Water mites may be found on aquatic animals, such as seals and otters.

Ticks are all external parasites (ectoparasites) of animals and feed solely on the blood of their unsuspecting hosts – they don't obtain food in any other way. Ticks parasitize mammals

Silk

There are several groups of mites in the Acariformes (Prostigmata) that use silk in several ways. Some mites use silk in the most ancestral way, to make silk mats on which to lay their eggs and as moulting cocoons for larvae. Astigmatids and oribatids have lost the ability to make silk.

In several groups of Prostigmata, e.g. Eriophyidae, Eupodidae and Cheyletidae, the silk comes from within the mouth, opening from modified salivary glands. However, some eriophyids produce silk from a spinneret (hollow seta) at the tip of each palp. The silk gland is large and extends from within the body.

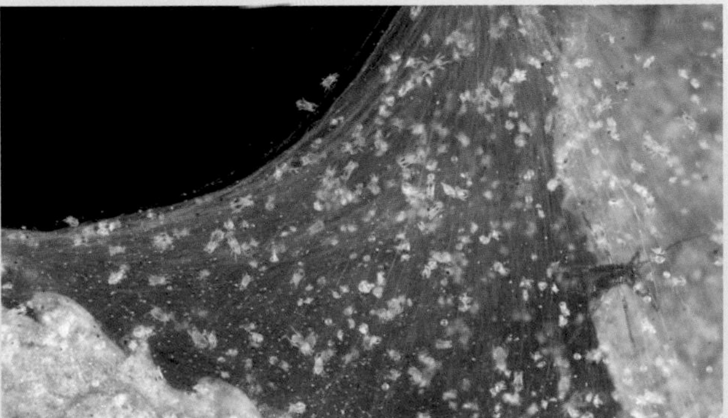

ABOVE Mites in the Tetranychidae (called spider mites because of the copious silk they produce), use silk throughout their life-cycle as protection for eggs, chambers for moulting, aerial dispersal, droplines and silken webs for shelter. Some predatory prostigmatids also use silk in prey capture.

in particular, but they can also be found on birds and reptiles. Rodents are the most popular hosts, having more species of tick than any other animals. Runners-up include rabbits and hares. Hard ticks normally have three hosts within their life-cycle, depending on the developmental stage, and these are often of different species. Most soft ticks live in the burrows and nests of their hosts. They usually have a single host, whose blood they suck as it hibernates. Free-living parasitic mites are also found in mammal nests. For example, blood-sucking mites in the genera *Haemogamasus* and *Androlaelaps* in the family Laelapidae (Mesostigmata) live in mammal nests and only go onto the host at feeding time.

Mites can be found as ectoparasites on birds, as parasites on feathers, and as free-living parasites in the nests of birds.

Ectoparasites: Many species in the order Astigmata infest birds and feed on their bodily fluids. For example, some species in the family Hypoderatidae, e.g. *Hypodectes propus*, are subcutaneous in their early life stages, and probably absorb liquid nutrients through their exoskeleton, because they lack mouthparts. Ectoparasitic mites can cause many problems for their hosts, such as reduced growth rates in juvenile pigeons, competition for food in hummingbirds and reduced virility in male domestic chickens.

Feathers: Specialized feather mites and ticks are found across nearly all orders, such as Ixodida, Mesostigmata, Prostigmata and Astigmata. However, the term 'feather mite' is usually used for those in the Astigmata and there are more than 2,000 species across the three superfamilies Analgoidea, Pterolichoidea and Freyanoidea. Feather mites are found on nearly every order of birds and have colonized all parts of the bird integument – on the surface and inside of feathers, and under the skin at the base of feathers. These plumage dwellers may be short term or long term. They may use the feathers only as a dwelling place while they suck the bird's blood, feed on skin fragments or imbibe feather oils, which is harmless to feathers. However, there are a few species that eat the 'pith' of feathers, causing quite a lot of structural damage. Feather mites have evolved the ability to steer clear of feathers that are about to be moulted, hence avoiding the problem of losing their home when the feather is shed.

Nests: Many mites reside in birds' nests, between feeding bouts on the nest inhabitants. A lot of these species aren't specific to birds, as they can be found in other animals' nests and also in human housing too – these include dust mites in the family Pyroglyphidae. Ubiquitous nest mites include *Dermanyssus gallinae* (Mesostigmata) and *D. hirundinis.*

Most soft ticks that parasitize birds do so during the host's nesting season, and live in the bird's nest. However, with domestic birds such as chickens living in a chicken house, their hosts are effectively nesting at all times of the year. Amphibians and fish don't have as large a diversity of acarine parasites as mammals and birds. Reptiles make very good hosts for a large number of endoparasitic and ectoparasitic mites. Ticks heavily parasitize terrestrial reptiles, but marine iguanas and snakes also succumb. Fish are lucky, as they appear to be generally avoided by acarine attack. When their normal food supplies become depleted, some fungivore and algivore mites may then feed on fish skin. However, it may also be the case that most associations between fish and mites just happen by chance.

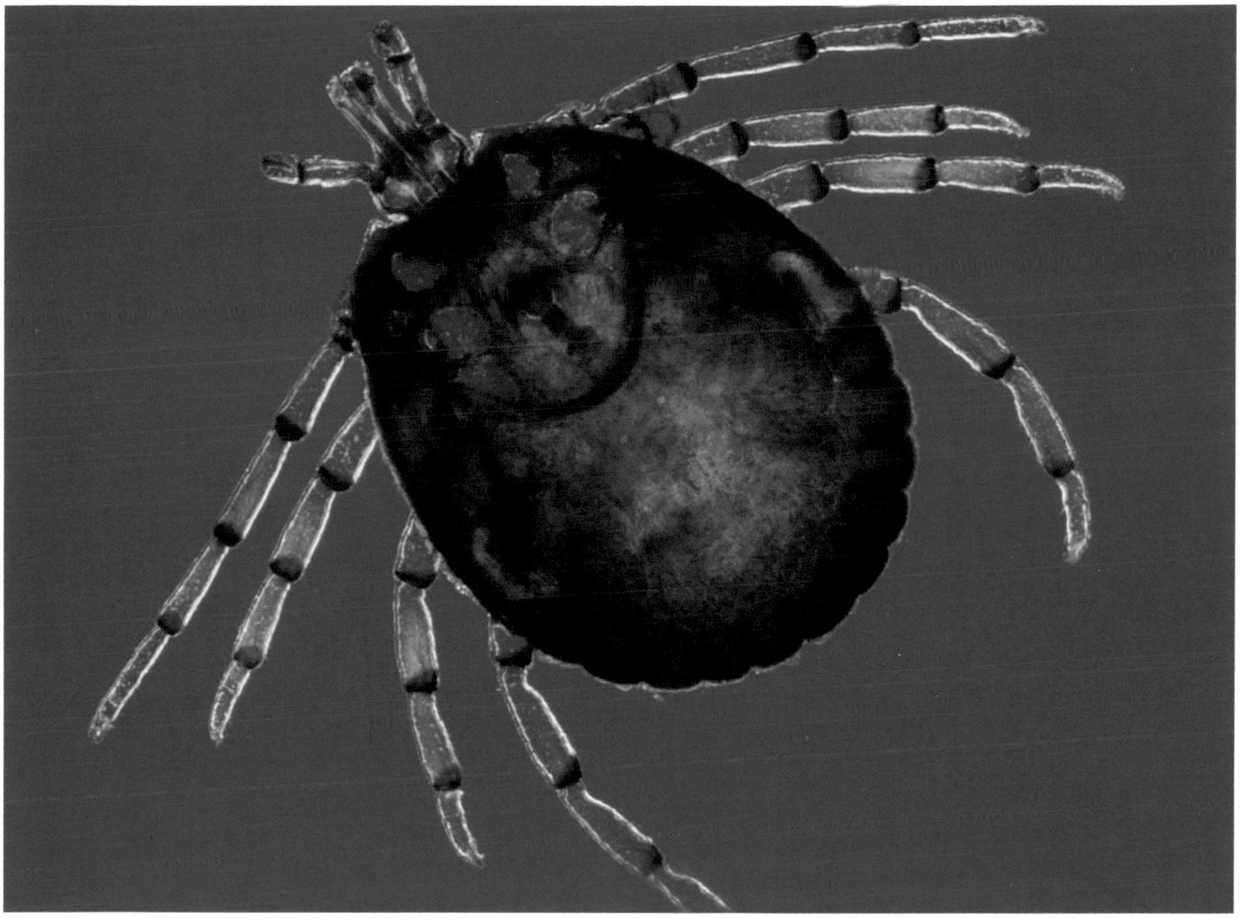

PARASITISM OF INVERTEBRATES Mites are enthusiastic parasites of arthropods, especially insects. In fact, mites have been recorded as parasitizing all of the main orders of arachnids and insects. Some insects are attacked at certain developmental stages only, e.g. thrips pupating in soil, and the eggs and larvae of fungus gnats, while other groups are attacked at any stage. The parasitization of insects by mites has many negative effects, such as reduced lifespan, increased susceptibility to predation and reduced productivity in honey bees. The crab-like mite *Varroa destructor* parasitizes honey bees (*Apis mellifera*). *Varroa* mites feed on the haemolymph of developing honey-bee larvae, pupae, and adults. They kill the bees outright, or cause them to have malformed bodies, legs and wings. *Varroa* mites can cause the loss of most of an affected colony, if left unchecked.

Ticks are also known to parasitize each other, known as hyperparasitism. In general, it tends to be males that parasitize females, and they have a particular liking for engorged individuals. However, males are less damaging to the ticks that they target; females hyperparasitized by other females have a high chance of dying, whereas female ticks parasitized by males have a much greater chance of survival.

Mites in the superfamily Halacaroidea are parasitic on crustacea and can be found in the gill chambers of lobsters and crayfish. They also parasitize molluscs, such as freshwater mussels, and feed on the mucus and blood of their host. Other invertebrate hosts include sea urchins, hydroids, slugs, millipedes and centipedes, and terrestrial hermit crabs and opiliones.

ABOVE Among the amphibians, salamanders and frogs, for example, may suffer intradermal parasitization by mites, while the respiratory passages of frogs and toads can also suffer mite infestation. *Ornithodoros* and *Amblyomma* ticks love to feed on amphibians. This tick was removed from a giant toad.

PARASITES LIVE IN UNUSUAL PLACES Mites parasitize many species of animals, but are not restricted to the same locations on each of these hosts. They are found in what we would consider to be the most unusual places, depending on the species. The following is a brief list of 'highlights' (after Walter and Proctor, 1999): 1) Although rather distasteful to us, an anus is a good place to be if you are a mite e.g. *Paraspinturnix globosus* lives in the anal canal of hibernating *Myotis* bats. These mites may be parasitic, although they might just be over-wintering. At the other end of the alimentary canal, mite infestation can cause the deterioration of the teeth of long-nosed bats. 2) Mites in the genus *Opsonyssus* are found on the eyeballs of fruit bats, and are possibly parasitic. 3) Adventurous mites dwell in the respiratory systems of aquatic mammals e.g. *Orthohalarachne* resides in the nasal passages and lungs of seals. Such parasitism can hinder the host's breathing. 4) The mite species *Pneumophagus bubonis* lives in the lungs of the great horned owl. 5) Astigmatid mites in the family Knemidokoptidae often burrow into the legs of domestic poultry, producing a swelling and scaling of the skin in a condition called 'scaly leg'. 6) The mite species *Vatacarus ipoides* lives in the tracheal passages of sea snakes. 7) Mites in the genus *Dicrocheles* live in the ears of noctuid moths. Shrewdly, these mites only park themselves in one of the two ears. This allows the encumbered moth to hear bats coming, so preventing an early demise of both moth *and* mite! 8) Not even the age-old enemy of gardeners is safe, as *Ricardoella limacum* is to be found in the pulmonary chambers of slugs. 9) Perhaps even more bizarrely, *Enterohalacarus minutipalpis* is found in the digestive systems of sea urchins, and is possibly parasitic.

BELOW The sheep tick, *Ixodes ricinus*, is an opportunistic or non-specific parasite – it feeds on anything that wanders past.

PARASITISM OF LIVING PLANTS Many mites are phytophegous, they feed on plants and damage plants with their cell-sucking, leaf-mining and gall-forming habits, and it seems that all parts are vulnerable. For example, leaves and stems are attacked by eriophyoid mites; fruit is attacked by tetranychid mites; tubers and roots are bored into by some oribatids and astigmatids. Mites are also particularly problematic because they can transmit plant pathogens, and are therefore the scourge of both crop and ornamental plants. Not only terrestrial plants are parasitized, aquatic plants are attacked too, e.g. duckweed and water hyacinth are parasitized by Hydrozetidae.

HOST SPECIFICITY The term 'host specificity' refers to how specific a parasite is to a host. Some tick species feed on anything that wanders past, so it is known as an opportunistic or non-specific parasite. Species that feed on a small range of hosts, such as bats, are considered as having moderate host specificity. The most discerning of tick species feed on just one host species, and therefore has strict host specificity. (This is not to say, of course, that a host only has one species of tick – indeed, they may have many different species feeding on them at any one time.) Such host-specific ticks tend to be found among animals in breeding colonies or roosts, where large numbers of the same species are available to feast upon.

SEARCHING FOR A HOST To feed, ticks must first find themselves a host. Endophilic ticks are those that hide away in burrows or nests, and wait for their future host to arrive home. Exophilic ticks, on the other hand, actively search for their hosts. To do this, they find a good vantage point (usually a plant stem), where they can wait for unsuspecting passing animals. There they cling, with their first pair of legs outstretched, in a pose called questing. Such ticks are extremely sensitive (because of the Haller's organ on the first pair of legs) to carbon dioxide, body odours and heat, which are emitted from passing animals. Physical disturbance of the vegetation also attracts them. As an animal brushes past, the alert tick clings on to the fur or hair, and then wanders over the host's body (sometimes for several hours) until it finds a suitable place for feeding. The patient ticks may have to wait for over a year for a host to come along!

Food by predation

IN SOIL There are huge numbers of predatory mites in the soil on the prowl for prey such as nematodes, collembolans and thrips. The main soil predators are in the Mesostigmata (Parasitina and Dermanyssina), and in the Prostigmata. Search behaviours by these soil mites are very variable – they may be classified as ambush predators, pursuit predators or saltatory (stop-and-start) search predators.

Ambush predators: Soil mite ambush predators are common in the Prostigmata. For example, mites in the genus *Neocaeculus* (from Australia) wait extremely patiently for up to several days, until a prey item wanders passed. The mite then springs into action, bringing its first pair of legs – specially adapted with spines on the inner surface – quickly down onto the prey. Certain mites in the subfamily Cunaxinae (Prostigmata) ambush prey in a similar way, although they grab with their spiny palps.

Pursuit predators: Mesostigmatids rush around at speed in pursuit of their prey, poking into every pore to look for food. In this way, they are able to find non-mobile prey, such as nematodes, that tend to clump together. These aggressive mites can attack prey larger than they are, holding on tenaciously to bring it down. There are also rapid pursuit predators within the Prostigmata, e.g. *Coeleoscirus simplex* is a cunaxid that is always on the move for food. Once it has tracked down its hapless prey, the mite circles like a shark around a wounded fish. It then tires the prey out by poking it with its chelicerae, before finally grabbing it and sucking out its juices. Whirligig mites in the family Anystidae whirl across the ground looking for prey.

Saltatory (stop-and-start) search predators: Many mite species (especially those in the Prostigmata) combine the two search behaviours above. They repeatedly stop and start, in a behaviour pattern called saltatory search. Mites in the genera *Veigaia* and *Polyaspis*, for example, deliberately search out their prey by walking around, but often sit and wait when they can't immediately locate what they're looking for.

ON VEGETATION There are many mites that are found on plants, mainly in the Prostigmata (e.g. Anystidae, Cheyletidae, Erythraeidae and Bdellidae), but also from the Mesostigmata, chiefly in the Phytoseiidae. Quite a few species feed on pollen, fungi and algae, but some are

RIGHT Egg predators such as mites in the family Erythraeidae often consume large insect eggs, while smaller eggs (including those of mites) are often consumed by mites in the Stigmaeidae. Here, a predatory mite (*Amblyseius finlandicus*) with prey mite egg (*Phytoseiidae*).

predators. As in the soil, these foliar predators have different ways of capturing prey. Foliar predators are mainly pursuit predators, e.g. Phytoseiidae (Mesostigmata). Ambush predators are also very common, e.g. Cheyletidae (Prostigmata). These mites ambush small arthropods that run along the leaf stalk.

IN WATER Some water mites eat only insect eggs, e.g. *Hydryphantes* (Hydryphantidae). Others have a much wider diet, e.g. *Limnesia* (Limnesiidae), readily consumes insect eggs and larvae, copepods, cladocerans, other mites and isopods. Insect pupae, small crustaceans such as ostracods, mussels, other mites, nematodes and oligochaetes also make up the diets of water mites. The palps of water mites are structured differently depending on what they predate – so, for example, mites that eat insect larvae tend to have sharp spines on their linear palps in order to trap prey, whereas egg-eating mites in the Hydrachnidae, Hydrodromidae and Hydryphantidae have a palpal claw.

These are some of the ways in which aquatic mites catch their prey. 1) Chemical attraction: Egg eaters are likely to be attracted by chemicals on the surface of the jelly coating of insect eggs. 2) Strong swimming: Strong swimming mites, e.g. in the family Eylaidae, are able to capture small crustaceans in the plankton. 3) 'Bump and grab': Mites that are smaller and slower can ably catch crustaceans that passively sink through the water column after they have been disturbed. 4) 'Net stance': Species in the family Unionicolidae capture their small crustacean prey by positioning themselves in a 'net stance'. They do this by placing their very spiny first pair of legs, and their less spiny second pair, outstretched off the substrate, thereby creating a network of spines. They orientate towards the swimming vibrations of prey and then grab them. 5) Serendipity: Mosquito larvae and pupae are often encountered by chance, and are then attacked. Often, a number of opportunistic mites take advantage of long and thin larvae, and share a group dinner. 6) Sit-and-wait: Some species use a sit-and-wait strategy, which mirrors that of their terrestrial counterparts, e.g. the mite *Limnochares aquatica* preys on chironomid larvae. It sits and waits until a larva pops its head out of its tube and then grabs the larva's head capsule and fixes on. 7) Burrowing: e.g. *Hygrobates trigonicus* burrows in lake bottoms in order to extract chironomid larvae buried in the sediment.

Other means of feeding

It's not always too easy to classify mites into separate feeding groups, because distinctions can be blurred, and some may fit into more than one group. Therefore, scavengers, detritivores, microbivores, fungivores and algivores are discussed here together.

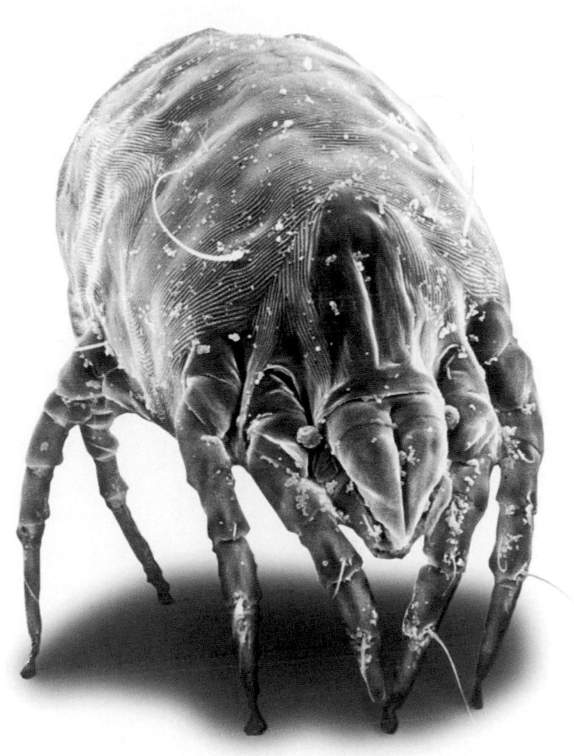

ABOVE Mite scavengers include dust mites such as *Dermatophagoides* sp., which feeds on human dander (tiny scales shed from skin). An adult human sheds enough skin cells in one day to maintain a population of 10,000 *D. pteronyssinus* for six months! Moulds also supplement their diet in the domestic environment.

SCAVENGERS A scavenger is an organism that feeds on dead or decaying matter. Mite scavengers feed on dead animal and plant material. Rotting fruit is eaten by mites brought in phoretically by beetles and flies, as well as by soil-dwelling mites. Mite species that feed on decaying plant matter usually have huge chelicerae.

DETRITIVORES AND MICROBIVORES A detritivore is an organism, such as an earthworm, that consumes decomposing organic particles and derives nutrition primarily from microbes on the particles. This is a more specialized way of obtaining food than a scavenger. A microbivore is an organism that feeds on microbes, which are micro-organisms. Detritivore–microbivores may be omnivore, fungivore, herbivore or predator, depending on the habitat, the stage in the life-cycle, and the time of year. Depending on the order, detritivore–microbivores ingest cell contents and cell walls. Prostigmatid microbivores don't ingest cell walls – they pierce the particles and suck out the cell contents. The large species of soil mites tend to consume detritus. Histiostomatid astigmatids filter yeasts, bacteria and very small microbes using their brush-like palps and chelicerae.

FUNGIVORES AND ALGIVORES A fungivore is an organism that feeds on fungi, while algivores feed on algae. Mites are a good size to be able to ingest fungal hyphae and spores. Oribatids and astigmatids that live in the soil feed on hyphae, spores and strands of algae. A number of oribatids use an articulating digit on their chelicerae to snip hyphae into pieces. Some mite species feed on mycorrhizal fungi that colonize plant roots and the soil. Large soil species tend to consume detritus, medium species prefer fungal hyphae and small species go for fungal spores. Feather mites may feed on fungal spores that grow on the feathers, which benefits the host bird. Not all algivores are content with a diet of pure algae. They may go for small invertebrates such as Collembola, given the chance. Aquatic mites in the subfamily Rhombognathidae (Halacaroidea) consume algal cells and have long, pointed chelicerae that can pierce the cells to get to the contents.

Feeding and digestion

FEEDING IN MITES Mites demonstrate both external and internal digestion. External digestion in mites e.g. Prostigmata, occurs just as for other arachnids. Internal digestion takes two forms – intake of fluids and intake of particles. Mites ingest fluid food after piercing the prey with sharp chelicerae, and draining the juice, e.g. Hydrachnidae pierce insect eggs with claws on the palps. Mites that ingest solid particles include those that bite off particles, e.g.

soil-inhabiting Astigmata and Oribatida. Other mites filter fine particles with modified palps and chelicerae, e.g. Hisiostomatidee (Astigmata). Most aquatic mites also don't use preoral digestion, because the digestive enzymes would be diluted in the watery habitat and become ineffective – instead, they consume particles, which are then internally digested. An exception is *Piona alpicola* (Pionidae), which feeds on *Daphnia* (water fleas) by injecting digestive fluids through a hole in the carapace and then sucking up the juices.

FEEDING IN IXODID TICKS Once a host has been 'boarded', the tick needs to select a suitable site for attachment, where blood capillaries are very close to the skin surface, such as behind a dog's ear. The tick is careful not to alert the host to its presence, so that it is not groomed off. It therefore punctures the skin painlessly. Substances in the salivary fluids cause irritation of the skin and result in the bite site enlarging, allowing the tick to gradually insert the hypostome. As well as enzymes, the salivary glands produce immuno-suppressive substances, anti-inflammatory substances, vasodilators (which increase blood flow) and anticoagulants. These overcome the host's defence system, to allow successful feeding. Additionally in many species, the salivary glands produce a kind of cement, which anchors the mouthparts to the host.

Ingestion is the act of taking food into the mouth. In ticks, blood is sucked up a groove in the hypostome and into the gut. The nutrients from the blood are concentrated by the salivary glands and excess fluid is returned via the bite site to the host's circulatory system. The food ingested is therefore essentially just the red blood cells.

Unless anything occurs to remove the tick, it remains attached until its meal is complete. The time taken to reach engorgement is very slow – most larvae take 2–6 days, and an adult female takes 6–15 days. Ingestion in ixodids has two phases: the initial, slow phase lasts about 7 days, and the weight increases about ten times; the following, rapid phase lasts only 12–24 hours, and results in a further tenfold weight increase. Within these phases, there are alternate periods of sucking and salivation, which are interspersed by resting. For adult females, the overall weight gain from unfed weight to fed weight is enormous – they can increase in weight up to 120 times! Female ticks engorge to such a size only if they have mated first – males never engorge.

RIGHT An enormously engorged female tick *Ixodes ricinus*.

Digestion occurs while the tick is still attached and continues after it has detached. It is in three phases. 1)The first continuous phase starts 24 hours after attachment, and continues until rapid engorgement. Nutrients gained in this phase are used for growth. 2) Delayed or reduced digestion is a resting phase during rapid engorgement. 3) The start of the second continuous phase is when the replete tick has finished feeding and detaches from its host. The tick finds a resting place in order to continue to digest its meal. Nutrients gained in this phase are used for egg production, so it does not occur in males and unmated females. After digestion is complete, larvae and nymphs moult to the next developmental stage and adult females lay eggs. Many ticks, once fed, don't need another meal for a whole year! In fact, ixodids spend over 90% of their time off the host, in the shelter of the host, or in vegetation.

FEEDING IN ARGASID TICKS This is somewhat different to that in ixodids: the feeding period is much shorter; the skin of the host is penetrated rapidly; the hypostome is usually rather weak although the chelicerae may be well developed; the salivary glands don't produce attachment cement; and weight increase is only up to 12 times. There are three phases of feeding. In the first phase, which lasts for about two days, no actual digestion occurs – instead, excess ions and water are removed from the blood and excreted mainly through the coxal glands. The next phase involves rapid digestion, and continues until the tick oviposits. The third phase is that of slow digestion – although it involves a low level of digestive activity, it enables the tick to survive for long periods without food.

Predators and parasites

Mites are eaten by a variety of organisms, including fellow arachnids such as predatory mites, spiders, pseudoscorpions and opilionids, as well as ants, ground beetles, diplurans and centipedes. Mites are found in huge numbers in the soil, and therefore make up a large proportion of their predators' diets. House dust mite predators include silverfish, dust lice, pseudoscorpions and other predatory mites. Vertebrate predators of mites include certain birds such as oxpeckers, small lizards, frogs and toads. In fact, mites often make up the largest proportion of an amphibian's diet.

The biggest killer of ticks is starvation. However, there are many organisms that predate ticks – some of which could be used in biological control. Birds such as chickens and crows are big fans of ticks – especially engorged ticks that have fallen to the ground. Cattle egrets are able to pluck off feeding ticks *in situ* on their hosts. When engorged ticks are searching for a place on the ground to lay eggs or to moult, they are vulnerable to predation from mice and rats, shrews and lizards. Indeed, an engorged female tick makes a fabulous meal – big, packed full of nourishment, and helpless because it has no defences and is too sluggish to escape! Arthropod predators include spiders (wolf spiders prey on ixodids, while jumping spiders prey on argasids), ground beetles and ants. Even though mites are very much smaller than adult ticks, some species in the family Anystidae do kill tick larvae, e.g. *Boophilus microplus*.

Organisms that are parasitic on acarines include flies and wasps. For example, the chalcid wasp *Ixodiphagus hookeri* lays its eggs on larval and nymphal ticks and the adults emerge about one-and-a-half months after the tick has engorged. This species parasitizes the tick species *Haemaphysalis punctata* in Spain, and *Ixodes ricinus* in Germany. Acarine parasites also include endoparasitic nematodes – the hosts can die rapidly from septicaemia caused by bacteria that the nematodes carry.

RIGHT Box mites are able to withdraw their legs into a cavity and lower their carapace like the hood on a pram, thereby turning themselves into an enclosed egg-shape, with no weaknesses in their armour.

Defensive behaviour and aggression

Mites have evolved several types of defence to avoid predation. Some can readily jump, using enlarged femora – a catapult mechanism similar to that in springtails – or by hydraulically extending all four pairs of legs. Several species of mites (and ticks) produce chemical secretions to deter predators. Species in the Opilioacaridae are known to shed their legs if captured by predators. Others play dead, by pulling their legs into recesses in the exoskeleton, and lying motionless. Many species of oribatids produce cerotegument (secretions from the integument). In some species, these are sticky and attract soil particles, which act as camouflage.

The most common defence is armour plating. Mites in the Mesostigmata often have one or two shields and several plates protecting the body, along the lines of plate armour worn by medieval knights. The beetle-like oribatid and uropodid mites exemplify this kind of armour plating superbly (see p.162 left). Many uropodids are nick-named turtle mites, because they can draw their legs into recesses in their armour. They are also sometimes referred to as panzer mites, resembling tiny acarine tanks trundling through an invertebrate battlefield! However, this armour plating isn't always effective. Indeed, there are certain beetles, e.g. in the family Scydmaenidae, which specialize in feeding on highly sclerotized mites, using their mandibles like a can opener on a tin of beans! Certain ants can be just as tenacious, easily ripping holes in the armour of the hapless mites. Some mites have gone down a different line of defence; they possess long spiny setae that they erect in times of attack, a bit like a miniature porcupine!

Water mites are rarely preyed upon, which is interesting considering that they are relatively abundant. It has been found that there is a positive correlation between the production of foul secretions by water mites, and their bright red colour. After a few bad experiences, fish learn that such mites are distasteful and avoid eating them. This is a good example of warning colouration.

Ticks certainly don't have as many defences against predators as other mites. Many species *do* have some type of armour plating, although it certainly isn't as protective as that worn by the oribatids. *Dermacentor variabilis* has large wax glands that produce a defence secretion, emitted during an attack from predators. This secretion acts as an alarm pheromone, alerting conspecific ticks in close vicinity.

Medical and veterinary importance of mites

Although acarines don't produce venom like spiders and scorpions, ticks and other mites are still of great medical and veterinary significance. This is because they are important vectors of disease, and they themselves can cause allergic reactions. It is perhaps shocking to discover just how many diseases and conditions ticks and other mites can cause. Disease organisms (pathogens) exist in a disease cycle, which often involves a vector and one or more hosts. A disease organism doesn't spend equal amounts of time in the hosts and the vector. It remains for most of the time in a long-term reservoir, which may be the vector or the host.

Medical importance of mites

Mites have quite a few adverse effects on humans. For example, mites can cause allergies such as asthma and dermatitis, and diseases such as scrub typhus, scabies and gastric disorders. An allergy is an abnormally high sensitivity of the body, caused by a disorder of the immune system. Allergens are substances that can cause an allergic reaction, and include pollen, micro-organisms, and certain substances in food – but these substances don't cause allergies in everyone.

DUST MITES Dust mites are in the family Pyroglyphidae (Astigmata). There are 13 species of pyroglyphids, found in house dust around the world – three species are the most common, comprising up to 90% of the house dust mite fauna: *Dermatophagoides pteronyssinus, D. farinae* and *Euroglyphus maynei*. However, mites from other families are found too: storage mites (families Acaridae, Chortoglyphidae and Glycyphagidae), predatory mites (family Cheyletidae), and the glistening mites (family Tarsonemidae).

As the name suggests, dust mites are usually found in dust. The abundance of mites within house dust is influenced by a number of factors, such as temperature and humidity. For a house dust mite, mattresses and carpets provide ideal environmental conditions in which to live and breed, with a copious supply of food, such as flakes of dead human skin. House dust mites are found in most homes in humid areas, while far fewer occur in homes at higher altitudes and in drier climates (and in lower population densities). The mites relocate to new areas by hitching a lift in crevices in furniture, in the clothing of humans, and on pets. The acarine detective Dr Matthew J. Colloff, from CSIRO, Australia, discovered ten species of mites in dust from passenger trains in Glasgow, UK, which could only have been transported there from homes via people's pets and clothing.

Dust mites are the most common source of allergens for people with allergies. It has been discovered that the faeces of these small creatures contain a particular protein, which is the principal cause of allergic reactions. When people vacuum, make their beds and dust their houses, they cause the microscopic faecal pellets to take to the air. These are then inhaled into the lungs and the mite protein causes an allergic reaction in some people. The faecal pellets may also sensitize the airway of an asthmatic, so that a subsequent problem such as a virus or other allergen is more likely to provoke a reaction. Signs and symptoms of dust mite allergy may include sneezing, runny or stuffiness of the nose, blocked up ears and respiratory problems such as asthma and eczema.

BELOW Asthma is often caused by mites. There are several species involved, including the flour mite or grain mite, *Acarus siro* (shown here) and the straw mite or grainstack mite, *Tyroglyphus longior* (both Astigmata, Acaridae).

DERMATITIS Dermatitis is inflammation of the skin, caused by an allergy. Almost any rash can be thought of as dermatitis, not just eczema. Dermatitis in humans can be caused by many species of mites, even those that usually infest only animals. The mites are transmitted to humans as they handle infested animals.

GROCER'S ITCH Prolonged contact with food infested by the mite species *Lepidoglyphus destructor* can cause a chronic, very itchy dermatitis called 'grocer's itch'. The materials these mites feed on include dried fruits, grain, copra (the dried white flesh of the coconut from which coconut oil is extracted) and cheese. Grocer's itch manifests itself as a rash or raw and crusted area on the knees and wrists, scalp, forehead and creases of the elbows. (It is also called 'baker's itch', although baker's itch in fact refers to several different inflammatory dermatoses of the hands, especially chronic fungal nail infection.)

CHIGGERS The name 'chigger' refers to the larval stage of mites in the family Trombiculidae (and sometimes to those in the Trombidiidae, although strictly speaking they are trombiculids). Although these beasts are very small, 0.4 mm (1/75 in) in length, they can cause severely itchy dermatitis in their host. A common harvest mite in North America is *Trombicula alfreddugesi*. In Australia and Asia, chiggers may carry scrub typhus. Chiggers are usually found in vegetation, such as tall grasses. They climb up the vegetation in order to transfer onto a host, and when the weather is dry, they hide beneath vegetation and in shady areas.

Chiggers attach themselves to the skin of their hosts, which may be domestic or wild animals (including birds and reptiles), or humans. They target areas where the skin is wrinkled or thin, e.g. in the elbow crease, or pressed against tight clothing e.g. under the waistband of a skirt. They pierce the skin, inject digestive enzymes into the wound, and then suck up the broken-down skin cells. They don't burrow into the skin (like scabies mites) and they don't suck blood. Their bites produce small, hardened welts, which are extremely itchy. The welts are *not* caused by chigger eggs being laid under the skin. Itching from a chigger bite may take several hours to develop and it may be a week before it goes.

OPPOSITE The storage mite *Lepidoglyphus destructor* is found in stores of grain and flour and feeds mainly on fungi. The faeces of mites such as these can provoke allergic reactions in affected people.

LEFT A common chigger is the harvest mite, *Neotrombicula autumnalis*, which is active in the UK during late autumn.

SCRUB TYPHUS Scrub typhus is a disease caused by *Orientia tsutsugamushi*. This zoonotic pathogen (a pathogen that normally exists in animals but that can infect humans) is one of a diverse group of intracellular bacteria known as rickettsiae, which includes the genera *Anaplasma*, *Coxiella* and *Orientia*. Some researchers class these as being intermediate between viruses and bacteria. They cause infections that spread in the blood to many organs and can result in fatal complications. They are found in ticks, other mites, fleas, lice and mammals.

In the case of scrub typhus, the disease is spread by several species of chiggers in the genus *Leptotrombidium*, including *L.* (*Trombicula*) *akamushi* and *L. deliense* (Trombidiidae). *Orientia tsutsugamushi* is found throughout the mite's body, but is particularly concentrated in the salivary glands. When the mite feeds on hosts, such as rodents (which are the secondary reservoirs) and humans, the *O. tsutsugamushi* is transmitted. Scrub typhus was first officially described in 1899 as being transmitted by a mite from Japan and Southeast Asia, where it typically occurs. It is also found in India and northern Australia. It got its name because infected rodents mainly live in scrub vegetation. However, it can also be found in more unexpected areas such as rainforests, and beaches.

Depending on the geographic area and the strain of the rickettsiae, mortality is between 1% and 60%. Death can occur from the primary infection or from secondary complications, such as circulatory failure. Most fatalities occur by the end of the second week of infection. If the patient doesn't die, symptoms may last for more than two weeks if the patient does not receive treatment – with treatment, the patient recovers within 36 hours. The devastating effects of scrub typhus were seen during the Second World War, when it swept through troops stationed in rural or jungle areas in the Pacific, killing and debilitating thousands.

SCABIES Mange is a generalized term used to refer to a skin disease caused by a number of species of mites in the families Sarcoptidae, Demodicidae and Psoroptidae, which infest the skin of more than 100 animal species – including humans. The mite *Sarcoptes scabiei* in the family Sarcoptidae (Astigmata, see p.162 bottom right) causes mange in humans, which is therefore called scabies. However, *S. scabiei* can also infest other mammals, such as horses, sheep and cattle, when the disease is then referred to as sarcoptic mange or canine scabies.

The small (less than 0.3 mm or 1/100 in), globular and semi-transparent *S. scabiei* mites attach themselves to the skin of the host by suckers on their ambulacra. They then cut channels into the horny outer layers of skin, using their claw-like chelicerae and the spines on the tarsi of their first two pairs of legs, and burrow in. In addition to the effects of the toxins they release, their burrowing actions cause bacterial infections. Scabies and mange mites are easily spread between people and animals, although it usually requires prolonged direct contact between humans for the mites to be passed on. They can also live for several days off the host, which makes the chance of them infesting a new host rather high. Scabies is still a widespread disease around the world.

GASTRIC DISORDERS It is not just the respiratory system and skin that can be adversely affected by mites in food. Eating food that is infested by mites is certainly not something to be recommended, as it may cause gastric disorders, such as extreme diarrhoea. Two mite species that are both from the family Acaridae (Astigmata) – *Acarus siro* (see p.185, the flour mite, and *Tyroglyphus longior*, the straw mite or grainstack mite – are often implicated in infesting food stuffs and causing gastric disorders. Believe it or not, however, there is a famous German cheese called Altenburger Milkenkase that is intentionally loaded with thousands of mites in order to give it its distinctive sharp flavour. Anyone wishing to taste this cheese requires a hardy constitution!

Veterinary importance of mites

MITES THAT AFFECT PETS The most common ectoparasitic mites on pets are *Sarcoptes scabiei* (which causes sarcoptic mange or canine scabies), *Demodex* (causing demodectic mange) and ear mites. There are several other mite species that affect pets, such as *Notoedres cati*, which causes notoedric mange in cats, and *Dermoglyphus passerinus*, which is responsible for feather loss in caged birds.

Sarcoptic mange or canine scabies: *Sarcoptes scabiei* can affect both humans and animals. Sarcoptic mange may infect all breeds and ages of dogs (when the disease is called canine scabies), but may also infect foxes and cats. Signs of infestation include severe itching (due to allergic reaction) and hair loss (due to scratching).

Demodectic mange: Demodectic mange is also known as puppy mange, red mange, and follicular mange. It is most commonly caused by *Demodex canis*, but can also be caused by *D. gatoi*, and *D. injai*. The mites live in the dog's hair follicles, and usually affects young dogs most severely. All dogs have these mites, but if the dog becomes ill or its immune system is suppressed, then the mites may cause a problem.

Ear mites: Ear mites in the family Psoroptidae (Astigmata), infest the ears of pets such as dogs, cats, rabbits, ferrets and horses. Signs and symptoms of *Otodectes cynotis* infestation include an allergic reaction resulting in intense itching of the ear, head shaking, ulceration and sometimes convulsions. The ear canal may become congested with inflammatory secretions, and thick, reddish-brown or black crusts may appear on the outer ear. The species of ear mite that infests rabbits and horses is *Psoroptes cuniculi*, which live on the skin surface under the edge of scabs (created by their feeding on exudates leaking from the skin because of the mite's abrasive claws and chelicerae). The mites also feed on loose skin. They are usually found in the ear, but are also be found on the entire head and body if the animal is heavily infested.

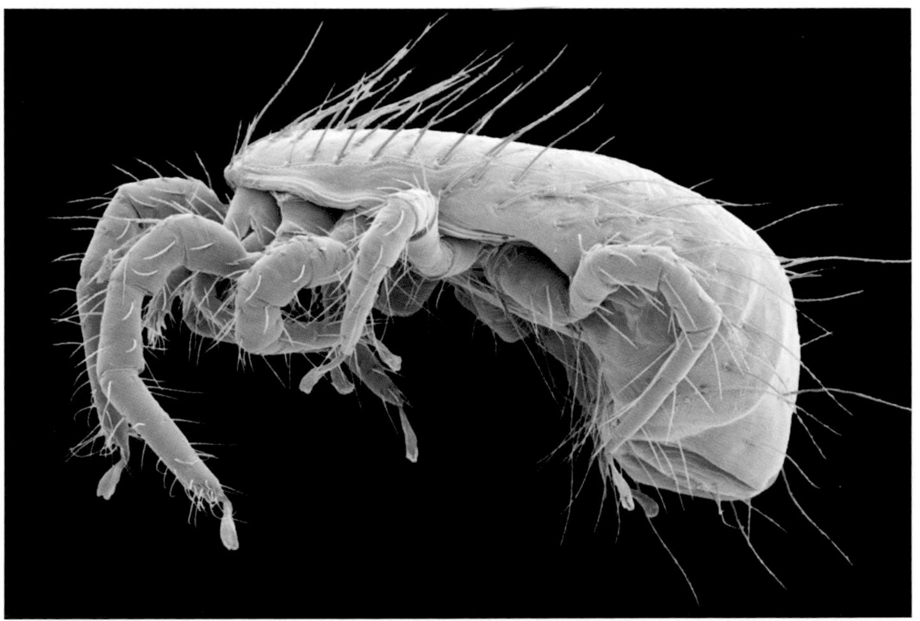

LEFT *Otodectes cynotis* infests dogs, cats and ferrets. The mites are usually found in the outer ear, but heavy infestations may spread to the inner ear canal. Housed in their auditory sanctuary, the mites feed on skin debris, blood and lymph.

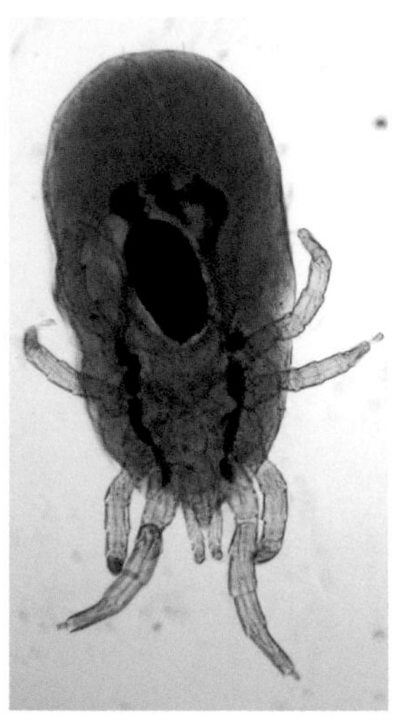

ABOVE The mites *Dermanyssus gallinae* (shown here) and *Ornithonyssus sylviarum* (Mesostigmata, Macronyssidae) affect domestic birds such as chickens. They produce a combination of effects, such as reduced egg production, reduced fertility, anaemia, skin irritation, and even death.

MITES THAT AFFECT LIVESTOCK There are many species of mites that affect livestock. Their effects include scaly leg in poultry, respiratory problems in birds and mange in sheep and domestic cattle. There are also mites that cause a combination of ill effects in domestic birds as follows:

1) Scaly leg: The mite species *Knemidokoptes mutans* causes 'scaly leg' in poultry, such as chickens, turkeys, pheasants and partridges. The legs and feet become affected, and have a whitish, scaly, crusty and thickened look. Toes can drop off if left untreated.

2) Respiratory problems: The mite species *Cytodites nudus*, known as the air sac mite, causes severe respiratory problems and death in birds such as chickens, ducks and turkeys, and in caged birds such as canaries.

3) Mange: It's not just pets that can be affected by mange – sheep are infested by the mite species *Psorergates ovis*, and domestic cattle are infested by *P. bos*. An infestation, can cause cattle to develop pruritis (skin itching), lesions, alopecia and weight loss. The mite species *Chorioptes bovis* also causes mange of cattle and sheep. *Psorobia ovis* mites are found under the skin of sheep, which is covered by fleece, but also affect the face and legs. The sheep may chew its fleece because of the itchiness.

Medical and veterinary importance of ticks

Ticks transmit a greater variety and number of disease organisms than any other arthropod worldwide, and tick-borne diseases are a real problem to livestock farming around the world. Their economic impact includes loss of production due to mortality and restrictions placed on moving animals around. Not all of the 880 tick species are disease vectors – a maximum of 200 species carry disease organisms and, not surprisingly, these have a wide range of hosts. The Ixodidae is the most important tick family in terms of its medical and veterinary importance – it is also the most dominant in terms of species. Ticks are particularly effective vectors of disease for numerous reasons. Those with a wide host range can transmit disease organisms from one host to another. They attach themselves firmly to the host and often stay there for long periods, until they are fully engorged – this allows them plenty of time to pick up or transmit disease organisms. The tick's salivary glands are the key structures in disease transmission because they return excess fluid back to the host, and this fluid contains disease pathogens from the tick. The host doesn't have to be directly bitten by the tick, however, for contamination to take place – it may be indirectly contaminated by the tick's secretions, waste products or squashed body.

Ticks are rather resistant to starvation and are long-lived, so they make very good long-term reservoirs for disease organisms. Ticks have an extremely high reproductive rate, with females of some species being capable of laying up to 10,000 eggs! Ticks often suppress the host's immune response, which makes the host more susceptible to the disease the tick is transmitting. As ticks are often attached to birds, they can be very widely distributed and are therefore able to transmit diseases to totally new areas.

Medical importance

Ticks are of enormous medical significance, because they transmit a great many diseases, including bacterial diseases such as Lyme disease, Boutonneuse fever and Q fever. They also spread viral diseases such as tick-borne encephalitis – there are over 500 known arthropod-borne viruses, known as arboviruses, and ticks transmit more than 100 of them. Protozoan diseases,

such as human babesiosis, are also passed on by ticks. (A protozoan is a single-celled, usually microscopic, organism, such as an amoeba, ciliate, flagellate or sporozoan, which contains a distinct membrane-bound nucleus.) Ticks also cause other conditions such as tick paralysis.

LYME DISEASE The infamous Lyme disease is considered to be 'the most serious arthropod-borne human disease of Europe and North America' (Hillyard, *Ticks of North West Europe*). It is also found in other parts of the world including South Africa, Australia and Asia. Lyme disease symptoms were first noted in Europe in 1883, but it took until 1975 for all the complex symptoms to be finally recognized as a single disease.

Spirochaetes are bacteria that are spiral-shaped like fusilli pasta. They are found in water, and blood, e.g. *Borrelia* species. There are several species of *Borrelia* around the world, spread by ticks in the genus *Ornithodorus*, which cause tick-borne fevers. Lyme disease is caused by the spirochaete *Borrelia burgdorferi*. There are three different strains, and each has its own characteristic symptoms. There are quite a few tick vectors that transmit *B. burgdorferi* – they include several species in the genus *Ixodes*, such as *I. ricinus* and *I. gibbosus* in Europe, and *I. pacificus* and *I. scapularis* in North America. Other ixodid vector species in Europe include *Haemaphysalis punctata*, *Dermacentor reticulatus* and the argasid *Argas reflexus*.

BOUTONNEUSE FEVER This exotic sounding condition (also called Marseilles fever and Mediterranean fever) is endemic in southern Europe, from where it is spreading northwards, and is widespread in Asia and Africa. It affects humans, and 2.5% of cases are fatal. The signs and symptoms include a black sore at the bite site that looks like a button, a rash on the soles of the feet and the palms of the hands, aching muscles and joints, and fever. It is caused by *Rickettsia conorii* bacteria and the main tick culprit for transmission in Europe is *Rhipicephalus sanguineus*, although some *Dermacentor* and *Ixodes* species also act as vectors. Dogs are one of the prime reservoirs of *R. conorii*. Humans are therefore at risk from transmission of *R. conorii* from pet dogs that have acquired infected ticks.

Q FEVER Q here stands for Queensland, which is where this disease was first seen in 1935. It is now found around the world. It affects livestock (especially sheep, but also goats and cattle) and humans who are in contact with livestock. It is caused by the rickettsia bacterium *Coxiella bourneti* and many tick species act as vectors (both Argasidae and Ixodidae). Worryingly, the rickettsia doesn't need an arthropod vector in order to remain viable.

TICK-BORNE ENCEPHALITIS This disease affects only humans and has a mortality rate of between 1% and 5%. The signs and symptoms are fever, headache, paralysis and coma. Tick-borne encephalitis is caused by a *Flavovirus*. There are a number of vectors of the tick-borne encephalitis *Flavovirus* including several species in the genera *Ixodes*, *Dermacentor* and *Haemaphysalis*.

HUMAN BABESIOSIS It is usually only elderly people or those who have problems with their spleens who contract human babesiosis, and most patients suffer only mild symptoms. However, it recently became evident that healthy people may contract it too. Human babesiosis is caused by the protozoans *Babesia divergens* and *B. microti*, via the main tick vector *Ixodes ricinus*. The signs and symptoms are similar to those of malaria.

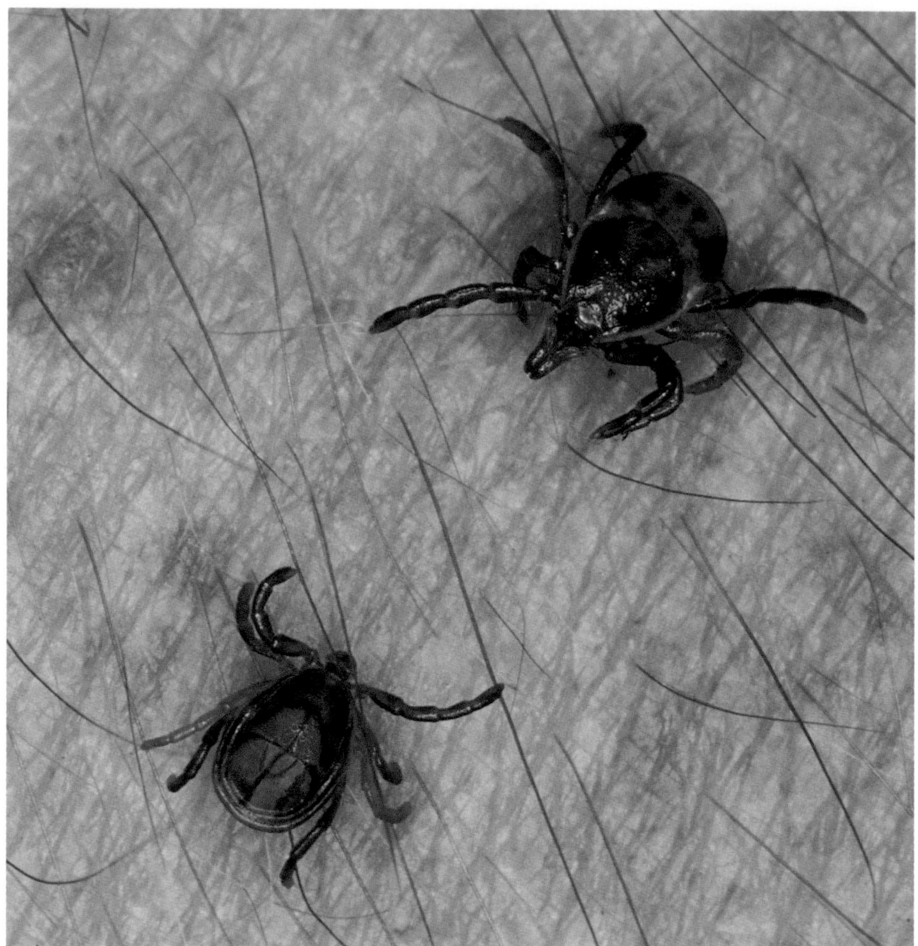

RIGHT There are around 50 species of ticks worldwide (in the Ixodidae and Argasidae) that cause tick paralysis. They include species in *Ixodes* (here *I. holocylus*), *Rhipicephalus*, *Haemaphysalis*, *Hyalomma* and *Argas*. *Dermacenter andersani* is the most important cause of tick paralysis in North America.

TICK PARALYSIS – TOXICOSIS Tick paralysis is caused by toxins in the tick's saliva being injected into the host's bloodstream during feeding. The condition is therefore a toxicosis rather than a disease. Although it is not properly understood exactly how these toxins work, it *is* known that they inhibit the transmission of nerve impulses across neuromuscular junctions. They are therefore neurotoxins – as are some spider venoms. Tick paralysis affects humans and domestic animals.

Veterinary importance of ticks

In terms of veterinary importance, ticks transmit bacterial diseases such as tick-borne fever, viral diseases such as Louping ill, and protozoan diseases such as redwater fever. Ticks also cause tick paralysis, tick worry exsangunation, and transmit fungi and nematodes. Here, we look at just a small selection of diseases and other conditions.

TICK-BORNE FEVER Tick-borne fever is caused by the rickettsia bacterium *Anaplasma phagocytophilum*, which is 'the most common tick-borne pathogen of veterinary importance in the UK' (Brodie *et al*, 1986). It is also found in northern Europe, India and South Africa. *Anaplasma phagocytophilum* is found in many mammalian hosts, ranging from wild deer and rodents, through to sheep, cows, horses and dogs. The common *Ixodes ricinus* is the vector

in Europe, with *I. pacificus* and *I. scapularis* transmitting the disease in the USA. Severe cases of tick-borne fever can result in bad losses of sheep stock. Even in low-level cases, there is a drop in productivity due to reduced weight loss in lambs.

LOUPING ILL The strange name of this disease comes from the Scottish word for leap – 'loup'. It describes the effect of the extreme muscular twitching that the disease causes in sheep. A high mortality rate is likely to occur in livestock with no immunity. This disease can also affect cattle, and sometimes humans working in close proximity to infected stock, such as vets and abattoir workers. It is caused by a *Flavovirus* and transmitted mainly by *Ixodes ricinus*. It is found in the UK and northern Europe.

REDWATER FEVER The name of this condition is taken from one of the signs of infection of the disease. Haemoglobin is found in the urine of affected cattle, making it dark red. Other signs are anaemia and fever, and infection can result in large losses of livestock. Death occurs due to the destruction of the red blood cells. Redwater fever is caused by *Babesia divergens* and is transmitted, as so many tick-borne diseases are, by *Ixodes ricinus*. Cattle living on highly tick-populated areas can acquire immunity. Redwater fever is found in the UK and is endemic to Europe.

TICK WORRY AND EXSANGUINATION Apart from the diseases ticks transmit, their blood-sucking presence on an animal brings its own problems, classed under the rather understated name of tick worry. Such problems include anaemia in animals with heavy tick infestations, skin irritation leading to open wounds, impairment of liver function, weakened immune system, and hair loss due to the scratching of irritated skin. Exsanguination results in high mortality from sheer blood loss recorded in chickens and other smaller farm animals.

NEMATODE AND FUNGI TRANSMISSION As if all the above diseases and conditions weren't enough, ticks are also known to transmit nematodes that live under the skin, as well as fungi. The fungus – called *Dermatophilus congolensis* ('dermatophilus' means 'having a preference for skin') – causes a skin condition that affects both humans and animals worldwide.

Commercial and agricultural importance

The commercial and agricultural importance of mites is a very large subject area. The following sections outline some of the commercial and agricultural practises affected by mites.

Plant parasites

Mites can be extremely problematic as plant parasites. They can transmit plant pathogens, as well as damaging plants with their cell-sucking, leaf-mining and gall-forming habits, and by feeding on the growing points of the plant. They can therefore be the scourge of both crop plant and ornamental plant growers.

False spider mites (family Tenuipalpidae): These have been reported for several decades as transmitting virus-like diseases of plants. The most economically important genus of tenuipalpid mites is *Brevipalpus*. For example, *B. californicus*, *B. obovatus* and *B. phoenicus* are vectors of the viral diseases *Lingustrum* ring spot, orchid fleck, leprosis (on citrus), coffee ring spot (on coffee), green spot (on passion fruit) and zonate chlorosis. These diseases have caused major economic

Crop pests

The Prostigmata include a great many plant parasites. The superfamily Eriophyoidea has thousands of species, and many of these are important as crop pests, because of their role as vectors of plant viruses and through the direct damage they cause. The Tetranychoidae contains around 1,600 species known as spider mites which usually live on the undersides of leaves, and feed on plant cells; the puncturing of the leaves to feed can cause a lot of damage. These little beasts (less than 1 mm (1/32 in) in size) often produce silk webbing as protection, which gives them their common name. Red spider mites (family Tetranychidae) feed on sap and plant tissue. They are major pests of grapes worldwide, reducing the harvest because of damage to the leaves and the fruit itself. Certain spider mites, such as the Pacific spider mite, *Tetranychus pacificus*, attack a broad range of plants, and are almost immune to chemical miticides because they hide beneath thick silk sheets, produced by their large colonies. The fruit tree red spider mite, *Panonychus ulmi*, attacks cultivated plants such as plums, pears and apples.

ABOVE The glasshouse red spider mite, *Tetranychus urticae*, is probably the best known red spider mite and is found in tropical, warm temperate zones and glasshouses.

losses through damage to globally important crops, such as citrus and coffee, in the Minas Gerais State in Brazil.

Gall mites (family Eriophyidae): Both immatures and adults feed on plant tissues, thereby causing serious damage to crops, e.g. *Cecidophyopsis ribis* is a particular problem in blackcurrants. Some species cause the plant to produce galls, which can be very conspicuous.

Tarsonemid mites (family Tarsonemidae): These mites usually feed on young plant tissues such as buds, and so cause distortion, reduced growth and eventually the death of new flowers and leaves. Strawberry crops are attacked by tarsonemids.

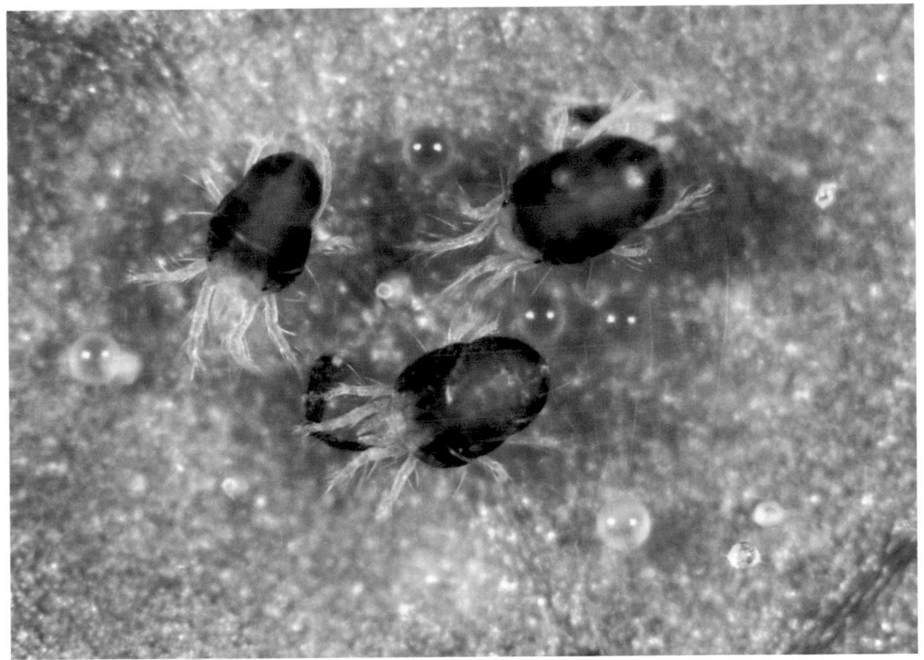

LEFT Greenhouse and house plants are often attacked by the ubiquitous red spider mite *Tetranychus urticae* (Tetranychidae). The predatory mite *Phytoseiulus persimilis* feeds on spider mites and successfully controls *T. urticae* on vegetables. Biological control using *P. persimilis* on the ornamental plant *Gerbera jamesonii* works because *T. urticae* (shown here) only attacks the leaves, and the leaves of *Gerbera jamesonii* are not harvested, only the flowers.

ORNAMENTAL PLANT PESTS As the name suggests, the Fuchsia gall mite *Aculops fuchsiae* (Eriophyidae) feasts on *Fuchsia* species and causes serious damage. The mites are extremely small, approximately 0.20–0.25 mm (1⁄100 – 1⁄125 in), so the first sign of infestation is when the *Fuchsia* is already showing symptoms including grossly deformed flowers and leaves that may have galls. All new growth may be suppressed as the infestation develops. *Aculops fuchsiae* is ranked as a major pest in Europe and is an EU quarantine listed pest. Mites in the family Tarsonemidae attack indoor and outdoor plants, such as azaleas, begonias, cyclamens and chrysanthemums.

Biological control agents

Mites can be used as biological control agents on other mites, on weeds and on nematode crop pests. Many species of mites prey on other mites, and these are found mainly in the Mesostigmata, but also in the Prostigmata.

Mites in the family Phytoseiidae (Mesostigmata) form one of the most important groups in biological control, and are often used to control spider mites in the family Tetranychidae (Prostigmata). Phytoseiid mites have recently been shown to sense their prey from long distances. They are also able to respond to special plant chemicals released when the plant is attacked by herbivorous invertebrates. For example, the control of the two-spotted spider mite *Tetranychus urticae* (Tetranychidae, Prostigmata) on edible crops in glasshouses has recently been undertaken using two interacting species of predatory mite – *Amblyseius californicus* and *Phytoseiulus persimilis*, both phytoseiids. By introducing a combination of predators that are particularly adapted to a certain part of the life-cycle of the pest, it is thought that the control of *Tetranychus urticae* will be improved. *Galendromus occidentalis* and *Typhlodromus pyri* (Phytoseiidae) have been successfully used in deciduous orchards in Australia to control *Tetranychus urticae*. Australian phytoseiids also provide good control of pest mites in horticultural crops such as grapes and lychees.

ABOVE Predatory mite (*Phytoseiulus persimilis*) feeding on two spotted spider mite (*Tetranychus urticae*) eggs.

It is not only the phytoseiids that are used as biological control agents, although most research effort on biological control by mites has been invested in this family. There is an increase in other mesostigmatid families as biocontrol agents too, e.g. Laelapidae, Macrochelidae, Ascidae. Soil-dwelling mesostigmatids can be of commercial importance as biological control agents because they feed on non-mobile prey such as nematodes.

Mites in the Prostigmata are also used in the control of other mites on plants. For example, it has been recently discovered that grapes in California that are attacked early in the season by the Willamette spider mite *Eotetranychus willamettei* (Tetranychidae), which is not considered a serious crop pest, are resistant to attack by the Pacific spider mite *Tetranychus pacificus* – which *is* a serious crop pest – later in the season. Therefore, instead of using miticides, Willamette spider mites are released early in the season instead.

CONTROL OF INSECTS Scale insects (Hemiptera, Diaspididae) are the most serious pests of many ornamental plants. They suck fluids from plant stems and leaves, as well as the roots. Plants that are heavily infested look rather unhealthy, with little new growth. Heavy infestations cause the leaves to go yellow and drop off, branches die and ultimately the whole plant dies. However, these terrible pests can be kept under control by mites in the genus *Hemisarcoptes* (Astigmata), which predate upon scale insects. Other mites also predate upon pest insect species too.

CONTROL OF WEEDS The use of mites for biological control of weeds was first discovered in Australia in the 1920s, when *Tetranychus desertorum* (Prostigmata) – although accidentally introduced – was used to control the spread of prickly pear (*Opuntia* spp.). It wasn't until the

1970s, however, that deliberate introduction of mites was used to control weeds. Mites can actually make good biological controllers of weeds, for several reasons: they act as vectors of plant diseases, they have very restricted host ranges so that non-target plants are not harmed, they disperse rapidly, they show rapid population increases, and they can be used to target particular stages in a plant's life-cycle. Two examples of mite species and the host plants they can be used to control are *Aculus hyperici* on St. John's wort, and *Tetranychus lintearius* on gorse. Although mites haven't yet been able to boast a major weed control success, they have in many cases made important contributions overall, and may do well in future control projects.

CONTROL OF NEMATODES There are several species of nematodes in the soil that feed on subterranean parts of plants. These can have a huge impact on plants grown commercially for food or garden centres. Mites that predate upon nematodes therefore have a commercial importance. One of the most important mite families in this respect is the Tydeidae (Prostigmata). Other families include Ascidae, Macrochelidae, Zerconidae and Uropodidae (all Mesostigmata).

Parasites of honey bees

Varroa destructor (see p.161 right), which parasitizes honey bees, is of great economic importance. It is one of the most serious pests known for honey bees and has been described by beekeepers as the modern black plague of bee hives, resulting in the loss of most affected bee colonies if left unchecked. The effects of *Varroa* include not only loss in honey and wax production, but – even more seriously – loss in crop productivity because of poor pollination. In fact, this small creature can cause industries and agriculture losses of millions of pounds. Since 1951, when *Varroa* was found in Singapore, it has spread rapidly. This has been due in part to the movement of infested colonies of bees for pollination and the importation of queen bees from infested areas.

The control of *Varroa* mites is complex. They're not at all easy to get rid of for good, instead, beekeepers aim to keep numbers below the 'damage threshold'. Integrated Pest Management is used, which is a combination of environmentally friendly treatments such as the use of essential oils, at appropriate times. Treatment choice depends on the time of year and on the number of mites present inside the hive and beekeepers use a year-round plan for effective control. In the past, *Varroa* mites were usually controlled with chemical treatments, but these resulted in the accumulation of residues in honey and wax. In addition, the mites developed resistance, which meant ever higher dosages had to be used. Ideally, control methods should not cause residual effects on honey and wax, should not result in mites developing resistance, and should be environmentally friendly.

Many beekeepers in the Netherlands use 'drone brood removal' as an environmentally friendly treatment. *Varroa* mites prefer to reproduce in drone broods. Drones are male bees that are characteristically stingless – they perform no work, apart from mating with the queen bee. By removing the source of infestation (i.e. the drone brood) from the hive, the number of remaining mites can be kept to a minimum. Fortunately, the removal of a drone brood doesn't deplete the numbers of drones required to fertilize the young queens because enough drones develop on the other brood combs. This very effective method is a great way of avoiding chemical residues in honey, and can be used in spring and summer, but should be combined with other treatments in autumn and winter.

Infesters of food products

Mites seem to get everywhere, and human food is no exception. They infest sugar, dried milk, cheese, cereals, flour, grain, dried fruit, dried vegetables, ham and pet foods. Food mites can be found anywhere where there is food, such as homes, and commercial premises such as bakers, grocers, graneries, mills, warehouses and so on. Many species of mites are found on stored foodstuffs even though they don't feed on the food themselves – they feed on the mites that feed on the food. For example, *Cheyletus malaccensis* (Prostigmata) feeds on food mites in the Acaridae (Astigmata). *C. eruditus* also feeds on acarids, and particularly likes *Acarus siro*. There are several species of mites that infest stored foods. Two of the most common are the grain mite *Acarus siro*, and the mould mite *Tyrophagus putrescentiae*, both in the family Acaridae (Astigmata).

ACARUS SIRO – THE GRAIN MITE This is the most important pest of storage premises. These mites are found most often on processed cereal products, such as flour, as well as on cheese, paper, tobacco, moulds, and nests of birds and animals. They often prefer a moist microhabitat. *Acarus siro* commonly occurs in extremely dense populations, e.g. 10,000 mites in just one 200 g sample. Such enormous numbers of mites can give the impression that an infested surface is moving. Infested areas are covered in cast skins, faeces and dead mites, and these can cause severe allergic reactions if inhaled. *Acarus siro* can produce such large populations because of its incredibly fast life-cycle. It is also very well adapted to unfavourable conditions. This is because its second nymphal stage can be replaced by a highly resistant 'hypopus', which is able to survive fumigation, insecticides and even several months without food. When the hypopus encounters favourable conditions again, it moults and resumes development. Not surprisingly, because of the hypopus stage, this mite is very difficult to eradicate.

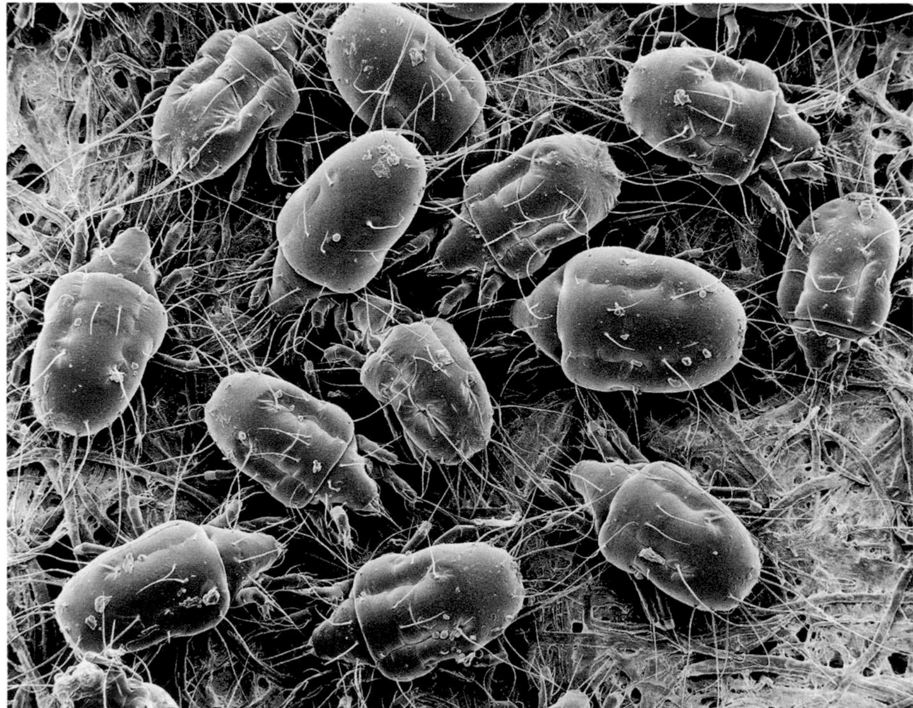

RIGHT *Tyrophagus putrescentiae* is also known as the cereal mite, cheese mite, Copra itch mite, forage mite and mill mite. It is found on food with a high protein and fat content, such as cheese, ham, dried eggs and nuts, and also on flour, barley, wheat, dried bananas and tobacco. It can cause dermatitis. This species is an extremely fast developer; it can complete a generation in only eight days if the conditions are right!

OTHER MITES Other astigmatids that infest foodstuffs include *T. longior* (found on barley, sugar beet, cucumber plants, tomatoes, beetroots), *T. perniciosus* (found in the dust of warehouses and granaries, as well as on barley and oats), and *T. similis* (found on vegetables such as spinach, and on mushrooms). *Tyrolichus casei*, found on cheese, is one of the most common food mites, and is known to cause dermatitis. There are also mites in the Prostigmata, such as *Tarsonemus granarius* (Tarsonemidae), that infest foodstuffs. These tarsonemids feed on fungi growing on stored grain and are restricted to granaries.

Environmental importance

Mites are often overlooked as bio-indicators, because they are almost impossible to identify in the field due to their size. As well as this, they are not often exploited as soil bio-indicators because of the way that soil is usually sampled. However, oribatid mites are actually great indicators of heavy metal pollution events, as they quickly retain heavy metals, and they live for a relatively long period of time. In fact, the relative proportion of parthenogenic oribatids increases due to heavy metal pollution and pesticide pollution. Oribatids have a specific preference for a certain pH, so that the absence or presence of particular species could act as a kind of 'litmus paper' for particular soils. They also indicate air pollution, usually decreasing in numbers with increasing pollution.

Water mites are not commonly used as bio-indicators for freshwater, possibly because many species of water mites are not terribly sensitive to contaminants. However, mites living in running water decline in numbers in polluted streams. In fact, organic pollution (such as sewage) can almost totally rid a stream of its acarine species richness. In a study undertaken on the Upper Danube, it was found that *Hygrobates fluviatilis* increasingly dominated the mite fauna with increasing pollution, making this species a good environmental indicator. Water mites can also indicate the presence of other organisms, e.g. *Unionicola crassipes* is only present when freshwater sponges are present, because the mite uses the sponge as an oviposition site.

Life history

Reproduction

Acarines are extremely diverse when it comes to reproduction, which is not surprising considering there are some 45,000 species. The various methods of sexual, and asexual, reproduction are briefly touched upon here, with a few examples.

PRE-COPULATION MATE GUARDING Males guard immature females and then mate with them once they have moulted into an adult – this is very common in mites. In the Mesostigmata, mate guarding is seen in several families, including the Macrochelidae, Laelapidae and Phytoseiidae. In the Astigmata, some male feather mites have suckers to get a grip on female tritonymphs. Other astigmatid families involved in pre-copulatory guarding are Chirodiscidae and Psoroptidae. In the Prostigmata, mate guarding occurs in families including the Cheyletidae, Tetranychidae and Eriophyidae. Females still in the process of moulting are carried around by males, who have modifications such as a sucker-like structure and sometimes a modified fourth pair of legs. In the Eriophyidae, there may be several males guarding immature females if there aren't enough to go round! In the Tarsonemidae, inactive

pre-adult females are kidnapped by eager adult males and stuck onto a sucker on the male's opisthosoma. Unfortunately, this system sometimes falls down, as developing males may be collected by mistake! Once the female has moulted into an adult, she is immediately impregnated.

POST-COPULATION MATE GUARDING: HAREMS Not just confined to certain cultures in the human world, harems are commonly found in other animal groups too – even within mites, although it is very rare. Harem-keeping is found in species in the water mite genus *Unionicola* (Unionicolidae), which live inside freshwater mussels. Harem-keeping is also found in spider mites (Tetranychidae).

MALE AGGRESSION Fighting for females seems to be the norm in some mite species, and has been well studied in the family Tetranychidae. Males have been seen to stab each other with their chelicerae, which can lead to the loss of haemolymph and ultimately death. They might also spit silk and use their front legs for fighting. Male water mites e.g. *Unionicola* may grapple with each other over females, trying to get hold of their opponent's leg in order to stab it. Males of *Caloglyphus berlesei* attack other males by grabbing them using a modified third pair of legs, and stabbing the exoskeleton with the fourth pair.

COURTSHIP There seems to be very little known about courtship in mites, apart from that demonstrated in water mites. For example, certain species of *Eylais* (Eylaidae) have a circular courtship dance. Many species of *Neumannia* (Unionicolidae), such as *N. papillator*, demonstrate very complex courtship dances, with many different leg movements. This courtship encourages the female to lunge and grab at the male as if he were prey (i.e. a copepod). Courtship that uses trickery in this way is unusual.

In ixodid ticks, courtship is a complex process involving a number of behaviours leading to copulation. In all but the genus *Ixodes*, ixodid species demonstrate a very similar pattern of courtship. Sex pheromones play a vital role in reproduction. The female releases attractant sex pheromones, and the male follows the chemical lure. When he finds the female, he begins courtship. She then releases mounting sex pheromones, which enable the male to recognize her as a potential mating partner. Some ticks produce a third type of pheromone, the genital sex pheromone, which the male needs to recognize before he can produce his spermatophore.

SPERMATOPHORE STRUCTURE In the acarines, spermatophores are extremely diverse, especially in groups such as the Hydracarina. Different spermatophore structures include: droplet spermatophore, e.g. Oribatida, Macrochelidae; highly ornamented, e.g. Halacaridae and Bdellidae; flask-shaped or globular spermatophore, e.g. some mesostigmatids; sticky coating on the outside of the spermatophore, e.g. Unionicolidae; blobs of gelatinous material containing miniature spermatophores, e.g. Pionidae (Hydracarina); stalked spermatophores, e.g. Nanorchestidae; long, sac-like spermatophore carried on the dorsal surface, e.g. Pterygosomatidae; spermatophores that look like lily pads, as they have a short stalk and a flattened and broad, horizontal head, e.g. Eriophyidae.

Ticks have the most complex spermatophores of all. These unique parcels of biological engineering take only 30 seconds to produce. The outer layer, called the ectospermatophore, is a sac-like, membranous structure, coagulated from proteins and polysaccharides, into which seminal fluid flows. In ixodids, an inner droplet called the endospermatophore is then exuded

Superorders and orders	Methods of transfer
Parasitiformes	
Ixodida – ticks	• spermatophore transfer
Holothyrida	• unknown sperm transfer
Mesostigmata	• spermatophore transfer
Acariformes	
Oribatida	• usually have disassociated transfer with spermatophores, many are parthenogenic, some deposit directly onto the female's genital opening
Astigmata	• direct sperm transfer with intromittent organ
Prostigmata	• the greatest diversity of transfer, includes spermatophore transfer and direct sperm transfer with intromittent organ
Opilioacariformes	
Opilioacarida	• unknown sperm transfer

LEFT Summary of the various methods of sperm transfer in the Acari, divided by order.

into the ectospermatophore, which contains proteins and a chemical called spermine, which is vital in fertilization. In argasids, two endospermatophores are produced.

SPERMATOPHORE TRANSFER IN MITES The transfer of the spermatophore varies between groups. There may be complete dissociation between the sexes, so the male leaves a spermatophore on the substrate for the female to find later, e.g. Rhagidiidae and Bdellidae. This is not necessarily as haphazard as it sounds, because in such species the males have evolved ways of enabling females to find their spermatophores. These include silk signalling threads (e.g. many Prostigmata and Oribatida), pheromones (e.g. Hydrachnidae, Limnesiidae and Hydrodromidae), contact chemicals on the spermatophores, and the placing of spermatophores in places where females go in the usual course of events (e.g. some gall mites and Hydrodromidae). Spermatophores may be produced when females are present, e.g. *Halacarellus basteri* (Halacaridae), and mites in the superfamily Eriophyoidea.

The male may place the spermatophore on the substrate and guides the female over them. Alternatively, the spermatophore is passed directly to the female without touching the substrate. For example, in all parasitiform species (except those which are asexual), the male takes the spermatophore from his genital opening using his chelicerae, and places it in the female's reproductive opening. The chelicerae are often unmodified. In some species, the male's palps and first pair of legs assist in removing the spermatophore from the genital opening, once it has formed. In *Neumannia* (Unionicolidae), once the female has grabbed the male, he deposits a mass of sticky spermatophores that stick themselves to the female's dorsal surface and end up getting pushed into the genital opening.

Many species have modified structures for spermatophore transfer. For example, males in the Dermanyssina (Mesostigmata) often possess a spermatodactyl, on the moveable digit of each chelicera. The structure of the spermatodactyl is species specific. Males in the suborder Parasitina (Mesostigmata) have a spermatotreme. In the Macrochelidae, males dip their arthrodial brush (a group of setae at the junction between the moveable and fixed digits of the chelicera) into a droplet of seminal fluid. He then places the brush next to the other chelicera,

where the seminal fluid moves by capillary action up the groove of the spermatodactyl to the tip, ready for transfer to the female. Some water mites have modified legs for holding and transferring the spermatophore, e.g. Pionidae have a modified third pair of legs. In fact, water mites indulge in all possible types of spermatophore transfer.

In a few cases, the genital openings of both the male and female are placed close together, and the spermatophore is deposited onto the female's genital opening, e.g. *Caminella peraphora* (Mesostigmata) and *Pilogalumna* (Oribatida). This is rare behaviour among arthropods. In most mites and ticks, fertilization takes place in the ovary after the spermatozoa have moved along the oviducts. However, in parasitiform and dermanyssine mites that have 'ribbon' sperm (the sperm cells are modified into ribbon-like structures), it seems that the spermatozoa actually penetrate the ovary walls!

SPERMATOPHORE TRANSFER IN TICKS In the argasid ticks, mating takes place off the host, because they have little association with the host apart from quick feeding bouts. In the Ixodidae, however, mating occurs either on or off the host – it depends on the species. Endophilic ticks in the genus *Ixodes* usually mate off the host, within their host's shelter, whereas exophilic *Ixodes* usually mate on their host.

During the insemination process, the male grabs the spermatophore using his chelicerae and places it onto the female's genital pore. The neck of the spermatophore then turns inside out into the pore. Inside the ectospermatophore, a chemical reaction causes carbon dioxide gas to be generated. Pressure is created, as under the cork of a champagne bottle, which forces the contents of the ectospermatophore into the endospermatophore. The single or double endospermatophores expand and either detach and are stored in the female's seminal receptacle (as in ixodids), or travel as little sperm-filled capsules into the uterus (in argasids). The ectospermatophore shrivels up and drops off the genital opening.

DIRECT TRANSFER This occurs via an organ called an aedeagus (penis in other arachnid groups). Astigmatids and some prostigmatids reproduce in this way. Males of some species have specialized structures that assist in copulation. For example, males in the Acaridae get a purchase on the exoskeleton of the female using suckers on the fourth pair of legs and near the anus. Feather mites have specialized hind legs and setae. To copulate, mites assume a variety of positions. For example, both male and female face the same direction in the Glycyphagidae, while male and female face in opposite directions in the Acaridae.

SPERM COMPETITION Sperm competition is defined as competition between spermatozoa of two or more males for the fertilization of an ovum. In many mites, the first male to mate with a female gets to fertilize most of her eggs – this appears to be the same in argasid ticks. However, in ixodid ticks, the story is rather different – it is the second male to mate with the female that has greater success in egg fertilization. He probably removes sperm deposited by the previous male's mating efforts, from the female's spermathecae. This is called 'last male precedence' and is also found in some species of Acaridae (Astigmata). Other males are successful in sperm competition simply because they have the largest spermatozoa. In some species, males that encounter the spermatophores of other males destroy them and then produce their own on the same site. Other species, such as *Allothrombium lerouxi* (Trombidiidae, Prostigmata), defend deposition sites from other males.

PARTHENOGENESIS Parthenogenesis, which is reproduction by virgin females, is very common in Acari and is found across most of the orders. Parthenogenesis has two main forms – arrhenotoky and thelytoky. The former is asexual production of males that are haploid, from unfertilized eggs (they have a single set of chromosomes – half the full set of genetic material) e.g. Mesostigmata and possibly in Endoestigmata, Astigmata. Thelytoky is asexual production of females that are diploid (two sets of chromosomes – the full set of genetic material), e.g. Ixodida, Oribatida, Mesostigmata, Endoestigmata, Prostigmata, Astigmata.

There doesn't seem to be a clear reason why some species have sexual reproduction, while closely related species are thelytokous. But thelytoky appears to be much more common in habitats with low biotic diversity, such as recently disturbed or deep soil, bogs and so on. Hundreds of species have evolved from thelytokous species, which is surprising – one might assume that such asexual lineages wouldn't have sufficient genetic variation and would die out.

Egg laying and parental care

EGG LAYING IN MITES Oviposition rarely occurs immediately after sperm transfer – it can vary between days, weeks and months afterwards, depending on the species. In many mites (e.g. Phytoseiidae), a single mating is apparently all that is required to fertilize all of the female's eggs, whereas in other species it is necessary for the female to mate more than once in order for oviposition to take place. Most mites don't have ovipositors, and lay their eggs onto the substrate. Mites that do have ovipositors, e.g. Oribatida, are able to insert their eggs into soil and leaf litter. Water mites, e.g. *Hydrachna*, are able to insert their eggs into the tissues of aquatic plants using their ovipositors.

Eggs are laid singly, or as a batch. For example, females in the suborder Dermanyssina (Mesostigmata) produce one egg at a time, which may weigh up to 40% of the mite's original body weight! They can lay up to eight of these massive eggs in one day. To lay such large eggs, the mother must stand at full leg extension. The eggs may be laid in one go, or there may be repeated egg laying, e.g. in *Acarus siro* a female may produce between 100 and 500 eggs over a period of 10–12 days.

The number of eggs and body size of mites are usually correlated, so miniature mites may only produce one egg per oviposition. Additionally, the size of host in parasitic water mites appears to affect the number of eggs produced, so that those that parasitize small hosts produce small clutch sizes (but with largish eggs), whereas those that parasitize large hosts produce larger clutch sizes (with smaller eggs). The majority of mites lay eggs in which the embryos are at an early developmental stage (oviparity), whereas in a few parasitic species the eggs develop inside the female and larvae are born (larviparity). In some Dermanyssoidea that are parasites of vertebrates protonymphs are born.

BELOW Mite eggs vary in shape – they may be spherical, bird-egg shaped, ornamented, or with strangely shaped structures such as this oribatid egg. They also vary in colour from white through to red.

Egg laying in ticks

After engorgement, an ixodid female drops off the host onto the ground. The time between engorgement and the start of oviposition is called the pre-oviposition period. It is variable in length, and depends mainly on the species and environmental temperature. Once the female is ready to lay her eggs, she searches out a sheltered microhabitat, such as under leaf litter. Argasid ticks, on the other hand, are much more modest on the egg-laying front. The female does not put 'all her eggs in one basket' – rather, she lays several small clutches of eggs (approximately 100 in a clutch) over a period of time (up to a maximum of six cycles) after each time she has fed. After an oviposition event, the female remains active – she is hungry, and so seeks a new host, feeds and oviposits again. She may have to wait a long time between each host. This pattern of several cycles over a period of time can spread the offspring of argasid females over many years, which is in sharp contrast to ixodids producing all their offspring in one massive hit.

LEFT A female *Ixodes holocyclus* tick lays up to 20,000 eggs in a single oviposition! She uses the nutrients from her huge blood meal to produce these eggs. Production of eggs reaches a peak up to 3–5 days after the start of oviposition, and 90% are produced by the tenth day. However, she does produce a few more eggs for up to another 10 days. By this stage, she is totally worn out, her mission is complete, and she dies.

PARENTAL CARE Acarines show little parental care compared to other arachnids. The most they do is hide or camouflage their eggs. Hiding consists of carefully laying the eggs in crevices (if they have a decent sized ovipositor), inserting them into plant stems (if they are aquatic), or laying them within webs made from silk. Camouflaging consists of covering the eggs with soil particles or a jelly-like substance.

Of course, there are always exceptions. For example, some females lay their eggs on the backs of males, e.g. *Damaeus verticillipes* (Oribatida). Some carry eggs attached to their legs, e.g. some Halacaridae. Some guard their eggs against predators, e.g. *Cheyletus eruditus*. A rather extreme form of parental care is found in red-legged earth mites; after the female has died in the spring, her eggs remain in diapause (suspended development) until the autumn comes, when they hatch.

Post-embryonic development

In general, the different developmental stages of acarines, after the egg are: a non-feeding six-legged prelarva; a six-legged larva which is active and usually able to feed; one to three eight-legged nymphs - a protonymph, a deutonymph, tritonymph and then the adult. There may be suppression and skipping of certain nymphal stages within the Acari, depending on the group. For example, the prelarva is only found in some acariformes (oribatids, astigmatids, prostigmatids). The deutonymph is suppressed in parasitic species (e.g. ixodid ticks), so the protonymph moults directly to tritonymph. The tritonymph is suppressed in certain acarines, e.g. Mesostigmata, Eriophyoidea and ixodid ticks.

HOMEOMORPHISM, HETEROMORPHISM AND HYPOPI With each subsequent moult into the next stage, various structures such as setae, sclerotized plates, lyrifissures and pores

appear. These structures make it relatively easy to recognize the different developmental stages. In general, immature stages resemble their respective adults (apart from lack of sclerotization, certain characteristics and genitalia) and this is termed homeomorphism. In a few mite groups, however, one developmental stage differs dramatically from the others, and this is termed heteromorphism.

A classic example of heteromorphism in mites is the hypopus. The hypopus replaces the deutonymph stage in some astigmatid families. Hypopi are produced when the surrounding environmental conditions are unfavourable. They are non-feeding, highly resilient forms which are resistant to insecticides, fumigation and low temperatures. They have no mouthparts (or they are much reduced) and can last for several months without feeding. They are either 'motile' (their four pairs of legs are functional), or they are 'inert' (their legs are non-functional). In order to gain access to new sources of food, motile hypopi use other organisms such as arthropods as transport. The hypopi often have ventral suckers to attach to such animals. They can also disperse on air currents. Inert hypopi, on the other hand, stay put inside their protonymphal exoskeleton until favourable conditions return, and then moult into a tritonymph. Because of the high resistance of hypopi to unfavourable conditions such as fumigation, species that have this form, such as the grain mite *Acarus siro*, are very difficult to eradicate. When environmental conditions do not require dispersal or diapause, the mites capable of forming a hypopus moult directly from the protonymph to the tritonymph stage.

DIFFERING DEVELOPMENTAL TIMES Many mite plant parasites characteristically have generation times of less than 10 days – some examples are described in the table. In some Macrochelidae, Phytoseiidae on plants and mesostigmatids in the soil, the amount of development each day may be 10–20% of the entire generation cycle. Although those mites that exploit ephemeral resources have very rapid developmental rates, they retain the full sequence of developmental stages.

Order, family or species	Developmental time	Economic importance
Prostigmata		
Polyphagotarsonemus latus (Tarsonemidae)	3 days from egg to adult	pests of economically important plants
Tetranychidae	10–14 days for a whole generation	
Astigmata		
Acarus siro (Acaridae)	9–11 days from egg to adult pests of foodstuffs	one of the most important
Mesostigmata		
Zygoseius furcifer (Pachylaelapidae)	108 hours (±4 hours) from egg to adult	predates nematodes
Macrocheles muscaedomesticae (Macrochelidae)	66–77 hours from egg to adult	none
Phytoseiidae	usually takes 8–10 days, some species may take only 5 days	often used as a biological control of mite plant parasites

LEFT Differing developmental times from egg to adult for a selection of mites. These have implications for control especially for economically important crops.

However, not all mites have such fast developmental rates. Mites that have a close interaction with their hosts' life-cycle may only pass through a single generation per year. Mites that have complicated life-cycles often have extended developmental and generation times.

DIFFERENT DEVELOPMENTAL STAGES IN TICKS Ixodid ticks have four developmental stages – the egg, and three parasitic stages (the larva, nymph and adult). Moulting in the larval and nymphal stages occurs after the blood meal has been digested. Argasids are different to ixodids, because they have more than one nymphal stage – there may be up to eight instars, depending on the species, availability of nutrients and the tick's gender.

LIFE-CYCLE OF IXODIDAE The life-cycle of ticks in the family Ixodidae is remarkably similar throughout: mating occurs on the host (except in *Ixodes*), the female engorges on the host and then drops off, and oviposition commences in a sheltered site. Ixodids have three-host, two-host or one-host life-cycles. A three-host life-cycle is where each separate stage – larva, nymph and adult – has a different host. In contrast, in a few species, the fed immature stages remain and develop on the host. Depending on whether later stages drop off and seek another host, or if they remain on the initial host, they are called 'two-host' or 'one-host' ticks. Hard ticks spend more than 90% of their life off the host. Under cover of deep vegetation, or in the shelter of their hosts, they spend long periods of time resting and developing. The life-cycle takes 1–6 years, depending on environmental conditions, but it is usually around 2–3 years. In a laboratory setting, however, it may take a mere 3 months.

There are several variations of this pattern. In two-host ticks, such as *Hyalomma anatolicum excavatum*, the larvae remain on the host once they have fed. They moult *in situ* and then reattach as an unfed nymph. They only detach once they have engorged and will moult to the adult stage off the host.

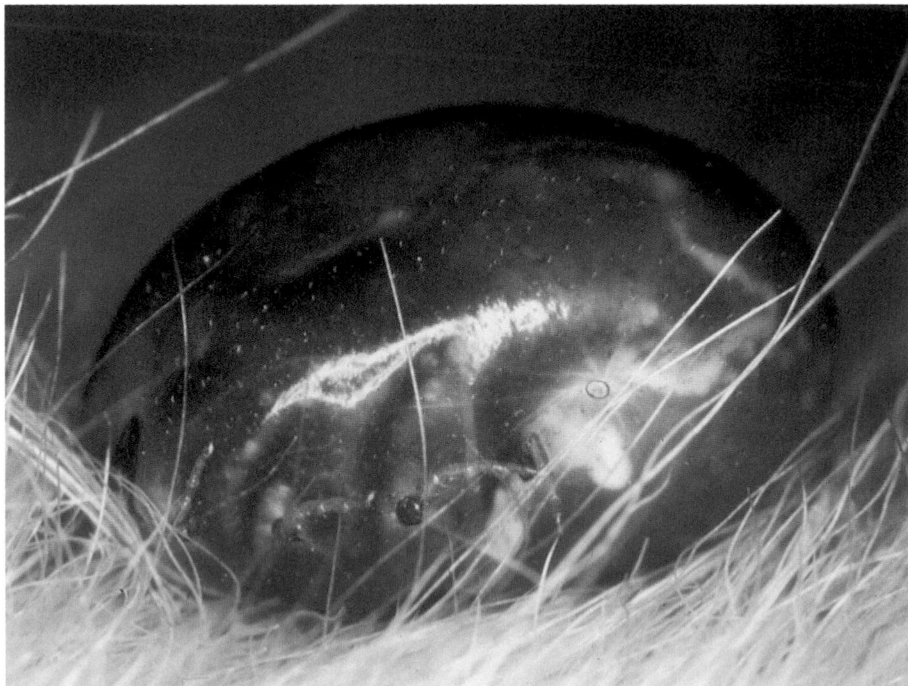

RIGHT In one-host ticks, such as *Boophilus microplus*, all the stages of the life-cycle remain on the host. Even once moulted to the adult stage, both sexes stay on the host to feed and mate. Only the engorged females finally detach in order to oviposit.

LIFE-CYCLE OF ARGASIDAE In contrast to ixodid ticks, argasids demonstrate a much greater diversity in their life-cycles. Nearly all species in the Argasidae have a multi-host feeding cycle, with many more hosts than the maximum three-host cycle of ixodids. This is because they feed regularly for very short periods (up to 30 minutes), and then drop off their host. There are exceptions to this typical argasid life-cycle. For example, the larvae of certain bat parasites, e.g. many *Ornithodoros* species, stay attached for several days, as if they were ixodid ticks. Slow-feeding larvae are also common in species that parasitize birds. Because argasids have so many nymphal stages, they have much longer life-cycles than ixodids. Additionally, their life-cycle may be extended for some years, because they are able to resist starvation for long periods of time during their development.

9 Opiliones

Harvestmen, harvest spiders, daddy long legs spiders

OPPOSITE The harvestman, *Pristocnemis pustulatus*, is only found on some mountain chains in southeastern Brazil, and is extremely sensitive to disturbances in the forest structure. Like other diurnal harvestmen the individulas are brightly coloured.

Harvestmen, daddy long legs and harvest spiders are all common names for the diverse order Opiliones (previously Phalangida). In France, they are known as reapers. In folklore, opilionids have a connection with the harvest season, possibly because some species are abundant at that time of the year. In Greece, and in the UK historically, they have been called shepherd spiders – and 'opilio' means sheep master, in Latin. In the past, European shepherds used stilts in order to have a better view of their flocks, so perhaps these long-legged creatures were reminiscent of the shepherds, and this is how the order got its name.

A typical opilionid species has extremely long, fragile-looking legs, with a small, rounded body, hence the common name 'daddy long legs'. However, some opilionids are bandy-legged, chunky and squat, with shortish legs and raptorial palps. Opilionids have no silk glands, and no venom glands, but possess odoriferous glands to repel predators. Many of their common names around the world refer to the horrible smell of these glands, e.g. the Argentinean name 'chichina' means stink bug in Spanish. Other special characteristics of the order include a penis in males and an ovipositor in females, and certain species have an omnivorous diet. Opilionids range in size from around 1 mm to 20 mm (1⁄25 in to ¾ in) in body length. The largest species is *Trogulus torosus* (from Europe) measuring 22 mm (1 in) in body length. One of the smallest species is *Crosbycus dasycnemus* (from North America and East Asia), which is less than 1 mm (1⁄25 in) in body length.

Classification

There are approximately 6,411 described species, making it the third largest arachnid order (below acarines and spiders). However, it's thought that this number may exceed 10,000 in reality. There are 45 families grouped into four suborders – Eupnoi, Dyspnoi, Laniatores and Cyphophthalmi. Eupnoi has six families and 1,800 species, Dyspnoi has seven families and 375 species, Laniatores has 26 families with 4,085 species and Cyphophthalmi has six families and 143 species.

The relationship among the opilionid higher ranks and the status of the order as a whole is one of the most challenging to arachnologists; it has been difficult for scientists to come up with a cladistic hypothesis with which everyone agrees. However, recent studies combining morphological characters with molecular evidence are progressing the understanding of the phylogenetic placement of the Opiliones. In some classifications, the suborders Dyspnoi and Eupnoi don't have suborder status, rather they are part of the suborder Palpatores, which

means to feel one's way, and refers to the way that opilionids use their palps. In other classifications, the Dyspnoi are grouped with either the Laniatores or Eupnoi. One of the most recently proposed hypotheses was published in 2002 by Giribet, Edgecombe, Wheeler & Babbitt. It places the Opiliones as the sister-group of the clade comprising pseudoscorpions and solifugids. This is in contrast to the hypotheses of Weygoldt and Paulus (1979) and van der Hammen (1986) (see p.8).

Diversity

Opilionids are actually much more varied in shape and size than one would imagine, and they differ quite radically between the four suborders. The great majority in the Eupnoi have a small, leathery body and long, slender legs, which look as if they could break under the slightest pressure. They have prominent eyes, and chelicerae like pincers. They are agile, fast and live in vegetation, and range in size from 1 mm to 12 mm (1/25 in to ¾ in) in body length. These are the 'daddy long legs' of the opilionid world e.g. *Phalangium opilio*, which has a holarctic distribution (it is found throughout the northern continents of the world).

Opilionids in Dyspnoi are completely different; many are short-legged and sluggish, although there is a great diversity of morphology and sizes – they range from less than 1 mm (1/25 in) right up to 22 mm (1 in) long. They often live in the soil and leaf litter and some are cave dwellers. An example of a Dyspnoi species is *Ischyropsalis kollari*.

BELOW This harvestman, *Leiobunum rotundum*, clearly shows the form of the majority of opiliones in the Eupnoi suborder- a small body and long slender legs.

LEFT Most opilionids in the Dyspnoi are short-legged but this *Ortholasma pictipes* (Nemastomatidae) shows there is variation within the suborder.

'Laniatores' means to tear in pieces, in Latin, and refers to the large claws on the raptorial spiny palps. These are bandy-legged, chunky and squat-looking opilionids, which primarily live in the tropics in leaf mould, beneath stones and under rotten logs. They are also known as short-legged harvestmen, although their legs may also be very long. In fact, Laniatores is the most diverse suborder. They are very heavily sclerotized, and range in size from 0.6 mm to 16 mm (2/100 in to ¾ in) long.

Cyphophthalmi are, in general, small, mite-like opilionids, which are heavily sclerotized. Although their scientific name means humped eye (ophthalmic means related to the eye), nearly all cyphophthalmids, except those in the family Stylocellidae, have no eyes. The name actually refers to odoriferous glands located on tubercles, which look like eyes. They have short legs, and are between 1 mm and 7 mm (1/25 in and ¼ in) long. They live in soil and leaf mould. An example of a Cyphophthalmi species is *Pettalus thwaitesi* (see right).

ABOVE *Pettalus thwaitesi* (Cyphophthalmi) is brown and mite-like and quite inconspicuous in the soil.

LEFT This opilionid in the family Stygnommatidae (Laniatores) has the large raptorial spiny palps which is indicative of the suborder.

Colouration

There are many species of opilionids that have inconspicuous liveries of browns and blacks, and troglobitic species that are a dull yellowish colour due to lack of pigment. However, within the four suborders, there are species with some amazing and unexpected colours, as outlined in the table.

RIGHT The range of colourings within the suborders of the Opiliones.

Suborder and family	Colour varieties
Cyphophthalmi	
Sironidae	some species are red, orange, yellow or black
Stylocellidae	some species are black, brown, orangey red, maroon
Troglosironidae	*Troglosiro raveni* shows distinctive brown and black pattern
Eupnoi	
Monoscutidae	some Megalopsalis species are jet black – males of *M. inconstans* are jet black with bright orange patches on the carapace
Neopilionidae	a few species have blue pigment (rare)
Phalangiidae	some species have greenish colour, or reddish pink
Sclerosomatidae	very variable colouration – some species have golden, silvery, copper metallic coloured patches, spots or stripes that vary in size and colour, or striking combinations of black, yellow, orange and green
Dyspnoi	
Ceratolasmatidae	genus *Ceratolasma* has a striking pattern with white and brown parts
Nemastomatidae	several genera have a common and rather striking pattern of silvery white or golden spots on a black dorsum
Laniatores	
Agoristenidae	species colouration varies from yellow to dark brown, and may have yellow stripes or green and white patches
Cosmetidae	in some species, the scutum and legs are reddish to blackish brown, with varied patterns of white, yellow or green; *Rhaucus vulneratus* (Latin species name translates as 'wounded') has four parallel blood-red stripes, made of wax produced by glands on the exoskeleton
Gonyleptidae	extremely variable – some species have very colourful patterns, in black, white, orange, yellow, yellowish-green or green, while *Gertia hatschbachi* has white markings on its legs and dorsal surface
Podoctidae	some species are a deep green, with legs ringed in yellow and black

RIGHT Cosmetidae sp. wandering at night on vegetation, Costa Rica.

Anatomy

External anatomy

At first glance, the body of an opilionid appears to consist of a one-piece, raisin-like structure. However, closer inspection reveals that, as in all arachnids, the body consists of two parts – the prosoma and opisthosoma. They give the appearance of being one because they are broadly joined. This is in stark contrast to a spider's tiny-waisted pedicel.

EXOSKELETON The exoskeleton varies in thickness, depending on the group of opilionids. Many Eupnoi and Dyspnoi have a soft, leathery and thin exoskeleton, whereas Laniatores and Cyphophthalmi are generally heavily sclerotized. The exoskeleton is covered with spines, setae, tubercles, denticles, spicules, and so on. Tubercles are hollow outgrowths of the cuticle, denticles are solid outgrowths of the cuticle, and spicules are very small denticles. All these structures together are referred to as an opilionid's 'armature'. Most spines are in contact with nerve endings. There is a generous sprinkling of lyriform organs on the body, legs (especially near the joints) and palps. These nerve connections are stimulated by movements of the body such as those caused by walking, vibrations through the air (airborne sounds), and vibrations through the substrate. Opilionids use sensory hairs on their palps and legs, called sensilla trichodea, to detect mechanical stimuli caused by moving prey, for example. The exoskeleton also possesses chemosensitive setae, but there are no trichobothria, which are so common in other orders such as scorpions and spiders. In some species, the patterning of yellow and white spots and stripes on the exoskeleton fluoresces under UV light, as does the whole exoskeleton of other species. However, the intensity of fluoresence is not nearly as impressive as in scorpions. As with scorpions, the function of fluorescence is unknown.

LEFT Species in the Gonyleptidae are particularly well endowed with very large spines e.g. *Glysterus* sp. from Costa Rica.

RIGHT Dorsal view (top) and ventral view (bottom) of the body of a typical opilionid.

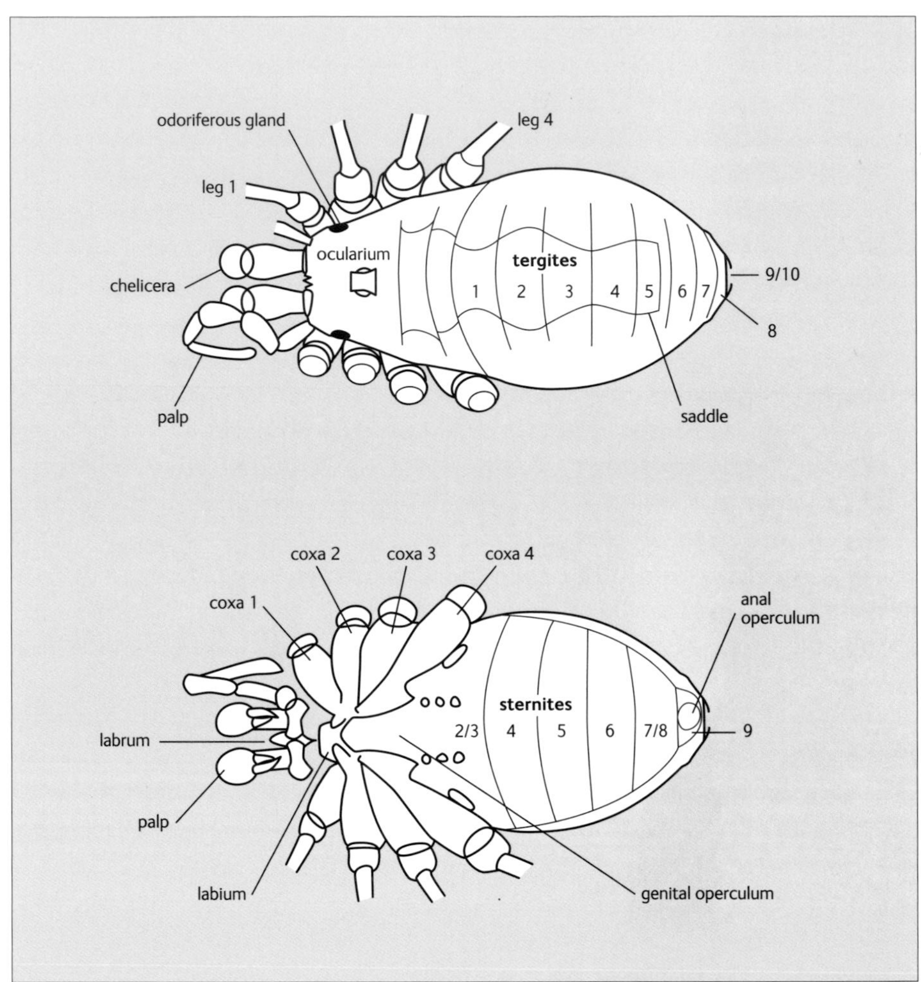

PROSOMA The prosoma is composed of six segments and each has a pair of appendages attached – the chelicerae on the first, the palps on the second, and four pairs of legs on the third, fourth, fifth and sixth segments. Dorsally, the segments are fused and form a carapace. In many families (e.g. Trogulidae, Cyphophthalmi and Laniatores), the carapace is joined to some or all of the tergites on the opisthosoma, forming a single shield called a scutum. In many species, this dorsal shield is present in males, but not in females. In the Trogulidae, there is a distinctive 'hood' structure, which projects forward from the base of the eye tubercle and overhangs the chelicerae.

EYES Opilionids usually have one pair of eyes, although those in the Cyphophthalmi have no eyes (apart from those in the family Stylocellidae). A typical eye is made up of transparent cells, called lentigen cells, sandwiched between a lens formed by a thickening of the cuticle, and the retina. It is believed that these eyes can do no more than distinguish between different light intensities, although they might possibly detect movement too, because animals in captivity appear to react to movement outside their confines by running around or bobbing their bodies up and down. The structural complexity and size of the median ocelli are very variable among opilionids, and are related to the amount of light available within the habitat.

So, species living in a darkened habitat have a reduced number of lentigen cells, for example In all but the Trogulidae, opilionid eyes are sited either side of a mound-like structure called an ocularium, which is found dorsally on the midline of the carapace. There are often tubercles, denticles or spicules on top of the ocularium. The size of the ocularium is related to the habitat in which the species is found. So, species that live above ground (e.g. *Mitopus morio*, which has a cosmopolitan distribution) have a well-developed ocularium, whereas species that live in darker habitats, such as in deep leaf litter (e.g. *Nemastoma bimaculatum*, from northern to southern Europe) have a reduced ocularium. In troglobitic or subterranean species, the eyes have non-functional retinas, or are totally absent.

FEEDING APPENDAGES Opilionid chelicerae have three segments, unlike the chelicerae of other arachnids, which usually have two. The smallest segment takes the form of a hinged claw (apotele) working against the fixed digit to produce a pincer, which is used to manipulate and kill prey, and to dismember food. In males, the chelicerae may be a lot larger than those of the female, and may have prominent extensions (apophyses). A cheliceral gland is found only in adult males of the Sabaconidae and the Nemastomatidae. This gland produces a fluid that is transferred to the female as a nuptial gift. The coxae of the palps and those of the anterior legs assist in masticating the food. This is known as coxisternal feeding, and also occurs in scorpions.

STRIDULATORY RIDGES As with some spiders and amblypygids, some groups of opilionids stridulate. The noise is produced by several parallel ridges (located on the inner surface of the chelicerae, or on the inner surface of the femora of the palps) rubbing against other ridges or some sort of scraper on the outside surface of the chelicerae, or on the cuticular outgrowths of the ocularium.

PALPS The palps are leg-like, and have six segments: the coxa, trochanter, femur, patella, tibia and tarsus. They may possess setae, spines, denticles and tubercles, and usually have a sensory function. The palps of opilionids are rather varied between the different families – they may be slender (e.g. Nemastomatidae), shaped like a boxing glove (e.g. Sabaconidae), or enormously large. Some opilionids (e.g. Laniatores) have palps equipped with large claws on the tip. These are either smooth (as in the Phalangiidae), or with teeth (as in the Leiobunidae). These claws are raptorial, and are adapted to grasp prey, in much the same way as the chelicerae. The coxae bear gnathobases ('gnatho' indicates the jaws), which form part of the complex mouth; they are used to grind food. The palps are used as tactile structures, for manipulating food and for grasping mates.

LEGS The legs of most opilionids have the same divisions as a typical arachnid leg, but the tarsus is divided into the metatarsus (basitarsus) and tarsus (telotarsus). Most of the leg segments possess hairs, tubercles, denticles and spines. There are two groups of lyriform organs, lying either side of the femur–trochanter joint, where autotomy takes place. There are groups of slit sensillae, and campaniform (bell-shaped) organs, which are mechanoreceptors. These organs enable the owner to recognize others of the same species. The tarsus is often divided into many segments called tarsomeres; there may be up to 100 of these articulating subdivisions. Such a long and very flexible tarsus acts a bit like a cowboy's lasso, allowing the opilionid to grasp wispy substrata. In general, there is a single claw on the tarsus, except in the 3rd and 4th pairs of legs in Grassatores (an infra order of Laniatores).

Long legs

How does a small body support the long legs so characteristic of opilionids? The coxae of most opilionids fan out and are attached to the ventral surface of the prosoma; this structure is vital for support. Articulation is at the joint of the trochanter–femur. One might imagine that very long legs would be a problem to extend, but in fact normal hydrostatic pressures in the haemolymph (compared to other arachnids) are totally sufficient for extending the legs when they are off the ground. Another potential problem with such long legs is the possibility of getting them tangled; however, each pair of legs is different in length to the others, so overlap is avoided.

ABOVE An extremely long-legged opilionid (Danum Valley, Sabah)

Many species of opilionids have the longest legs of all arachnids compared to their body size (with a few exceptions, such as spiders in the family Pholcidae). The legs are sometimes as much as 40 times the length of the body! Long legs are certainly a good way of getting around e.g. *Leiobunum rotundum* (from the UK, Canary Islands and Africa) in the family Sclerosomatidae (Eupnoi) strides over tall vegetation that other invertebrates would find impossible to cross quickly. Such species usually have a greater number of tarsal segments. Animals with shorter legs, as in the suborder Dyspnoi, are found in leaf litter and under stones, where they lead a rather more sluggish life.

When a group of opilionids are at rest, their legs touch those of their neighbours. This greatly increases the size of their sensory range, much in the same way as a web does for a

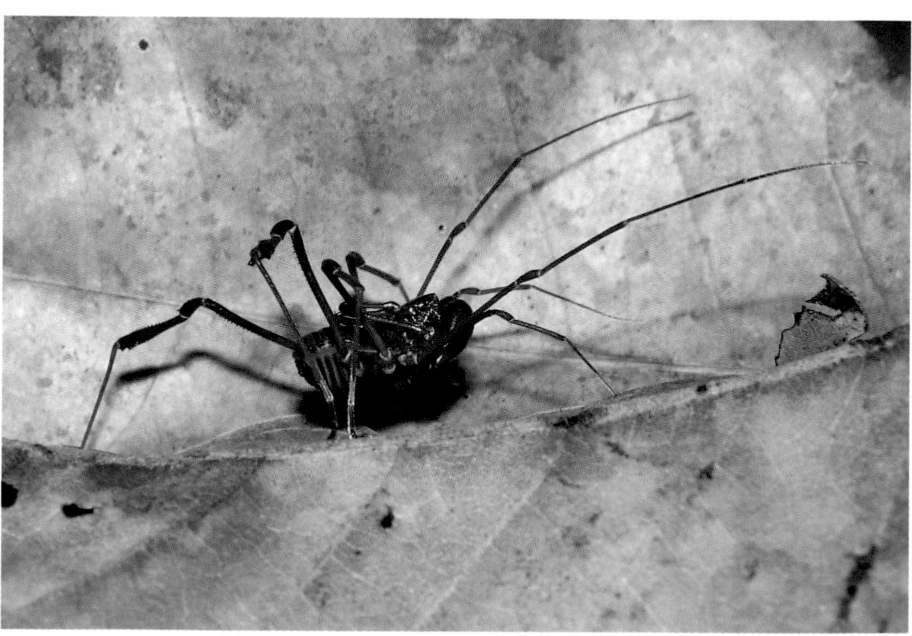

RIGHT As with amblypygids and uropygids, opilionids use a pair of their legs as sensory structures in place of antennae, unlike amblypygids and uropygids, they use their longer second pair of legs (instead of their first). The opilionid's second pair of legs touches the surroundings as the animal moves around, as seen with this species in the Stygynidae (Laniatores).

spider. Even if only a few individuals on the outside of the group are alerted, the entire group very rapidly becomes aware of the possible threat because an alarm 'message' is transmitted through the leg contacts. Eupnoi opilionids have the rather strange habit of bobbing up and down if disturbed. Although this behaviour is believed by some to be involved in killing prey, it is also thought to be a way of warming up the leg muscles and getting fresh air into the tracheae, so that the animal can move quickly away. It might be a form of defence, too.

ODORIFEROUS GLANDS Like uropygids and schizomids, opilionids have odoriferous glands, also known as stink or repugnatorial glands. Whereas uropygids and schizomids have anal glands at the base of the flagellum, the odoriferous glands of an opilionid are located on the dorso-lateral surface of the prosoma, next to the coxae of the second legs. These glands produce and expel a fine spray or droplet of defensive fluid, described as smelling like walnuts or horseradish! In Cyphophthalmi, the odoriferous glands are located on tubercles. Among the Laniatores and Cyphophthalmi, they are more developed (and therefore more visible) than in Dyspnoi and Eupnoi. In Cyphophthalmi the glands are controlled by muscles, while in the other groups there is a varying degree of musculature, which may even be absent.

OPISTHOSOMA Opilionids have a segmented opisthosoma, with ten usually distinctive segments marked by grooves or tubercles. The opisthosomal tergites can be fused together, along with the prosoma, to form a shield-like scutum. Each segment has a dorsal tergite and ventral sternite, although the sternites are not directly below the tergites. Sternite 10 is the anal cover, surrounded by sternite 9. The opisthosoma may have a pattern of markings on the dorsal surface, called the saddle, which may be continuous with patterning on the prosoma. The dorsal surface may be fairly smooth, or have tubercles; the ventral surface is usually smooth, but in some genera (e.g. *Sabacon*), it is hairy.

RESPIRATORY STRUCTURES Opilionids usually have an extensive tracheal system, opening through a pair of spiracles on the second or third opisthosomal sternite, just behind the fourth coxae. They possess no book lungs. The tracheal supply to muscles in slow-moving species is less developed than in more active species. Oxygen enters by diffusion. The spiracles have a number of different forms, depending on the family. The openings vary in shape, and may be exposed on the surface, or hidden. Some opilionid species have clusters of finely spaced spines (of different lengths depending on the species) on the inside rim of the spiracles. These form a mesh that may reduce water loss. In contrast, the spiracles of species in the Sclerosomatidae are open or weakly guarded. Oxygen is able to reach to the tips of the tarsi of the very long legs of phalangiid and sclerosomatid opilionids because of a pair of 'accessory' spiracles on the tibiae of all four pairs of legs – these allow oxygen to enter the tracheae at this point, to supplement the supply to the extremities.

SEXUAL DIMORPHISM As with all arachnids, males are usually a bit smaller than females, although they have longer legs. Males are more heavily sclerotized, so cuticular armature is usually more pronounced. A cheliceral gland is found only in adult males of the Sabaconidae and the Nemastomatidae. In several species, the armature of the chelicerae, palps, fourth pair of legs and mesotergum may be modified in males, and have distinctive shapes, swellings and apophyses.

EXTERNAL GENITALIA When not in use, the genitalia lie behind the genital operculum, which is quite far forward on the ventral side of the body, and is usually located between the leg coxae. In Eupnoi, it is even further forward, immediately behind the labium, just behind the mouth, and is derived from the second and third sternites.

Internal anatomy

Relatively little is known about the internal structure of opilionids. Researchers have done intensive studies on only a few species, mainly *Phalangium opilio*, so the descriptions that follow are based largely on this species.

CENTRAL NERVOUS SYSTEM The CNS of opilionids is centralized around the oesophagus and consists of a large consolidated neural mass. The supraesophageal ganglion consists of two main parts, the protocerebrum and the deutocerebrum. The protocerebrum is hooked up to the eyes by the optic nerves, and its main job is to receive and then process visual information. The deutocerebrum is hooked up to the chelicerae by well-developed cheliceral nerves, as well as being associated with the pharynx and oesophagus. The subesophageal ganglion is associated with the palps, legs and opisthosoma. The heart is innervated by a cardiac ganglion that runs along the mid-dorsal surface.

CIRCULATORY SYSTEM The heart is short and tubular, and located dorsally in the first three opisthosomal segments. It only has two pairs of ostia, in contrast to a spider's heart, which may have up to 18 ostia. The anterior aorta follows the dorsal surface of the midgut, until it reaches the oesophagus, where it empties into a perineural channel that surrounds the central nervous system and the major nerves. The posterior aorta passes along the median dorsal surface of the midgut to the posterior of the body. The circulatory system of an opilionid is unusual in having haemocyanin, the oxygen-binding respiratory pigment found in the haemolymph of arachnids with book lungs.

RESPIRATORY SYSTEM Opilionids don't have book lungs, but possess tracheae, though the tracheal system doesn't have extensive arterial branching. There are two 'stem' tracheae, starting at the two spiracles, and branching into secondary tracheae. The tracheae are located mainly in the prosoma, where the nervous system, parts of the gut, the genital organs and the leg muscles are located. From each spiracle, there is a single main trunk which projects upward and forward into the prosoma. It gradually narrows before ending in the chelicera. The trunk sends off lateral branches along its path to the appendages.

DIGESTIVE AND EXCRETORY SYSTEMS What sets opilionids apart from many arachnids is that they don't have a sucking stomach. It is therefore necessary for them to chew their food. The pharynx is a tubular structure lined with a thin layer of cuticle, so it doesn't absorb liquids. It becomes the oesophagus at the point of penetration of the central nervous system.

The oesophagus joins the vastly expanded midgut, which has a lining of epithelial cells. It is the largest organ in opilionids and performs many vital functions including digestion, nutrient absorption, and water uptake. It is also used as a larder to store glycogen and lipids that can be used during over-wintering. Digested nutrients are absorbed in the anterior midgut (lined with digestive cells), while indigestible food is passed to the posterior midgut, which is lined with

transport cells and secretion cells. The main jobs of the posterior midgut are reabsorpion of water, via the transport cells, and compression of waste to produce faecal pellets. The hindgut, or proctodeum, links the posterior midgut to the anus. Its walls are lined with thin chitin. It can be shortened and dilated by different sets of muscles, which, when contracted, push the faecal pellet out through the anus.

Opilionids don't have Malpighian tubules – coxal glands and nephrocytes are responsible for excretion. Water and ions are regulated by the coxal glands, which open on the border of the third coxae from which coxal fluid exits.

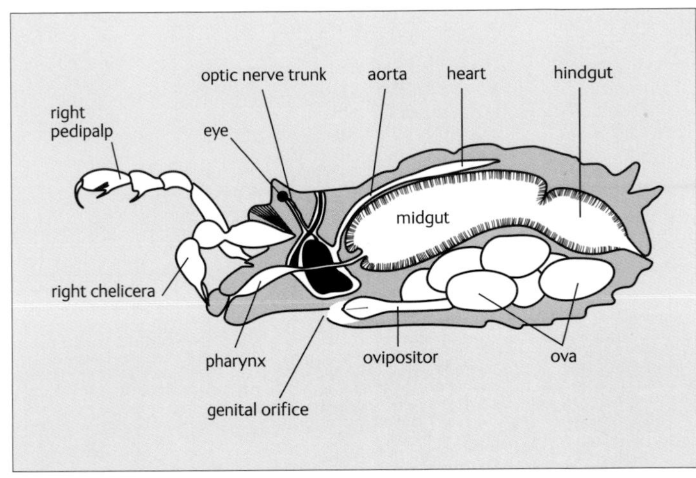

ABOVE Cross-section of generalized gut of female opilionid.

INTERNAL GENITALIA Opilionids are very unusual among the arachnids, because there is direct sperm transfer. Male opilionids have an intromittent organ for reproduction, usually called a penis, although the term 'spermatopositor' is used in the Cyphophthalmi. In contrast, the penis of Dyspnoi, Eupnoi and Laniatores consists of two main parts – a long shaft (corpus) and a short, round-headed tip called the glans, which is either operated by muscular or hydraulic action, depending on the opilionid group. In Eupnoi, Dyspnoi and Insidiatores (a group within Laniatores) the tip of the penis is moved relative to the shaft by a long tendon connected to a muscle bundle at the enlarged base. The shaft is therefore sclerotized to prevent it from collapsing during muscle contraction. In the Grassatores (a group that comprises the superfamilies Gonyleptoidea and Oncopodoidea in the Laniatores) the tip of the penis is erected by haemolymph pressure. There are therefore no muscles and the penis shaft is only weakly sclerotized.

In females, the ovipositor consists of a long tube made from up to 40 membranous rings in most species in Eupnoi and all species in Cyphophthalmi, but is short and unsegmented in Dyspnoi and Laniatores. In most species in the Cyphophthalmi, the tip of the ovipositor has a forked structure, called a furca, with the genital opening between the two processes. On each of the two processes is a sense organ. The furca may help females find spermatophores and may also have a role in the selection of appropriate oviposition sites.

The basic structures of male and female reproductive systems are similar in structure (the eversible organ is the penis or the ovipositor). The female has a horseshoe-shaped ovary, and the anterior ends are joined to an oviduct (gonoduct), which continues forward as the uterus. This then becomes the vagina, which is lined with cuticle. The vagina moves forward through the ovipositor during egg laying. When not extended, the ovipositor is protected within a chamber that has flexible cuticular walls. The posterior end of the ovipositor attaches to the posterior end of the chamber. Eversion of the ovipositor (being turned outwards) occurs when muscles on the walls of the pregenital chamber expand its lumen. Increase in haemolymph pressure also contributes to the eversion process. Just inside the genital opening is a pair of seminal receptacles, which store the sperm until the female is ready to fertilize her eggs. The ovipositors of some Laniatores have four lobes at the tip, instead of two.

The male has a horseshoe-shaped testis, and the anterior ends are joined to a sperm duct, which continues forward as the vas deferens. This then becomes the ejaculatory duct lined

with cuticle, which moves forward through the penis during insemination. The ejaculatory duct is connected to the expulsion organ at the base of the penis. This is attached in turn to the seminal reservoir, a closely spiralled structure where the semen is stored. In Dyspnoi, Eupnoi and Laniatores, sperm is pushed through the elongate ejaculatory duct by the action of the expulsion organ. In the Cyphopthalmi, the expulsion organ is absent. When not extended, the penis is retracted behind the genital operculum, and is protected within a chamber with flexible cuticular walls. The posterior end of the penis attaches to the posterior end of the chamber. Eversion occurs when muscles on the walls of the pregenital chamber expand its lumen, though haemolymph pressure is also involved.

OPPOSITE Opilionids do have a wide distribution across the globe, as illustrated by the ubiquitous *Mitopus morio* (Phalangiidae, Eupnoi). It has a very wide range – Europe, Asia (north of the Himalayas) North Africa, and most of the Arctic.

Distribution

Opilionids are found primarily in the humid tropics, where there is the greatest diversity. Such diversity decreases sharply in drier and cooler climates, e.g. there are only 25 species in the whole of the UK. *Phalangium opilio* (Phalangiidae) is thought to have a wide distribution as a result of human activities – urbanisation, transport and agriculture. It is found in North America and New Zealand, as well as in the Palaearctic; its extended range is mainly in disturbed habitats. There are, of course, opilionids with much more limited distribution, e.g. the family Troglosironidae is endemic to New Caledonia. In fact, many species have limited distribution ranges.

The Eupnoi are widely distributed in both hemispheres. The Dyspnoi are mainly found in the northern hemisphere, extending down into Asia. Only a few families in the suborder Laniatores – the Cladonychiidae, Phalangodidae and Travuniidae – are found in Europe. The rest are found mainly in tropical and temperate regions of the southern hemisphere, such as Southeast Asia, Africa, Australia, Central and South America. Although the suborder Cyphophthalmi has only 130 species, these are found on all continents and major islands, except Antarctica.

Habitats

Humidity and high temperatures

As a whole, opilionids are found in habitats across a range of humidities, from 50% to 100%. Certain species have preferences e.g. *Oligolophus hansenii* lives on branches at 50–60% humidity, whereas *Nemastoma bimaculatum* lives in the ground layer at 85–100% humidity. The need for a humid habitat is an important limiting factor to their distribution, which is why opilionids are much more common around dark, shady areas next to water. Some species have a tolerance to high temperature, and these are distributed at latitudes close to the equator. For example, the 'heat death point' (the highest temperature a species can be subjected to for one hour at high humidity and still be alive 10 hours later) of *Odiellus spinosus* (Phalangiidae, Eupnoi) is a scorching 45°C (113°F)! Many species are, in contrast, extremely cold tolerant, e.g. *Mitopus morio*, also a phalangid, can withstand the freezing tundra.

Multidudinous habitats

Opilionids are found in a multitude of habitats, such as woodlands, heaths, meadows, bogs, rocky landscapes and gardens. The species *Trachyrhinus marmoratus* (Sclerosomatidae, Eupnoi) is even found in the Chihuahuan Desert. In these habitats, they are found in leaf litter,

RIGHT Opilionid running in vegetation, Surrey, UK.

soil, moss, under rocks and stones, on tree trunks, in clumps of grass, or running on vegetation. Woodlands have more than double the number of species of open habitats. Species have been found at 4,000 m (13,000 ft) above sea level in the Himalayas but, as is the case with other arachnids, species richness decreases with increasing altitude.

All four suborders of opilionids have species that are troglobitic, especially Laniatores (e.g. most of the species in the genus *Texella*, from Texas), and show signs of 'regressive' evolution, such as blindness, de-pigmentation and elongation of appendages. Certain species are troglophiles so they can be found both in caves and in habitats above ground, e.g. the gonyleptid *Goniosoma spelaeum* shelters in a cave during the day, but follows a fixed trail out of the cave and up the same tree almost every night to forage. Some opilionids may use caves during part of their life-cycle, e.g. some species in the genus *Gyas* (Phalangiidae), from eastern Europe, over-winter in caves to avoid the snow. These are known as trogloxenes.

Some have very specific habitats, e.g. Trogulidae are only found under stones and in leaf litter on limestone. This may be because their snail prey requires calcium in the soil for shell formation. In contrast, other opilionids have colonised a wide range of habitats, e.g. *Mitopus morio* is found in tundra, as well as in broad-leaved woodland. Some opilionid species, such as *Phalangium opilio*, have a preference for human-made habitats. They may be found in gardens and on waste ground, possibly distributed by accidental introductions, e.g. eggs may be laid in the soil of exotic plants, and end up in gardens in temperate regions.

On the horizon

Within a habitat, opilionid species are often characteristic of the different horizons that exist there. In the layer above the ground layer (which is usually about 10 cm (4 in) above the soil surface), the field or herbaceous layer (including the surface of leaf litter and heather), species such as *Mitopus morio* and the long-legged beast *Leiobunum rotundum* are found. In general, short-legged species are associated with dense litter, whereas long-legged species are usually

LEFT In the shrub and tree layers are species that are also at home in the field layer, such as *Leiobunum rotundum* (shown here), as well as branch-dwelling species, such as *Oligolophus hansenii*.

BELOW In the ground layer of a typical UK woodland habitat, short-legged beasts such as *Trogulus tricarinatus* are found.

found in widely spaced foliage. However, the relationship between leg length and habitat isn't always straightforward – this 'rule of thumb' seems to apply for some European and North American species, but not for most tropical Laniatores.

The developmental stage of an opilionid may also determine its position in a horizon. As juveniles are more prone to dehydration than adults, they tend to be found in the ground or herbaceous horizons, where it is more humid. As adults, they may well move to a higher horizon. To complicate matters, adults may undergo a daily vertical migration, through different levels of the vegetation – they will therefore be found in different places at different times of the day.

General biology and behaviour

Food preferences

Opilionids are very enthusiastic feeders, and many are omnivorous. They also scavenge upon an even greater selection of dead organisms, which may be difficult to deal with when alive because of their tough exoskeletons, such as decomposing centipedes and beetles. Most surprisingly, even dead vertebrates are on the menu – several species of Phalangiidae have been seen to feed on mammal, bird, frog and fish remains. Among the arachnids, scavenging on dead prey is almost unique to opilionids; it is only found additionally in some mite species. The wide-ranging diets of opilionids are sometimes supplemented further by fruit, vegetables

ABOVE Opilionids catch and kill the majority of their food, which includes snails, leeches, earthworms, woodlice, collembolans, insects such as moths, caddis flies, flies and mayflies, arachnids such as other opilionids (including their own young) and spiders. *Nelima elegans* with prey, western North Carolina, USA.

and fungi. Feasting on fungi is rare among arachnids, but is common in insects. Opilionids have even been observed to eat bird droppings. In captivity, they feed on freshly killed insects, as well as raw meat, wholemeal flour, marshmallow and even peanut butter. Opilionids in the family Phalangiidae that were fed on vegetables and bread, but no animal matter, for a long period were found to get constipation!

Although the broad culinary spectrum described above might suggest that all opilionids eat almost anything, in fact it applies only at higher taxonomic levels. Most species focus on a very narrow range of prey types – normally small and lightly sclerotized invertebrates, which are easily grabbed and eaten. Large prey, and prey that has some form of chemical defence, is usually avoided. Some species are particularly narrow in their gastronomic tastes, e.g. certain Trogulidae are specialist feeders on snails and slugs.

Because of their feeding preferences, some opilionids have proved to be extremely useful in biological control. For example, a study in 2001 estimated that the opilionid species *Phalangium opilio* in New Zealand was responsible for 31.5% of total arthropod predation on cabbage white butterflies (*Pieris rapae*) in cabbage crops. This was much higher than for a species of lycosid spider, which managed only 2% of the total predation.

Prey capture

Since opilionids have a wide range of food choices, it stands to reason that there are different ways that they find their food. They capture live prey, but they also forage for immobile food items.

FINDING LIVE FOOD Opilionids are usually nocturnal predators that hide away during the daytime. There are a number of diurnal species in the families Sclerosomatidae (Leiobuninae, Gagrellinae), Phalangiidae and Gonyleptidae, which feed during the daytime. For example, British phalangiids seem to be active during the day only when there is a high humidity and diurnal activities are more frequent when the animals are particularly hungry.

In general, opilionids are ambush predators that stay in a 'sit-and-wait' position while hunting, rather than being active hunters, e.g. individuals of *Goniosoma spelaeum* from Brazil have been observed in ambush position. They stay completely still for hours, with all their legs stretched out like the spokes of a wheel. It is thought that this arrangement increases the chance of detecting prey, by creating a larger area of possible contact. Other individuals of the same species have been observed suspended in the air, stretched out between two leaves. They attach themselves by their first, third and fourth pairs of legs, leaving their second pair of legs free to constantly move around. Such an ambush position is a good way of intercepting flying insects. However, there are exceptions to the ambush strategy, e.g. *Mitopus glacialis* (Eupnoi) actively searches for and captures collembolans. In general, the eyes are not used in prey detection. The cave-dwelling species *Megalopsalis tumida* and *Hendea myersi cavernicola* hunt glow worms and use their well-developed eyes to detect the light source from their prey. Additionally, the non-cavernicolous species *Caddo agilis* (Caddidae) has large eyes that may be used to detect prey.

LEFT *Protolophus singularis* feeding on a cranefly at night, the most common time for opilionids to prey.

Chemoreception plays a part in prey detection with the use of the palps and the first two pairs of legs. The most important chemoreceptors in opilionids are the sensilla chaetica, which are sensory setae located on first and second pairs of legs. A good example of chemoreception is found in species in the family Trogulidae, which are able to follow fresh snail mucus trails. There may be some kind of airborne chemoreception involved in prey capture, because of the extremely rapid response shown by some captive opilionids when food has been placed in their container. Such chemosensory capabilities are more likely to assist opilionids when they scavenge rather than when they go out hunting.

FORAGING AND KLEPTOPARASITIC BEHAVIOUR Because many opilionid species have a large range of possible food types, they need a mixed strategy of hunting and active foraging. Opilionids regularly feed upon discarded carcasses of bees and moths that have been captured and partially eaten by crab spiders. But they are not always satisfied with leftovers – a *Phalangium opilio* individual was observed trying to steal a dead honey bee (*Apis mellifera*) from an adult female crab spider (*Misumena vatia*), before the spider had even started its meal! The opilionid was very insistent, making several attempts to grab it from the spider, but to no avail. Such kleptoparasitic behaviour has been seen in a species of Gonyleptidae too. The gonyleptid bravely stole a moth from a spider (Ctenidae) that was hunting on a tree trunk. Cosmetids are also known to steal prey from pholcid webs. However, this behaviour is rare and the incidences observed may have been cases of opportunistic theft. Sometimes the theft of food from conspecifics results in several individuals sharing the same piece of food. Some species share food regularly, e.g. in *Pachyloidellus goliath* food sharing is commonplace.

CAPTURING PREY Opiliones are not able to subdue their prey with venom, as do most spiders and scorpions; instead, they immobilize the prey using their chelicerae, palps and legs. This may take up to half an hour in opilionids with long legs and small bodies. In phalangiids, the first pair of legs and palps holds the prey, if it is still struggling. Certain genera, such as *Megabunus*, have large spines on the ventral side of their palpal femurs which help to restrain prey.

The most common method of prey capture is to grab the prey with the cheliceral claws (e.g. in the Cyphophthalmi), but other methods include suddenly drawing all legs inwards and grabbing with the palps (e.g. in certain Phalangiidae and Sclerosomatidae). Juveniles and smaller species in the Dyspnoi have adhesive hairs on their palps to capture prey, which is then passed to the chelicerae. With some species, physical contact with the prey is required for detection and capture to take place. *Mitopus morio*, for example, will only go in for the kill if contact with a leg has occurred. The first pair of legs is more important than the second pair for identifying food. In a study of the hunting abilities of *Leiobunum nigripes* and *Leiobunum vittatum*, flies were eaten significantly sooner after contact with either of the first pair of legs, than after contact with the second pair. However, many species go in for the kill without the need for any touching, so these must have mechanoreceptors that operate from a distance.

Feeding and digestion

DISMEMBERING AND TEARING Once an opilionid has secured a meal, it often picks up and carries the food item away from the capture site, to a place where it will not be disturbed during feeding. Because the prey hasn't been dispatched by venom injection, it is usually eaten alive. Opilionids process an item of food by dismembering it, tearing it into smaller pieces by alternate movements of the chelicerae. Lightly sclerotized sections of the prey are attacked first. Snail-eating opilionids often break off pieces of the shell to get at the retracted snail inside. Others eat their way into the shell through the aperture. The palps, chelicerae and occasionally the first pair of legs are used to manipulate the food into the mouth.

Salivary juices are mixed with the food particles and taken up by the mouth. Opilionids are atypical feeders compared to other arachnids, because they ingest small particles of food (even small chitin pieces) that are not filtered out. Most digestion is therefore internal. Because of this, opilionids are likely to ingest pathogens and parasites that would normally get sifted out. This whole process can take between just a few minutes and several hours, during which time the second pair of legs is held out to the sides to monitor for any intrusion. Adult opilionids can withstand long periods of starvation. When given food again, they have a feeding frenzy and put on a vast amount of weight in one session, so that they have a highly distended abdomen.

RIGHT Tropical opilionid feeding on its prey (Jatun Sacha, Ecuador).

THE NEED FOR WATER Opilionids are more susceptible to dehydration than most other arachnids. This may be due to several things, such as their long legs resulting in a high surface to volume ratio, and the fact that they are unable to control the size of their spiracles because of lack of muscular control. Opilionids excrete solid guanine so water loss is minimal. Most opilionids also prefer damp and shaded areas, which minimizes evaporative water loss.

Opilionids do not need to drink water if they are living in moist conditions, but this changes if they find themselves in dry conditions. When its leg tarsi wilt like the leaves of an unwatered plant, it is a good indication that an opilionid is dehydrated. This unfortunate condition can, surprisingly, be fully reversed if the animal has a drink. When severely dehydrated, an opilionid will suck up freely available water for up to five minutes. In order to drink, some species touch the surface of a water body with the tips of their legs or palps. The water adheres to the dense hairs and the opilionid can bring minute water droplets to its mouth. Opilionids have been known to sup on over-ripe fruit, or chew on the edges of leaves or moss with their chelicerae to gain water. Opilionids also show behavioural adaptations for reducing water loss, e.g. those species living in the upper horizons of a habitat tend to be more active at night than during the day, when it is much cooler.

Predators and parasites

INVERTEBRATE PREDATORS Opilionids are not above a little cannibalism from time to time. This often takes place when an individual is at its most vulnerable during moulting. Immatures are cannibalized more often – probably because their softer bodies are easier to tear apart – though cannibalism among adults does happen on rare occasions. The eggs are most commonly cannibalized. In what is termed 'group cannibalism', several opilionids have been observed to attack and feast on an individual, although the behaviour was probably not co-ordinated between them. Several species have been seen eating already dead conspecifics (individuals belonging to the same species).

Centipedes, ground beetles, scorpions, flies, true bugs, ants, Mecoptera and snails such as *Neohelix albolabris*, all predate on opilionids, but spiders are probably the most important group of predators. Hunting spiders grab the opilionid, seize its legs and force them out of the way, before greedily tearing into the tissue of the prosoma. Web builders, such as *Theridion* species, eat opilionids too.

VERTEBRATE PREDATORS Vertebrate enemies include toads, frogs, salamanders, rodents such as mice, insectivorous mammals such as shrews (e.g. *Sorex araneus*, the Eurasian shrew), hedgehogs, badgers (e.g. *Meles meles*) and bats, and marsupials such as the gray four-eyed opossum (*Philander opossum*). Opilionids are particular culinary favourites of birds, such as meadow pipits. Even fish will eat them, but it is rare for opilionids to fall onto the surface of water.

PARASITES Opilionid parasites come from several different groups. Most parasites are not fatal to opilionids (such as mites), but some are (such as Diptera). Opilionids are often parasitized by ectoparasitic larval mites in the families Thrombidiidae, Trombiculidae and Erythraeidae. Erythraeids are small, red mites (mainly in the genus *Leptus*) that suck tissue fluids and haemolymph from their host. Heavy infestations occur in moist ground conditions. Those opilionid species that are very mobile are likely to be attacked by mites, possibly because they travel more. Some species of flies in the family Phoridae are parasitoids and lay eggs on opilionid hosts. The eggs develop internally and kill the hosts. Small flies in the family Ceratopogonidae are blood suckers.

Nematodes (roundworms) develop internally, but then emerge out of the anus and mouth, or burst out of the body cavity in order to moult to maturity, which is obviously fatal for the opilionid. Mermithid nematodes require damp microhabitats, where they deposit eggs. Some nematodes directly penetrate the exoskeleton in order to invade their opilionid host. Other parasites of opilionids include parasitic protozoa such as Microsporida and Sporozoasida, Cestoda (parasitic flatworms commonly known as tapeworms), and Trematoda (parasitic worms called flukes).

Defensive behaviour and aggression

Opilionids have no venom but, as an order, have a large repertoire of defence mechanisms that are chemical, morphological and behavioural.

CHEMICAL DEFENCES The best-known defence mechanism of an innocent-looking opilionid is the secretion of defensive fluid, which is either used as a chemical 'shield' in vapour or liquid form, or is directed towards the offending cause. A defensive 'shield' can be formed in several ways: liquid vapour is exhaled from the gland; a small drop of liquid is emitted, which evaporates; liquid is emitted as a fine spray, which covers the dorsal surface; liquid is emitted and then transferred to other parts of the body through a lateral or ventral groove in the exoskeleton, or through surface microsculpture.

The secretion can be actively directed to the offending cause in different ways too: liquid is squirted as a fine jet (as in Uropygids); in Gonyleptidae, the jet can reach up to 10 cm (4 in), aimed in any direction; the body is basted in defensive fluid, so that if a predator grabs the opilionid the fluid is transferred to the predator; the opilionid dabs a leg into a droplet of defensive fluid produced at the glands, and then brushes it against the offending cause.

BELOW Through a pair of scent glands, *Acutisoma longipes* releases a yellow liquid with a strong acidic odor, capable of fending off many of their predators.

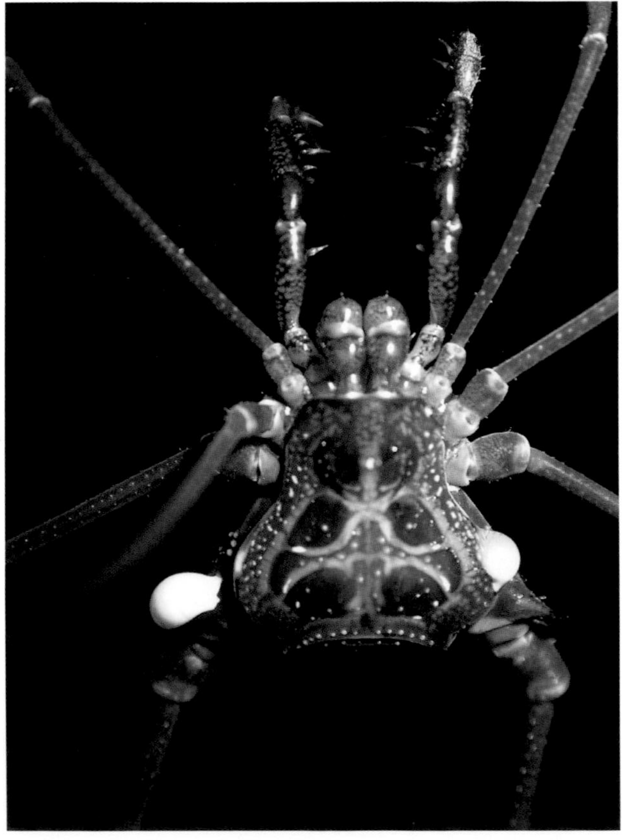

The chemical composition of the defensive chemical varies with the family. It may contain a cocktail of alcohols and acyclic ketones, naphthoquinones, terpenoids, phenols or alkylated benzoquinones. Most species use a mixture of compounds. Many opilionids mix their defensive compounds with aqueous enteric fluid (from the intestine). This is mainly to dilute their precious defensive compounds. Some defensive fluid compounds are antimicrobial, and therefore help to defend the opilionid against external pathogens.

Chemical defence certainly keeps ants, most spiders, and young birds at bay. It even causes dermal irritation in humans. The fluid is so distasteful that a frog will even spit out an opilionid after it has taken it into its mouth! However, some mammals and birds are very tenacious, and will repeatedly attack the opilionid until it has run dry of its defensive chemical. The opilionid is then consumed. Additionally, there are certain vertebrates, such as opossums, that are totally unaffected by the chemical.

LEFT Defence can be further enhanced by behaviour. For example, opilionids stay motionless for long periods of time, with their legs outstretched and their body close to the substrate, to avoid the attention of predators.

MORPHOLOGICAL DEFENCES Clearly, a hardened exoskeleton and spines will deter some predators. For example, species such as *Cobania picea*, *Discocyrtus* sp. and *Neosadocus* sp. in the family Gonyleptidae, have extremely large spines on their opisthosomas and on their fourth pair of legs (see p.213).

Cryptic colouration is widespread among opilionids. This form of defence enables the opilionid to avoid detection by predators. Colouration and patterning that disrupts the animal's outline can foil would-be predators, as do concealing mottled and dark colours. They also frequently spend time on substrate that enhances their concealment. Some species take camouflage one step further - trogulids (e.g. *Trogulus* spp.), nemastomatids and several other groups produce an adhesive secretion from glands on the exoskeleton, to which soil and other debris attach and camouflage the body. These opilionids resemble walking clods of earth!

Opilionids don't always blend in with their backgrounds, however. They may possess warning colouration (aposematic colouration) – conspicuous markings indicating to a predator that the animal is distasteful or venomous. Certain species in the Gonyleptidae have yellow and red, black and red, or black and yellow patterning, while some sclerosomatids are bright yellow or orange. These aposematically coloured beasts are usually diurnal and slow-moving.

Mimicry is rather under-studied in opilionids, but it is likely to be Müllerian mimicry – a form of protective mimicry in which two or more distasteful or harmful species closely resemble each other and are therefore avoided equally by all their natural predators.

BEHAVIOURAL DEFENCES Hiding away under stones and so on is called anachoresis, and greatly reduces the chance of contact with potential predators, so it can be considered a defence mechanism. Fleeing, too, is a simple but effective means of avoiding predation that long-legged opilionids employ. Feigning death is known as thanatosis, and is common in several families in the suborders Laniatores and Dyspnoi. To fool potential predators into thinking they have died, opilionids go rigid. Depending on the species, and indeed the stimulus, the opilionid may remain rigid from a minute or so, up to half an hour.

Sociality

Opilionids can be sociable beasts, unlike most of their arachnid relatives. Many species are highly tolerant of conspecifics. As already mentioned, many species form aggregations, where their legs extensively overlap. Laniatores and many Eupnoi form loose aggregations, where their bodies are orientated in different directions. In dense aggregations, which are only formed by Eupnoi, there is a much higher density of individuals, often forming several layers. The first layer holds onto the substrate, while the acrobatic individuals of the outer layers hold onto those rather burdened individuals beneath. Aggregations consist mainly of subadults and adults that are unrelated. This is possible because there is a lack of cannibalism among these stages. Up to five different species may be found in one aggregation, although it is much more common for one or two species to dominate. In the Laniatores, the number of individuals in an aggregation ranges from three to almost 200, but there may be up to 70,000 individuals in Eupnoi mass aggregations!

Apart from defence, the other reasons put forward to explain why opilionids aggregate are: reduced water loss, overwintering, heat production, optimal humidities and to improve success in mating. The ways in which aggregation enhances defence are discussed in the previous section. Overwintering opilionids congregate as temperature drop, and may stay together for several months until temperatures rise again. The numbers of individuals in such aggregations may reach 2,000! Forming aggregations probably reduces water loss for individuals that are closely grouped together because airflow is reduced, so less water is lost from the spiracles. It also seems that environmental factors encourage this sociality. For example, individuals often aggregate at sites close to water sources. If opilionids aggregate around the mating period, then their chances of meeting the opposite sex and mating are drastically improved. However, competition for mates is also increased, of course.

BELOW An aggregation of opilionids under a leaf, Tapanti National Park, Costa Rica.

The 'daddy long legs' opilionids in the suborder Eupnoi vibrate their bodies up and down extremely rapidly, in a behaviour called bobbing. This is thought to confuse predators as to where the body is, thereby deflecting an attack, perhaps to the legs, where it is likely to be less harmful. As is well known, opilionids are able to autotomize their legs if attacked, as a form of defence.

Although stridulation has not been well studied in opilionids, it is known that species in Laniatores and Dyspnoi possess stridulatory apparatus, consisting of sets of ridges or similar structures that are rubbed against one another. The ridges may be located on the inner surface of the palpal femurs, where they are rubbed against the denticles of the ocularium or the outer surface of the chelicerae. Otherwise, there may be stridulatory ridges on the second segment of the chelicerae. Although stridulation is mainly used for communication between conspecifics, it is very likely that it is also used in defence.

Many opilionid species aggregate in large groups – one of the main reasons for this is defence. The shared release of defensive chemicals is greatly amplified in a large group, which means that individuals can more easily repulse a potential predator. However, species that don't secrete defensive chemicals are occasionally found in aggregations where the majority of individuals *do* secrete – these are seemingly taking advantage of the protection provided by others. Being in a large group also means that an individual is less likely to be picked out by a predator, as well as being better able to detect and respond to a threat quickly. In this way, aggregation enhances the chances of an individual's safe escape.

Retaliation in opilionids may be of a chemical nature, as described above, or mechanical. Opilionids may use their chelicerae or palps to bite, pinch or spear a predator. Gonyleptids with sharp spines on their fourth pair of legs bring them together quickly, and if you were to have your skin trapped between then, the spines might even draw blood!

Grooming

Leg grooming is clearly very important to opilionids – they may suddenly interrupt vital activities such as mating, to indulge in it! The mouthparts act as the cleaning apparatus. In a process called 'leg threading' or 'mouthing', each leg in turn is pulled carefully through the mouthparts, with the help of the palps and chelicerae, working all the way along to the tip of the tarsus, and then suddenly released. Most time is spent cleaning the tarsi, especially those on the second pair of legs. Opilionids need to keep their legs immaculately clean, in order for their receptors to remain effective.

Autotomy and moulting

Compared with other arachnids, opilionids seem to shed their legs with almost gay abandon. If a creature is handled, limbs invariably part company from the body. If an opilionid inadvertently trails its legs in a small body of water, the weak surface tension may even be enough to pull a leg off! When a leg is autotomized, it parts company at the trochanter–femur joint and only a little droplet of haemolymph is lost. In a study in 1999 on two *Leiobunum* species, Cary Guffey (from Louisiana, USA) discovered that if an opilionid loses up to two legs, it is pretty much unaffected. If, however, it loses three or more legs, there is significant reduction of mobility and foraging ability. Although the loss of a number of legs has quite severe effects, surprisingly, opilionids do *not* regenerate them. At each moult, the legs of long-legged opilionids may extend in length up to 150% of the previous size! During and after

a moult, when its legs are still soft, it is impossible for an opilionid with long legs to remain upright. So, usually it hangs upside down, from beneath leaves or branches. This enables the animal to extract its legs from the old skin with the help of gravity, as well as to keep them extended during the time it takes for the exoskeleton to harden. Opilionids with much shorter legs (e.g. trogulids) need not go to this trouble.

Jerking movements of the legs and body help to split the exoskeleton during a moult. The femorae of the legs and the chelicerae then emerge rapidly, followed by the rest of the body. That's the easy bit – freeing the tibiae and tarsi from the inside of the old skin is more difficult. The palps and chelicerae are used to help out. The palps pull the legs towards the chelicerae, and then the chelicerae pull each leg in turn out of the old skin. Before hardening, the legs are cleaned by 'leg threading'.

The moulting rate decreases greatly if conditions are too dry – there must be a reasonable level of humidity. When a drought is broken by rain, opilionids will moult within a few hours.

Life history

Reproduction

Sexual reproduction in opilionids is usually therefore direct, with the male transferring sperm directly to the female using his penis; this increases insemination success. The Cyphophthalmi are are different, however, as it appears that they transfer sperm indirectly via spermatophores or 'sperm balls'. In some species of Grassatores (in the Laniatores), subadults are sexually mature and can reproduce. It is thought that they may have a unique life-cycle among arachnids, sharing their precocious maturity only with mayflies.

COURTSHIP Although there may be no real courtship, there are often quite complex social interactions before copulation in opilionids, which include the following. Males often first encounter females in mixed-sex clusters. Older males are seemingly more confident, and initiate mating activity before the younger ones. A female initially plays hard to get, and – unfortunately for the male – vigorously resists his amorous advances. He does not give up though, and continues to attempt copulation. Impressively, a male *Leiobunum longipes* was recorded in captivity attempting to copulate at least 29 times in just 2.5 hours. As the keen male's attempts increase in fervour, the female moves away from the group to a more secluded spot, where she finally allows herself to be approached. Males of some species present a nuptial gift for their prospective mate – this may consist of glandular secretions produced from the chelicerae.

Some opilionids are rather promiscuous. For example, large-bodied and powerful mature males of *Leiobunum calcar* and *L. vittatum* sometimes make a play for almost any mature female opilionid that crosses their paths. A female in this situation either turns away quickly to avoid his 'embrace', or lifts her opisthosoma and lowers her prosoma. A male of the same species will not intervene.

In same-species groups, males together will not compete if there is no female to fight over. However, this all changes when a female is added to the equation. If a group of males together encounter a receptive female, it may take up to half and hour to determine which of them will mate. Fighting males use dirty tactics – they try to pull each other's legs off! Males of *Platybunus bucephalus* (Phalangiidae) often fight for females during the reproductive season.

Being the loser is not a good option, as it is generally eaten by the winner.

When males are in charge of brooding eggs, the tables turn. The species *Zygopachylus albomarginis* (Gonyleptidae) from Panama is one of the few arachnids to demonstrate paternal care. Females of this species fight for the males and then court them. The males may be quite choosy and reject the females. Interestingly, sexual dimorphism in this species is very subtle.

ABOVE A copulating pair of *Acutisoma proximus*.

A STRAIGHTFORWARD ACT The act of copulation itself is quite straightforward. In all but the Cyphophthalmi, the pair faces each other, and the male grasps the female with his palps. At the same time, he extends his second pair of legs outwards to detect any interference from other males. The male extrudes his penis through his genital operculum. The female helpfully guides it into her genital opening, using her palps and chelicerae. Copulation takes between a few seconds and a few minutes – this depends on the species, and the female's mood. She usually ends the encounter by suddenly pressing down her prosoma. In the gonyleptid species *Acutisoma longipes*, the male intensively taps the female's body and hind legs during copulation, and may mate with her multiple times. In the Cyphophthalmi, the male mounts the female from behind and copulation is ventral surface to ventral surface.

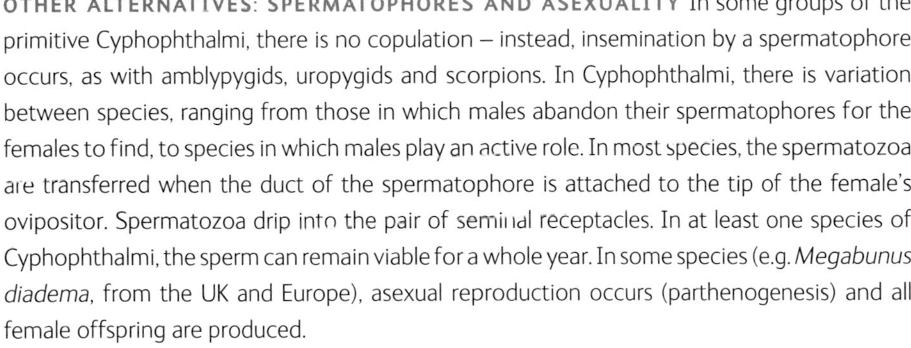

OTHER ALTERNATIVES: SPERMATOPHORES AND ASEXUALITY In some groups of the primitive Cyphophthalmi, there is no copulation – instead, insemination by a spermatophore occurs, as with amblypygids, uropygids and scorpions. In Cyphophthalmi, there is variation between species, ranging from those in which males abandon their spermatophores for the females to find, to species in which males play an active role. In most species, the spermatozoa are transferred when the duct of the spermatophore is attached to the tip of the female's ovipositor. Spermatozoa drip into the pair of seminal receptacles. In at least one species of Cyphophthalmi, the sperm can remain viable for a whole year. In some species (e.g. *Megabunus diadema*, from the UK and Europe), asexual reproduction occurs (parthenogenesis) and all female offspring are produced.

POST-COITAL BEHAVIOUR A copulating pair may mate several times with each other, or with different partners. A male must guard a female with whom he has just copulated, otherwise other males will try to mate with her too. If a superior mated male is challenged by another male, he will leave the female unguarded to chase away the intruder. While his 'back' is turned, another smaller male who would never challenge the superior male may cheekily nip in and mate with the female. When the superior male has dealt with the intruder, he deals with the smaller nuisance male, before mating with the female again.

After mating, a male *Leiobunum longipes* actively encourages the female to get into an ovipositing position, by spreading his legs over her body in an umbrella-like stance, and gently stroking her legs with his. Without such treatment, the female just wanders off and does not oviposit. Not all female opilionids require such encouragement, however. In *Acutisoma*

longipes, the female lays her eggs immediately after copulation. The male remains close to her as she oviposits, touching her with his legs. In general, mate guarding by the male can go on for more than 24 hours. He tries to mate with the female – sometimes he is successful. Male opilionids sometimes have harems of up to five females, which they guard and fight over vigorously.

Egg laying and parental care

A female opilionid lays eggs through her ovipositor, which is extruded from beneath the genital operculum, and can sometimes extend nearly twice the length of her body! The end portion of the organ waves backwards and forwards between the laying of each egg, in order to aid the movement of the next egg down the tube. Females in Eupnoi and Cyphophthalmi have long ovipositors, and so can insert them beneath the bark of trees, into dead wood, under stones, or into a damp medium such as soil or leaf litter, in order to hide the precious eggs. Species in the Trogulidae (Dyspnoi) are particularly inventive, because they often lay their eggs in the empty shells of snails that they probably ate earlier. They then cover the opening with a membrane to protect the eggs. They are called 'shell intruders' – as compared with Ischyropsalididae, which are known as 'shell breakers'.

RIGHT A female *Acutisoma proximum* guarding her eggs.

Females in Laniatores, on the other hand, have short ovipositors. They therefore have to lay their eggs on exposed surfaces such as tree trunks and rocks. They try to protect the eggs by covering them with debris, or laying then in shallow crevices, or even staying to take care of the eggs. This may help to camouflage the eggs, or prevent them from dehydration.

The eggs of opilionids are usually spherical and creamy white, yellow or pale green, and 0.5–1.0 mm (1⁄50–1⁄25 in) in size. They can be laid in batches of up to 120 eggs at a time, usually in late summer or early autumn. Female Cyphophthalmi are much less prolific than this, as they only lay 10 eggs per year and no more than 50 in their lifetime. Females of two different species of *Leiobunum* have been known to oviposit side by side in the same spot with no dispute – so perhaps such tolerance is widespread among other genera too.

EGG DEVELOPMENT Egg development generally takes 1–2 months, but may take from 20 days up to six months. The eggs darken during embryonic development, and just prior to hatching in the spring the occupant can be clearly seen, moving around through the nearly transparent egg case. In order to rupture the egg case, many young opilionids have a special structure equivalent to an egg 'tooth' in birds.

DIFFERENT FORMS OF PARENTAL CARE In the Opiliones, there are different forms of parental care, which are important in increasing reproductive success. Hiding or covering the eggs, or secreting a chemical or mucus onto the eggs, are all methods of protection that don't rely on one of the parents being around. These different protective mechanisms can occur separately or in combination. Egg covering can take the form of camouflage. For example, females of *Promitobates ornatus* (Gonyleptidae) found in southern Brazil lay a solitary egg on soil, which is then covered in soil particles. To make sure the egg is fully covered, the female rotates the egg before applying more soil. Mucus production protects eggs against dehydration, microbe attack and freezing.

Post-oviposition parental care consists of the adults staying around after the eggs have been laid, in order to protect them. This type of care can be separated into maternal egg guarding, maternal egg covering and guarding, and paternal care. Adult females of Laniatores and Dyspnoi have a short ovipositor, so the vast majority of species in these suborders lay their eggs on exposed surfaces. In these situations, post-oviposition parental care is highly likely to enhance the survival of the offspring. Post-ovipostion parental care is only found in the Laniatores, especially within the neotropical superfamily Gonyleptoidea. It is likely that parental care evolved in the Gonyleptoidea, because in the tropics there is a particularly high risk of ant and fungal attack on the eggs. All species in the family Gonyleptidae have a form of maternal care.

It is thought that intense predation on eggs by generalist predators provides a selection pressure favouring the evolution of parental care. In other words, those opilionids that protect their eggs have a greater number of offspring survive, which go on to produce more eggs than non-guarding opilionids, and so on. Guarding females (e.g. *Acutisoma proximum*, Gonyleptidae) always stay over their eggs, which they constantly monitor with their first pair of legs. Occasionally they use their second pair of legs and palps. They are not easily disturbed by light like other opilionids, but run away if touched.

There are several factors that make relatively long-term associations between offspring and parents possible in the Laniatores. For example, most species reproduce throughout the

year and can live for more than two years. In addition, Gonyleptids have evolved superior mechanical and chemical defence mechanisms, which make them better at repelling possible predators of their eggs. Cyphophthalmi and Eupnoi have not evolved post-oviposition parental care because females of these suborders have a long ovipositor that allows them to lay eggs into substrates that will protect them from parasites and predators.

PATERNAL CARE

Paternal investment once the eggs have been fertilized is unique to opilionids, among the arachnids, and is found only in Laniatores. Males of *Zygopachylus albomarginis* build mud nests on tree trunks. Once the mother has made her choice of nest, she will lay a modest 1 to 5 eggs and then leave. The male is then the one to guard the eggs against predators such as ants, and from cannibalism by conspecifics. He also removes fungi, to prevent the eggs from spoiling. The control of fungal attack on eggs appears to be unique to this species.

Males of *Ampheres leucopheus* (Gonyleptidae) from Brazil have been observed guarding egg batches attached to the undersides of leaves. The males appear to patrol the area around the eggs looking for possible predators. In *Leytpodoctis oviger* (Podoctidae) from the Philippines the female attaches 4–5 eggs onto the male's fourth femur, and he dutifully carries them around until they hatch. Arachnologists in Sao Paulo, Brazil, recently postulated that males of some opilionid species might accept eggs to guard from more than female, or from the same female at different times.

DIFFERENCES BETWEEN SOLE MATERNAL AND SOLE PATERNAL CARE There are differences between the ways in which females and males guard their egg batches. For example, females guard batches that contain eggs in only one stage of embryonic development, while males guard batches containing eggs in several stages of development, which are likely to come from several different egg-laying sessions.

The amount of time for which a female guards her eggs is no more than 60 days, depending on the species, whereas in species that have paternal care, females continue to add eggs to the batches, so the male's parental duties can last for up to eight months! During her egg guarding, a female doesn't leave the egg batch in order to forage. On the other hand, a guarding male may frequently leave his eggs and can be found up to 5 m (16½ ft) away. The male probably does this because he can not last for such long periods of time without foraging for food, or he may be on patrol for predators or looking for other mates.

There have also been cases of biparental care observed in a few species in the family Gonyleptidae. When the guarding female dies, or deserts her offspring, the male takes over her duties. Males have also been seen to defend territories where females are caring for their broods, or even defending their mates or offspring.

Post-embryonic development

Post-embryonic development in Opiliones is fairly homogeneous across different taxa. However, developmental time is very variable – from just a few months up to three years for completion.

NYMPHAL STAGES There are three main stages: the larva, nymph and adult. Not surprisingly, species with longer lifespans have a greater number of nymphal stages, e.g. *Phalangium opilio*

has 6–8 nymphal stages, whereas *Lophopilio palpinalis* has four. The number of nymphal stages can vary within the same species. In Laniatores, Dyspnoi and Eupnoi, the larva is quick off the mark, and undergoes its first moult within just one hour of hatching – it might not even bother to leave its egg case! However, with Cyphophthalmi, the larva is in a less mature state, and so may take up to seven days before moulting. There may be up to nine moults between new nymph and adult. There are around 6–20 days between each moult.

DIFFERENCES IN LIFE-CYCLES There is a wide range of life-cycle patterns among opilionids. Some species are stenochrone, or seasonal, while others are eurychrone – non-seasonal species where adults can be found the whole year round and have overlapping generations (e.g. most species in the neotropical Laniatores, and the Cyphophthalmi).

There are three main life-history patterns for British opilionids: species that are active all year with overlapping generations and where adults are always present in greater or smaller numbers; annual species that over-winter as nymphs from eggs that hatch in the autumn; and annual species that over-winter as eggs, hatch in spring and die in the autumn. However, other patterns of development can also be seen in some British species. There are also variations in life-cycles among opilionids that live in tropical and near-subtropical parts of the world. These are related to differences in environmental conditions such as litter decomposition changes, annual flooding regimes (e.g. in parts of the wet central Amazonian forest of Brazil) and temperature. Indeed, within any species, the duration of the larval, nymphal and adult phases is dependent on environmental factors – mainly temperature and humidity.

Longevity

Longevity within opilionids is rather variable. There are annual species with 2–3 months of nymphal stages and ephemeral adults, biennial adults, and perennial species with a year of nymphal stages and adults that live for more than three years. Within the Cyphophthalmi, there is particularly great variation. The general pattern is that Laniatores and Cyphophthalmi may live for several years as adults, whereas Eupnoi usually have a short lifespan of a few months.

10 Scorpiones

Scorpions

Scorpions are recognized the world over, by their distinctive stinging apparatus (called the telson) and characteristic claw-like 'pincers' (chelae - the tibia and tarsus parts of the palps). Myths and legends have perpetuated about scorpions through the ages. Unfortunately, they have nearly always been associated with evil, which is a great shame, because they are amazing animals that demonstrate a wonderful array of behaviours, such as intricate courtship and a high level of maternal care. Scorpions have occasionally been associated with good. For example, the ancient Egyptians venerated the scorpion, believing that it was sacred to Isis, the goddess of rebirth.

Scorpions are infamous as they produce venom. A few species, such as the aptly named 'death stalker' (*Leiurus quinquestriatus*), are lethally venomous to humans, but the vast majority are not harmful. Scorpions are unique in possessing a pair of comb-like structures called pectines, which are sensory in function. Nearly all species are solitary and nocturnal, and many live in burrows. Scorpions possess a very unusual and seemingly magical property – they fluoresce under UV light! This is due to the unique structure of their exoskeleton. They vary in size from the miniature *Typhlochactas mitchelli* from Mexico, at 8.5 mm (¼ in), to the enormous *Pandinus imperator* from central Africa, a specimen of which has been measured at just over 229 mm (9 in).

Classification and diversity

There are approximately 1,500 species of Scorpiones in 170 genera, and between 13 and 20 families. It is not possible to be more specific over the higher phylogeny and classification of the order, because scorpion workers are currently divided on the issue. Higher classification changes proposed by certain workers since 2001 have been rigorously rejected by others, and so it is necessary to work with these estimates for the present. Scorpions have had the same basic body plan for the last 425 million years. They have many morphological similarities, and generally look quite alike. In fact, they have been described as being morphologically 'conservative', although there *are* some striking differences, which are obvious even to the layperson.

Ecomorphotypes

Among scorpions, there is diversity in the size and shape of structures such as palps and opisthosomas because there are distinct ecomorphotypes within the order. Ecomorphology is the study of the relationship between the ecological niche of an individual and its morphological adaptations, i.e. how the body structures of an organism enable it to be adapted to its environment.

OPPOSITE Many parts of the scorpion body, particularly the palps, and the opisthosoma, possess various granules, ridges and keels (carinae), although these are not found in all species.

Diversity in shape

A great deal of diversity among scorpions is found in the size and shape of the palps, and the chelae on them. Some scorpions, such as those in the family Scorpionidae, boast very powerful, large and fearsome-looking palps (e.g. *Pandinus* sp. from Africa, and the genus *Heterometrus* from Southeast Asia). Scorpions in the family Scorpionidae are large and have a hostile look about them even though they are not, so people needlessly fear them. At the other end of the scale, are scorpions in the family Buthidae.

Additionally, there is diversity in the width of scorpion 'tails' between different families. However, the term 'tail' is a misnomer because, contrary to popular belief, scorpions *don't* have tails. What people know as a 'tail', is actually an extremely long and tapered opisthosoma. The common name used for buthids is 'thick tailed' scorpions. As the name suggests, many buthids possess a wide, and powerful opisthosoma, in striking contrast to those of liochelids such as *Hadogenes troglodytes* from southern Africa.

RIGHT *Pandinus* sp. (above) from Africa has large palps, whereas buthids, such as this *Parabuthus* sp. from Etosha National Park, Namibia (below), usually have very slim and delicate-looking palps, and a wide and powerful opisthosoma.

Scorpions can be divided into different ecomorphotypes depending on the substrates on which they live: rock, sand, or soil. Such scorpions are substratum specialists. Substratum specialization is defined by Dr Lorenzo Prendini at the American Museum of Natural History in New York, as 'the dependence of different species on substrata of specific hardness, texture or composition for survival. Accordingly, different substrata exert different selection pressures on the animals living on or in them, resulting in the evolution of specialized ecomorphological adaptations.' And this means that 'the more physically unfavourable the substratum, the more specialized the adaptations for colonization and survival'. So if you were to place a substratum specialist, e.g. psammophilic – sand adapted, onto a substrate to which it was not adapted, it would stand no chance of survival not least, because it couldn't move around. Substratum specialization therefore plays a central role in speciation and hence the diversity of scorpions.

The basic divisions of ecomorphotypes for scorpions are described below, to show how morphological diversity is inextricably linked to habitat. Pelophilic scorpions are those that spend almost their entire time in burrows in hard soil. Because their palps are such powerful weapons, they do not usually sting their prey at all, and so have a reduced telson. Apart from their palps, they may use their legs, chelicerae and opisthosoma to dig, and compact the walls

of their burrows. Therefore, their legs and opisthosomas tend to be short and robust. Pelophilic scorpions are less restricted in their range than lithophilic or psammophilic scorpions.

Lithophilic scorpions are those adapted to living in crevices and cracks in rocks. They spend the majority of their time away from the ground surface, and their specially adapted bodies allow them to move around in their confined world. Paired, highly curved lateral claws on the tarsi, and spine-like setae, give the legs a very firm grip on rocks, allowing these scorpions to move rapidly, even upside down! They also have extra large lateral eyes to help detect low light.

Psammophilic scorpions are those adapted to life in sand. They have streamlined opisthosomas that enable them to move through the sand easily. They also have enlarged setae in groups called bristlecombs, and long tarsal claws, which increase the surface area in contact with the ground. Like camels, these scorpions are able to walk on loose sand without sinking. These structures also enable the scorpion to burrow in loose sand.

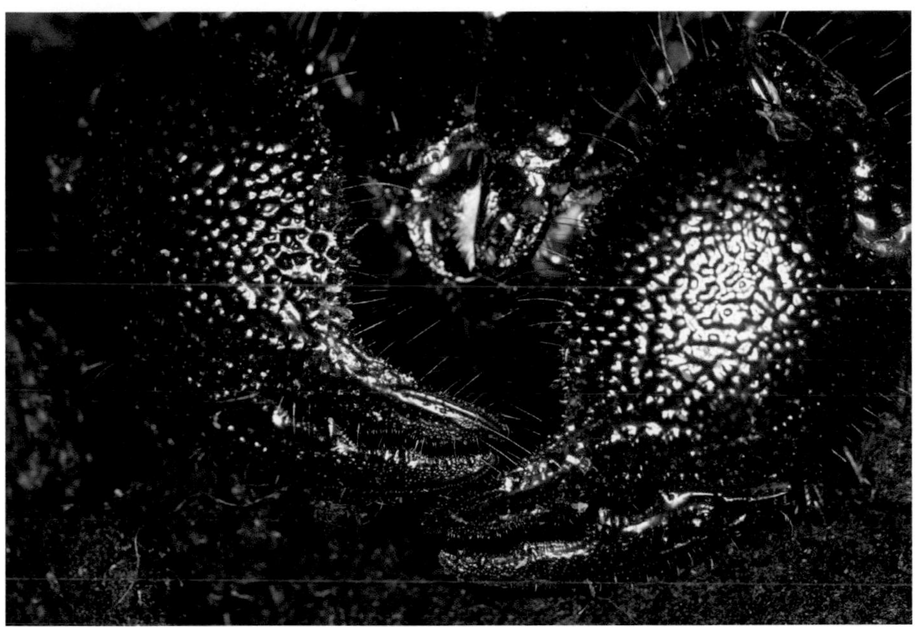

ABOVE LEFT Pelophilic scorpions use their large, spade-like chelae on their palps – which are rather like crab claws – to dig their burrows, to capture prey and for defence.

LEFT This flat rock scorpion *Hadogenes* sp. is dorso-ventrally flattened to enable it to live in both horizontal and vertical narrow rock crevices of outcrops and cliff faces.

Colouration

Many scorpions are yellow or brown, but there are some species whose colouration varies from this general pattern. For example, they may be dark green e.g. *Pandinus imperator*, dark brown or black. *Heteronebo bermudezi* is almost white, while *Hadrurus concolor* is rusty-red. Some may be mixed in colour – *Androctonus crassicauda* is dark brown or black, with yellowish or greenish tips to the legs and palps. Some colours are due to darkly sclerotized cuticle. However, many are the result of pigment in the hypodermis of the exoskeleton. In many troglobites (scorpions found exclusively in caves), there is a *lack* of pigment in the cuticle, making them very pale. Sand scorpions are often the colour of the substrate.

RIGHT Scorpions can also be patterned; *Centruroides vittatus* from the USA is an attractive example.

BELOW An example of a yellow scorpion, *Centruroides sculpturatus*.

Anatomy

External anatomy

The exoskeleton structure of a scorpion is basically the same as that of all other arthropods, although there are some unique features. For example, scorpion exoskeleton contains hyaline exocuticle, which is one of the very thin, uppermost layers.

HEAVY METAL REINFORCEMENT Small quantities of heavy metals are found in the bodies of some groups of organisms. Scorpion exoskeleton contains the highest metal concentrations (especially of zinc and manganese) of all the Arachnida, Chilopoda, Diplopoda, Insecta, Annelida and Crustacea. Zinc is concentrated in the tips and edges of structures that are prone to mechanical force and abrasion, such as the fangs of spiders, jaws of worms, and the chelicerae, palps, tarsal claws and telson of scorpions. It appears to produce increased resistance to wear (just as metal tips on the heels of shoes help prevent them from wearing out). Manganese is associated with rigidity, and tends to be found in the shafts of structures. Creatures with such metals in their bodies appear to assimilate large quantities from the environment in which they live.

GRANULAR AND HAIRY Although scorpion exoskeleton has a waxy layer, it is not smooth. Lithophilic scorpions have particularly raised carinae to protect the median eyes from being scratched by rock surfaces. Several genera of scorpions have ridges on the dorsal surface of the opisthosoma, which are referred to grandly as stridulatory 'organs'. Such scorpions stridulate by scraping the telson across these ridges to produce a noise that is even audible to humans. Scorpions are also covered in sensory hairs. Some take this to extremes, such as the aptly named *Hadrurus hirsutus* from southern North America and Mexico.

EXOSKELETON FLUORESCENCE Apart from some species of Solifugae and Opiliones, scorpions are the only arachnids to fluoresce under UV light. Even museum specimens that met their demise decades ago still demonstrate the same amazing property. The fluorescence is produced by organic compounds, such as coumarins (usually found in plants), located in the hyaline exocuticle. The intersegmental membranes do not posses hyaline exocuticle, so as a result they do not fluoresce. Fluorescence does not occur in juveniles or freshly moulted scorpions; it only occurs in hardened individuals. When a scorpion is moulting, its exuvia fluoresce brightly, but the animal itself does not. It takes around 48 hours for the freshly moulted animal to fluoresce again. Desert scorpions fluoresce more brightly than those from less arid areas, which implies that there is a possible difference between the structures of their exoskeletons.

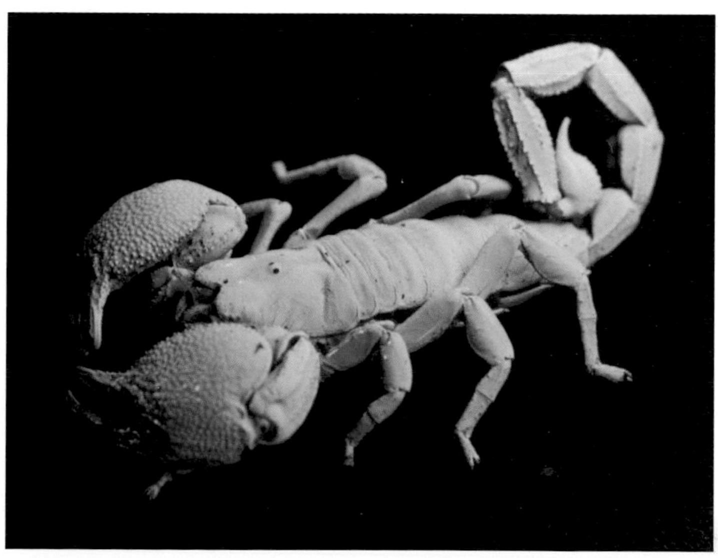

BELOW This *Pandinus imperator* specimen fluoresces vividly under UV light.

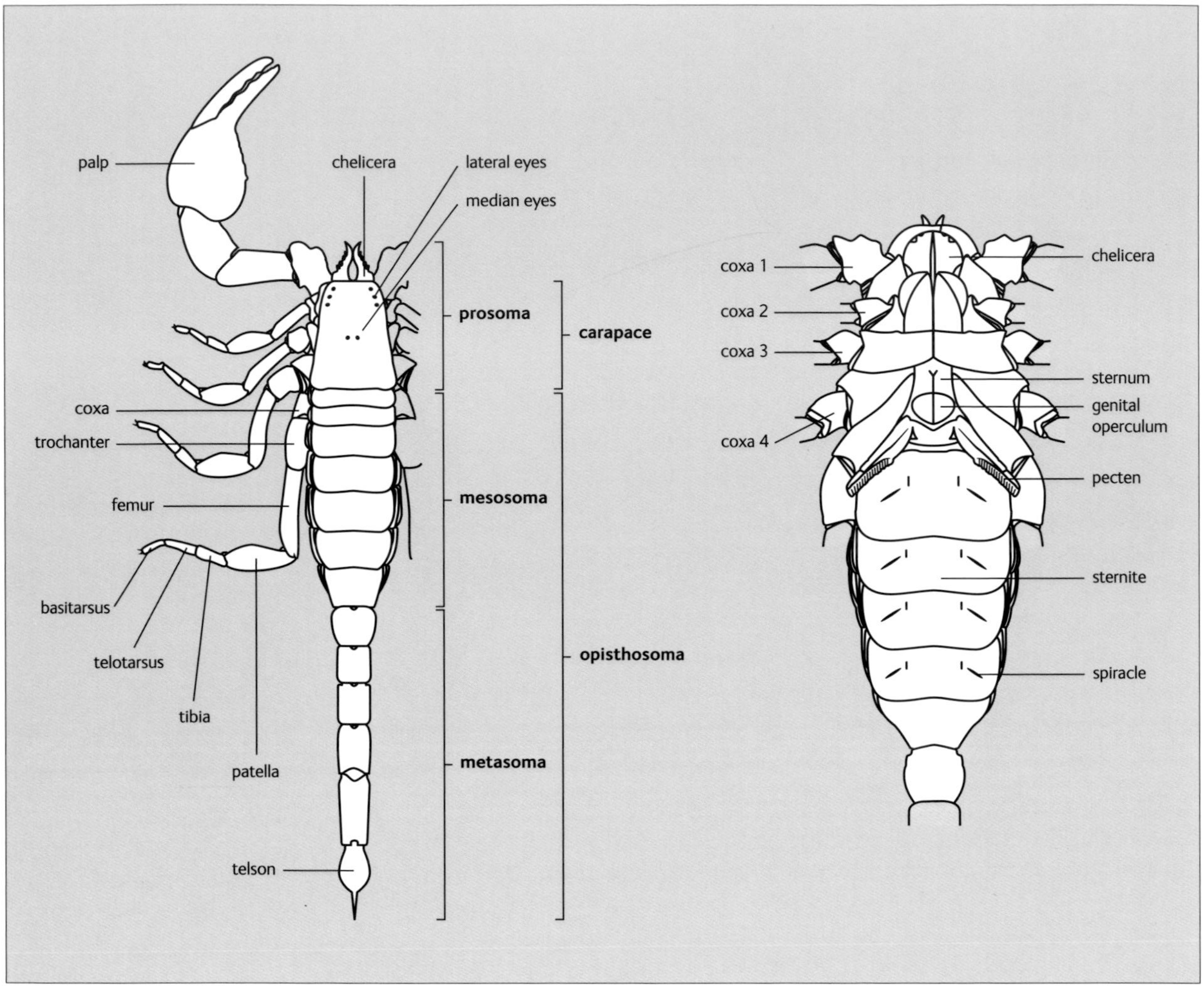

ABOVE Dorsal view (left) and ventral view (right) of an adult scorpion.

Although fluorescence was first reported in 1954, the supposed function of this phenomenon is still unproven. A few theories have been proposed. One suggests that it is a way for scorpions to recognize their own species, although this has been disputed – it is argued that because scorpions are colour blind (they have only one photoreceptor pigment), they can't see the fluorescence of their exoskeleton as a separate light source. Another theory, put forward in 2001 by Frost and colleagues at Marshall University in the USA, suggests that the fluorescent compounds could have served as a sun block in ancient scorpions.

The ancestors of scorpions were originally marine. During their move onto dry land, they would have been exposed to harmful UV rays. A compound in the exoskeleton that protected scorpions from the sun's irradiation would have been of great advantage, and it is well known that the earliest fossil scorpions had hyaline cuticle. Although this 'sun block' theory is supported by the discovery of coumarins that act as sun block in young plants, it does not explain why the fluorescent compounds are still selected for in nocturnal scorpions. Given that many minerals incidentally fluoresce under UV light, it is most likely that scorpion fluorescence has no real function, and is a purely incidental characteristic.

SENSORY RECEPTORS A scorpion's exoskeleton has various receptors, including chemoreceptors, mechanoreceptors and photoreceptors. These consist of different kinds of hairs, spines and sensory slit organs. The pectines, that distinctive pair of comb-like structures on the ventral surface, are probably the most important sensory organs a scorpion possesses, and these are discussed separately below.

Chemosensory hairs are found near the mouthparts and on the chelicerae, as well as in less obvious places such as on the appendages. They allow the scorpion to determine the chemical composition of different food sources, as well as other substances (such as pheromones), in the near vicinity. Such is the sensitivity of some of these chemosensory hairs that a female can use them to distinguish between insects as food, and her own young.

Mechanoreception is the best developed of all the scorpion sensory systems. Mechanoreceptors are very varied, and include slit sensory organs, sensory hairs, tarsal sense organs, joint receptors and metasomal muscle receptor organs. A slit sensilla, which is a slit-shaped thinning of the exoskeleton, is covered in membrane and hooked up to nerve cells. It detects the buckling of the exoskeleton caused by bending movements as a result of pressure.

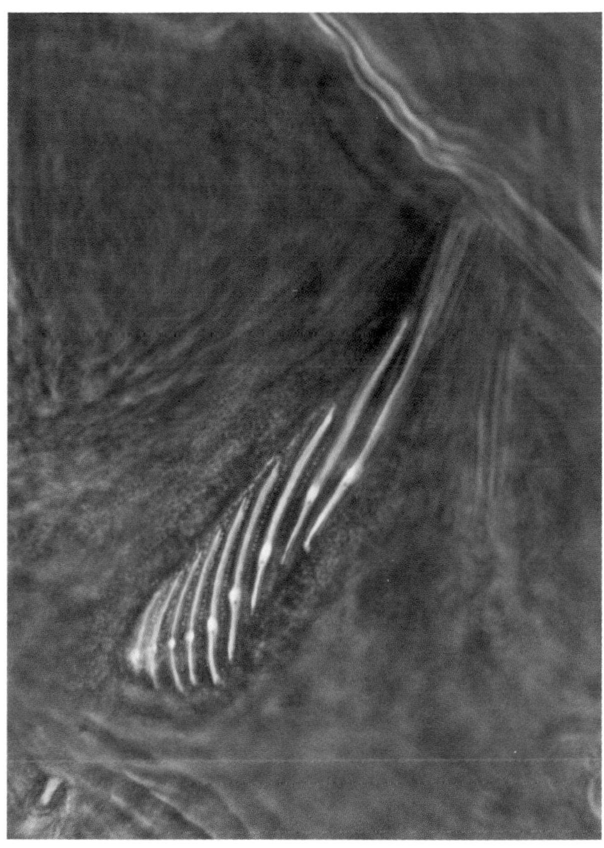

ABOVE Slit sensilla are found individually scattered, in vague groupings, or in small organized groups called lyriform organs. They are found on all parts of the body, including the telson and legs. In some species, they help to detect prey by picking up substrate vibrations.

Scorpions have numerous sensory hairs on their exoskeleton, many of which are mechanoreceptors. Some of these hairs are innervated (attached to nerves). Mechanical movement and air currents move these hairs and create electrical pulses. Trichobothria are long, thin sensory hairs which sit in sockets called bothria, and are also innervated. They are only found on the tibia, patella and femur of the palps, and are much less numerous than the tactile hairs. They are extremely sensitive because the very fine hair shaft is suspended in a membrane, rather than being firmly attached to the exoskeleton, and so the slightest air movements trigger nerve impulses. Interestingly, each trichobothrium has a directional sensitivity, which enables the scorpion to detect prey, as well as to locate other scorpions. Tarsal sense organs and joint receptors are found on the legs, while metasomal muscle receptor organs occur on the opisthosoma.

The main photoreceptors of a scorpion are the eyes. Amazingly, though, it has been discovered that the opisthosoma of scorpions possess metasomal ganglia which are photosensitive. It appears that the location of this photoreceptive activity in the opisthosoma varies between species. However, specific details are as yet unknown.

PROSOMA The prosoma has a carapace on the dorsal surface, which has various grooves extending from the eyes to the posterior edge of the carapace. There are also various keels that vary in structure between families. There is a small sternum on the ventral surface, which may be triangular (e.g. Buthidae), transverse (e.g. Bothriuridae) or pentagonal (e.g. Scorpionidae), depending on the family. This feature is therefore used in the identification of scorpions to family level.

Highly sensitive eyes

Scorpions have up to 12 eyes; there is usually one pair of median eyes on the dorsal surface of the carapace on a small ocular tubercle, and 0–5 smaller eyes in a row on the anterolateral edge either side of the carapace. Normally, there is the same number of eyes on either side but, bizarrely, there may sometimes be more on one side than on the other. The number of eyes on the anterolateral edge depends on the species. Scorpions that live in caves, not surprisingly, either have reduced eyes, or no eyes at all; this is also the case for some species that live in leaf litter.

A scorpion eye is made from modified, smooth and highly transparent cuticle, with a strong and regularly curved outer surface. The lens is large for an ocellus-type eye in a terrestrial invertebrate. It condenses light into a retina that has retinula units, made up of visual cells. The median eyes have vitreous humour, which separates the lens from the retina, but this is not present in the lateral eyes. The median and lateral eyes have very different roles. The large and prominent median eyes form a clear distortion-free image, and allow the scorpion to have good spatial awareness. Although the smaller lateral eyes form less accurate images, they are highly sensitive, and even at low light intensities they can detect differences in brightness. In lithophilic scorpions, the lateral eyes are particularly well developed in order to make the best of low light conditions found in rock crevices. In fact, the sensitivity of scorpion eyes ranks highly within the arthropods as a whole. Daylight or even moonlit nights are much too bright and would damage the eyes of nocturnal scorpions, so they only venture out at twilight and on moonless nights.

During the day, the eyes are protected by a shielding pigment that migrates in front of the photoreceptor cells, to prevent damage (much as sunglasses protect our own eyes). At night, the pigment withdraws to the base of the retinula cells. The movement of this pigment is controlled by the scorpion's circadian rhythm – the internal 'body clock' that regulates the approximate 24-hour cycle of biological processes in so many organisms.

BELOW Prosomal region of a scorpion showing the median and lateral eyes (Jatun Sacha, Ecuador).

CHELICERAE The chelicerae are three-segmented structures located between the bases of the palps. Each chelicera has a barrel-shaped segment with a fixed finger on the second segment, and a moveable finger. On the inner edge of the fixed finger is a fringe of thick hairs, which are sometimes a beautiful golden colour. These setae act like a food colander, straining food as it is ingested, and are also used to groom the legs and palps. The inner edges of the fingers also have various teeth and cuticular granules, which help to break down the exoskeleton of the prey. These structures are highly distinctive between different taxa, and so are useful in identification. The chelicerae have very strong adductor (closing) muscles, which help the scorpion hold on to its prey. There are chemoreceptors located on the chelicerae, which tell the scorpion whether prey is palatable or not.

CHARACTERISTIC PALPS The claw-like palps are one of the most characteristic features of a scorpion. They are very variable in size, in proportion to the size of the body, between different species. They can be slim (as in the Buthidae, see p.240 bottom), long and slim as in many lithophilic scorpions, or extremely chunky, crab-like and capable of powerful crushing, as in fossorial (burrowing) scorpions (see p.240 top). As with the palps of pseudoscorpions, scorpion palps have a 'hand' with one fixed digit (tibia) and one moveable digit (tarsus). Together the tibia and tarsus are known as the chela. Along the cutting inner surface of the digits, there are cuticular outgrowths, which can be granular, or tooth like. The palps have no abductor (opening) muscles – instead they are opened by hydraulic pressure (increase in haemolymph pressure).

The palps are used by the scorpion to capture food, to dig burrows, for defence, and in males to grasp the opposite sex during mating. In addition, they clearly have a sensory function, because they are the only appendages that possess trichobothria. These are found on the tibia, patella and femur. Their arrangement and number are important in the taxonomy of the order. The sensory importance of the palps is indicated by the way scorpions hold them outstretched, off the substrate, as they move around. Buthid scorpions (as well as some other taxa) have a small brush of setae on the ventral side of each palpal femur. When at rest, these scorpions use their palpal brush to remove dust from their median eyes.

LEGS Scorpions walk on all of their pairs of legs, unlike some other arachnids which have a pair of antenniform legs that have evolved as sensory structures only. Scorpion legs are covered in sensilla, so they are clearly used as sensory structures. Slit sensilla are found singly or in small groups on all segments of the legs, and are concentrated near the joints. One type of slit sensillum, the basitarsal compound slit sensillum is found in all scorpions and is sensitive to surface vibration, thereby enabling the scorpion to detect prey. Scorpions also have projections on their tarsi, thought to be sensory, and the legs have directionally sensitive sense cells called joint receptors. As well as being sensory structures, the legs in both sexes are used for digging, and in females they are used for catching young that are just being born. In many scorpions, the way that the legs are articulated, their direction of movement and the placement of special setae and claws show adaptation to a burrowing life.

OPISTHOSOMA The extremely long and tapered opisthosoma is divided into the broad, midbody region called the mesosoma, and the narrow, tail-like metasoma, with a conspicuous telson at the tip. Contrary to the case in many arachnid orders, the opisthosoma of a scorpion

is very conspicuously segmented. Additionally, it does not taper at the anterior, as with spiders, but is broadly joined to the prosoma by a wide pedicel. The intersegmental membranes are very conspicuous (see p.240), because they have quite a lot of stretch – particularly when an animal has fed. The opisthosoma is long, slim and dorso-ventrally flattened in lithophilic scorpions, streamlined in psammophilic scorpions, and heavy, robust and quite short in pelophilic scorpions.

OPPOSITE The metasoma of a 'yellow' scorpion *Buthus occitanus* (Spain).

The mesosoma has seven segments – the seventh narrows at the posterior, where it joins on to the metasoma. On the dorsal surface, each segment has a sclerotized plate (tergite). On the ventral surface, the sclerotized plates (sternites) cover segments 3 to 7. Segment 1 doesn't have a ventral plate, because it bears the genital aperture. Segment 2 bears the pectines. The spiracles of the book lungs are found on segments 3 to 6. Internally, most of the abdominal organs are found in the mesosoma, which varies in width between species. Scorpions that use their large, claw-like palps for defence often have a poorly developed mesosoma, while in many venomous scorpions, the musculature of the mesosoma is more developed.

The metasoma has five segments, which are basically rings of chitin covered with keels, setae and bristles, with no definite tergites and sternites. Each segment is longer than the previous one – segment 5 is always the longest, and this bears the anus and the characteristic telson at its end. There are paired innervated muscle fibres in the metasoma, which detect the bending and flexing of the metasoma and telson.

The telson comprises a bulb-shaped vesicle and a wicked-looking curved spine called an aculeus, with a tip used to pierce the prey. There is sometimes an accessory spine ventral to the aculeus, depending on the family. The vesicle is bulb-shaped because it contains not one, but two venom glands. The venom is secreted through ducts, which lead along the length of the aculeus, and exit through a small pore either side of the tip. Fortunately, it is easy to see when a scorpion is about to sting, because it curves its metasoma over the rest of its body in a characteristic arch before it strikes.

COMB-LIKE PECTINES Pectines are curious-looking appendages, unique to scorpions. They are comb-like structures located on the ventral surface just below the genital opening. They are therefore carried very close to the substrate, and are able to actively sweep and tap over the substrate as the scorpion walks. A pecten is an intricate structure, formed from a flattened, narrow and flexible strip made up of a number of thin lamellae. Attached to the lamellae are flattened teeth, which vary widely in number (between 5 and 40) in different species and even between the sexes, as males usually have more. The teeth are covered in thousands of micro and macrosetae, the most numerous being chemosensory sensilla that look much like inverted mushrooms, called peg sensilla. Each tooth has around 400 peg sensilla located on the margins.

The tip of each peg sensillum has a slit, opening into a fluid-filled chamber with 10–18 sensory cells. These cells are linked to dendrites (branches of nerve cells), which link to the nerves of the other peg sensilla, so that information is relayed across the entire surface of the pecten. The nerve connections form a ganglionic structure (a group of nerve cell bodies) within each pecten. This means that the sensory cells can pick up information from outside, and relay it to the central nervous system. Various other sensilla can also be found on the pectines. For example, stretch receptors, which are circular in shape, have been found on some burrowing scorpion species.

ABOVE Ventral surface of a scorpion, showing the pectines, genital operculum and leg coxae.

Pectines are the most densely innervated sensory organs in arthropods, and are used as contact chemoreceptors, as well as mechanoreceptors. They therefore play a crucial role in the reproductive biology of scorpions, as they enable males to identify and locate potential mates, because they are sensitive to pheromones that females leave behind on the substrate. They also differentiate between different substrates during spermatophore deposition. They are used by newborn scorpions to recognize their mother. In both males and females, they are important in feeding. For example, a hungry scorpion may walk over a dead insect, but will only go in reverse to pick up the prey when the pectines have made contact. They are also important in predator avoidance (when there are chemical traces left behind by predators) and in navigation. Some scorpion species also use their pectines to stridulate, by rubbing them against the sternites. Pectines are often greatly reduced in scorpions that only ever encounter sand (ultrapsammophiles), possibly indicating that they are of no real use for an animal that lives on one type of substrate.

Pectines are unique to Scorpiones, but Solifugae have structures with similar functions called malleoli, also found on the ventral surface of the animal. They too make contact with the substrate and are used as chemoreceptors.

EXTERNAL GENITALIA The genital aperture (known as the gonopore or gonotreme) of both male and female is situated on the first segment of the opisthosoma. It is covered by the operculum (see p.244 right), which is formed from either a pair of small plates that are partially or totally separate in the male, or a single plate in the female.

RESPIRATORY STRUCTURES Scorpions have four pairs of book lungs on the ventral surface on segments 3 to 6 of the mesosoma, which open to the outside by narrow slit-shaped, oval-shaped or circular apertures called spiracles (see p.244 right) set into the body cavity. This arrangement means that water loss by evaporation is kept to a minimum. Some species are able to control the size of the spiracle apertures through muscle control and this might be used to further restrict water loss.

SEXUAL DIMORPHISM In scorpions sexual dimorphism can be quite obvious or rather subtle, depending on the characteristics involved, and these can be grouped into several categories – body size and shape of body structures; presence of extra features; surface textures and greater development. As for arachnids generally, female scorpions are usually larger than males, and also tend to have more robust bodies. The shapes of body structures can vary quite considerably between the sexes. The metasoma is commonly elongated in males. In the case of *Hadogenes troglodytes*, the elongated metasoma of the male makes it one of the longest of all scorpions. The palps of male scorpions are usually longer and slimmer than in females. The telson can vary in shape too. The pectines of females are usually smaller than those of males. In the Bothriuridae, males have processes or depressions on the

palpal chelae, which hold the females' palpal fingers during courtship. Males are usually more granular on their bodies than females. In some species, there are heavy scallops on the inside cutting edge of the digits of the palpal chelae in males, but in females they are weakly formed.

Internal anatomy

CENTRAL NERVOUS SYSTEM Scorpions have a large cephalothoracic mass surrounding the gut in the prosoma, the dorsal portion of which is equivalent to the 'brain'. From the cephalothoracic mass, there are major nerves that run to the eyes, palps, chelicerae, legs, pectines, genital region, and the anterior part of the mesosoma. Additionally, there is a long nerve cord that runs almost to the tip of the telson. There are seven associated ganglia – three in the mesosoma and four in the metasoma – from which paired nerves branch. These run to the book lungs, and the various body segments. The very last ganglion gives rise to nerves to the telson, venom gland and aculeus.

CIRCULATORY SYSTEM A scorpion's heart consists of a tube, located in the mesosoma and extending as far as mesosomal segment 7. There are seven pairs of ostia, which open into the cavity and can be closed by valves.

RESPIRATORY SYSTEM Unlike spiders, which breathe passively, scorpions have active respiration. There are two groups of muscles opening each spiracle, expanding the atrial chamber to let air in, and then compressing air out. Rhythmic contractions push air into the pulmonary chamber.

DIGESTIVE SYSTEM A scorpion's digestive system is effectively split into three parts. The stomodeum consists of the mouth, pharynx and oesophagus; the mesenteron is the stomach, stomach glands, intestine and hepatopancreas; and the proctodeum comprises the hindgut

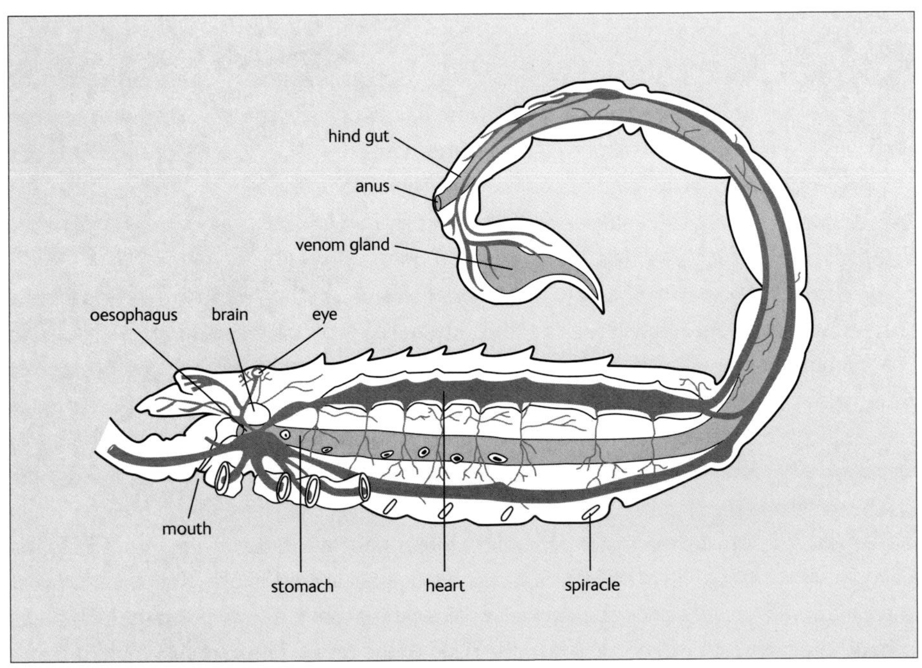

LEFT Cross-section of a generalized scorpion.

and anus. The preoral cavity is formed from the chelicerae, the coxae of the palps and the coxal processes of the first two pairs of legs (coxapophyses). The first pair of these coxapophyses has short bristles, used partly to filter incoming food. The second pair of coxapophyses has lateral channels that direct the liquid food to the mouth. The mouth is at the rear of the preoral cavity, and beneath the base of the labrum (upper lip).

The mouth leads into the pharynx, which is a small chamber used for sucking. The oesophagus is a thin tube that runs from the pharynx and is attached to the stomach, which is a dilated tube. The stomach glands are considered to be the anterior pair of the hepatopancreas glands. The hepatopancreas is the name for the five pairs of glands that lead off the anterior intestine, and accounts for around 20% of the total body mass in the mesosoma. It produces enzymes to break down food, and absorbs digested material, so it also acts as a food storage organ, by storing high concentrations of glycogen. The muscles act as a back-up larder by storing glycogen too. The hepatopancreas also discharges excretory products into the intestine. Attached to the posterior intestine is the hindgut, and then the anus, which opens immediately before the telson.

EXCRETORY SYSTEM The excretory system of a scorpion consists of the hepatopancreas, Malpighian tubules, coxal glands, nephrocytes and lymphatic glands. Malpighian tubules extend to the anterior end of the stomach, and into the hepatopancreatic lobes. The tubules absorb waste material, which is excreted. The coxal glands remove waste from the haemolymph and excrete it through the coxal gland ducts, which open on the fourth pair of coxae. Nephrocytes are large excretory cells that take up waste products, and are found in the ventral part of the prosoma and the mesosoma. Lastly, the lymphatic glands perform a major role in removing waste. They consist of cellular strands that run along the ventral nerve cord in the mesosoma.

Scorpions are able to conserve water by excreting very dry and almost insoluble waste products, such as guanine and uric acid. These compounds can be excreted in crystalline form without having to be dissolved in water, because they are non-toxic.

INTERNAL GENITALIA The female's reproductive apparatus is in the mesosoma. Through the genital aperture is the genital chamber, into which opens a pair of spermathecae. Each spermatheca runs into an oviduct, which is in turn attached to a system of ovarian tubes, forming a complex set of ovaries somewhat resembling a train line map (see p.13). The ovarian tubes are regularly interspersed with ovarian follicles (containing oocytes) that are attached to the tubing by a stalk, like grapes on a vine.

When a female has been inseminated, she can store the male's sperm in her seminal vesicles. Females from certain species in the families Diplocentridae and Buthidae are able to store sperm for long periods. At the appropriate time, spermatozoa leave the spermathecae, move along the oviducts and fertilize the oocytes. These develop *in situ* inside the ovarian follicles. Follicles may contain embryos at different stages of development. When an embryo has reached a certain stage of development, it breaks away from the stalk, moves into the ovarian tube and along to the genital operculum.

As with the female, the male's reproductive apparatus is located in the mesosoma. The paired testes consist of a system of tubules joined together at right-angles. At the anterior end, there are several paired glands, the seminal vesicles (where sperm is stored) and the long paired paraxial organs, which extend almost to the edge of the testes. The paraxial organs produce

both halves of the spermatophore capsule (hemispermatophores) – one in each organ – like two halves of a hollow chocolate Easter egg. Sperm is transferred from the seminal vesicles into the hemispermatophores, which are then 'glued' together by an exudate from the accessory glands. The ejaculatory sac presses the spermatophore capsule out of the gonopore.

VENOM GLANDS Scorpions have a pair of venom glands in the bulbous part of the telson. They are surrounded on the inner edge by muscles, which compress the glands against the exoskeleton in order to eject the venom. They have a duct, which leads to an exit aperture near the tip of the aculeus. Each gland has a lumen (the inner open space) surrounded by a single layer of folded, glandular secretory epithelium. The epithelium cells produce and secrete the venom into the lumen. The amount of folding of the epithelium is variable between different families and genera.

Distribution and habitats

Scorpions have a wide geographical distribution. They live on all major landmasses except Antarctica, and are found in the tropics, and in temperate zones. In addition to regions such as Africa and the Middle East, central Europe, and the Americas, scorpions are also found in more unexpected places such as Switzerland, limited to the southern and southeastern parts of the Alps. Two species found there are aptly named, as they belong to the subgenus *Alpiscorpius*. In France, a population of *Euscorpius italicus* has established in the Rhone Valley.

The surprising distributions of scorpions have occurred by accidental introduction through human activity. For example, the population of *E. italicus* in the Rhone Valley resulted from 19th century Italian traders producing and selling 'oil of scorpions' – a supposed panacea for ills. Clearly, the movements of human populations over the globe have greatly contributed to the present distribution of scorpions, as these animals have limited dispersal abilities of their own.

The most diverse communities of scorpions (up to 13 species) are found in deserts. They are also found in grasslands, savannah, temperate and tropical forests and rainforests. Although scorpions are not usually found at higher altitudes because of the cold, a few hardy species have been found at 3,500 m (11,400 ft) in Kashmir in areas of frozen snow, as well as at altitudes of 4,000 m (13,000 ft) in the Himalayas and 4,750 m (15, 584 ft) in the Andes! Certain species are troglobites, and one species has been found in caves at depths of more than 800 m (2,600 ft). Unusually for a troglobite, this animal is quite large (around 70 cm, or 28 in, long) and pigmented. Other species are troglophiles, or cave 'lovers', so although they are found in caves, they are also found in other favourable habitats. Several scorpion species live on the seashore, and in the intertidal zone. Some are adapted to being submerged daily,

BELOW Even in the wet and often cold climate of the UK, the European yellow-tailed scorpion *Euscorpius flavicaudis* has naturalized in parts of the Thames Estuary and has turned up elsewhere in southeast England. This scorpion is harmless, its sting provoking only a mild reaction.

Habitats

Most scorpions live on the ground, and some are arboreal. Ground dwellers often build burrows. They can be found under rocks, in crevices, under logs, and in leaf litter. Arboreal scorpions can be found under bark, in tree holes, in epiphytes, at the bases of leaves and so on. They are also found in fallen logs and dead vegetation. In fact, many buthid species specifically live under bark, and so are called bark scorpions. Some South American scorpion species live in bromeliads, and forage for prey near the water at the base of these plants. Certain species may forage on vegetation at night, but live in burrows in the ground during the day.

ABOVE AND BELOW A bark scorpion (above, Cockscombe Reserve, Belize), and a scorpion foraging in foliage at night (below, Jatun Sacha, Ecuador).

while others follow the waterline as the tide changes. A few scorpion species can live in many different microhabitats and habitats within their range. For example, the extremely versatile *Vaejovis janssi* (from Revillagigedo Islands, Mexico) is found on the beach, in rocky terrain, in vegetation, and in jungle.

General biology and behaviour

Venom

As with spiders, scorpions are infamous for producing venom. They use their venom to rapidly immobilize and kill their prey, as well as in offensive and defensive behaviour. Some species don't normally sting their prey at all, such as adults of *Pandinus imperator*, but they still produce venom.

ARE ALL SCORPIONS HIGHLY VENOMOUS TO HUMANS? All scorpions produce venom. However, the vast majority of scorpion stings are no more painful to humans than that of a honeybee. A few species possess venom that is highly potent to humans and are referred to as highly venomous. Venom is a complex cocktail of chemicals, whose components differ between species, so the effects can be extremely variable. It contains active neurotoxins, salts and various organic compounds. In some species, the venom also contains enzymes that increase the penetration of tissues.

Scorpion venoms are poisonous to a large range of animals. Vertebrate victims include rabbits, rodents, birds, fish and of course humans. Invertebrate victims include insects, arachnids and isopods. Some neurotoxins act selectively on insect targets, which has great implications for the control of agricultural pest insects. Other scorpion venoms act selectively on vertebrate targets. This does not mean that scorpions need to kill vertebrates in order to eat them; it could be that the venom has evolved to disable or kill vertebrate would-be predators.

Certain animals are resistant to scorpion venoms, e.g. the mongoose species *Herpestes edwardsi*. However, it is not clear whether this immunity is to venom from *all* scorpion species, or just from a small number. A scorpion is resistant to its own venom, but if stung by another scorpion – of the same species or a different species, it may well die.

There is a lack of importance of many species, but it is estimated that around 25 out of 1,500 species of scorpions are highly venomous, which is a very small percentage. Of this small number, many are in the family Buthidae; the following buthid genera all contain species that are highly venomous to humans: *Androctonus, Leiurus, Buthus, Mesobuthus, Parabuthus, Centruroides* and *Tityus*. However, within a genus, such as *Centruroides*, there may be both highly venomous and mildly venomous species. The lethal capacity of buthid venom is rather impressive, although only a small amount is injected, and the active neurotoxin makes up only around 5% of the whole venom, it can still be lethal.

WHAT ARE THE EFFECTS OF SCORPION VENOM ON HUMANS? The human mortality rate from scorpion envenomation is dropping, as a result of interventions such as antivenom production and scorpion control programmes. However, scorpion stings still cause hundreds of deaths each year around the world and are a significant cause of mortality in areas such as Mexico, the Middle East and North Africa. According to the Scorpion Systematics Research Group at the American Museum of Natural History in New York, around 100,000 scorpion stings occur in Mexico each year and as many as 800 people die as a result. Things may be even worse in North Africa and the Middle East.

Common name	Scientific name	Country
Arizona bark scorpion	*Centruroides sculpturatus*	USA and Mexico
Durango scorpion	*Centruroides suffusus*	Mexico
Mexican scorpion	*Centruroides noxius*	Mexico
Mexican scorpion	*Centruroides limpidus*	Mexico
Brazilian yellow scorpion	*Tityus serrulatus*	Brazil
São Paolo scorpion	*Tityus uniformis*	Brazil and Argentina
	Tityus bahiensis	Brazil
	Tityus stigmurus	Brazil
fat tailed scorpion	*Androctonus crassicauda*	Turkey, Egypt, Iraq and Israel
black fat tailed scorpion	*Androctonus bicolor*	Israel, Egypt, Lebanon, Libya, Jordan, Morocco, Algeria and Tunisia
fat tailed scorpion	*Androctonus australis*	Morocco, Algeria and Tunisia
fat tailed scorpion	*Androctonus mauretanicus*	Morocco
death stalker	*Leiurus quinquestriatus*	Turkey, Egypt, Israel, Lebanon, Libya, Jordan, Algeria and Tunisia
yellow thick tail	*Buthus occitanus*	Jordan, Algeria, Morocco and Tunisia
South African thick-tail scorpion	*Parabuthus* spp.	South Africa
Indian red scorpion	*Mesobuthus tamulus*	India

RIGHT Scorpions are a major health hazard in several parts of the world. The table gives an idea of the range of some of the most venomous scorpions.

BELOW Israeli deathstalker scorpion, *Leiurus quinquestriatus*, found in North Africa and the Arabian Peninsula.

DO BIGGER SCORPIONS PRODUCE STRONGER VENOM? As with spiders, size is not necessarily an indication of venom strength. Both small and large scorpions can be venomous. Examples of small scorpions that are dangerously venomous include species in the genus *Centruroides*, and the svelte and delicate-looking *Leiurus quinquestriatus*, which can really pack a lethal punch. Examples of large species include *Parabuthus villosus*, *P. granulatus* and several *Androctonus* species. Ironically, big and beefy *Pandinus imperator*, which is only mildly venomous and is harmless for healthy humans, is often mistakenly considered to be highly venomous.

Food preferences

Scorpions are generalist predators with a varied entomological diet, feeding on insects such as moths, butterfly larvae, termites, beetles, cockroaches, earwigs, bugs, wasps and ants. They consume other arachnids, mainly scorpions, Solifugae and spiders, and also enjoy chilopods, crustaceans such as woodlice, molluscs, and annelid worms. Small vertebrates, such as snakes and lizards, and harvest mice, are also on the menu (see p.238). However, their feeding patterns are seasonal, because of prey availability.

Although a few species are specialists – such as the intertidal scorpion *Serradigitus littoralis* (from Baja California, Mexico), which feeds almost exclusively on isopods living in the littoral zone – the choice of prey is usually just related to its size and palatability. Small scorpions go for small prey, and large scorpions go for large prey. Additionally, there is a lower size limit for large scorpions, because they can't manipulate very small prey. Age-specific differences in diet within a species also occur, as the smaller instars can't manage larger prey, and the larger instars can't manage smaller prey. Scorpions are important in food webs; they are the top insectivorous predators in several North American desert communities, and on islands with a poor vertebrate fauna.

A scorpion's chelicerae have chemoreceptors that detect just how palatable a potential dinner is going to be. So, stink bugs, which emit a noxious odour, are understandably rejected, except in times of food shortage. Prey that are difficult to process, such as insects with particularly thick exoskeletons, are avoided too, unless the scorpion is particularly hungry.

Prey capture

The vast majority of scorpions are nocturnal, and so capture their prey at night. There are two main tactics that scorpions use to locate their prey – they are either 'sit-and-wait' predators that stay in a motionless posture in the mouth of their burrow or on vegetation, or they actively forage as they move rapidly around. These behaviours are often species specific, but may also depend on sex, age and developmental stage.

Sit-and-wait predators catch the prey in their large palps, and usually consume it at the site of capture, although prey is sometimes taken back to the burrow and eaten there, or even taken up into foliage to be consumed. This strategy of staying totally motionless until the prey comes to the scorpion is energy efficient, because no energy is expended in hunting, and it avoids the scorpion itself becoming vulnerable to foraging predators.

Active foraging is less common, and expends greater energy. Amazingly, some scorpion species forage in the heat of the day. Foraging in vegetation is common in small species, probably because prey is more abundant among plants than on the ground (see p.254 bottom). Even those that live in burrows on the ground may forage on vegetation.

Although scorpions have very sensitive eyes, like most arachnids, these are not important in prey location. Instead, prey is located by the armoury of sensilla on the scorpion's body and appendages, and by the pectines. Prey location is very sophisticated, and very accurate. The trichobothria on the palps are sensitive to air pressure changes, and can detect the slightest movement of prey, which triggers an immediate response. They are so sensitive that scorpions can detect and capture flying insects! Prey more than 30 cm (11¾ in) away is detected by a combination of sensilla. Prey foolhardy enough to come closer is sensed by the tarsal sensilla, located and grabbed in one hit. Scorpions are able to discriminate between suitable prey items and potentially harmful creatures, e.g. young *Pandinus* are attracted to fluttering moths and termites, but keep well away from buzzing bees. The scorpion is able to discriminate in this way by detecting substrate vibrations with its sensilla.

Once the prey is caught by the palps it is not necessarily stung. If it is relatively small compared to the size of the palps, then the scorpion does not sting it. However, if the prey poses a potential problem – if it is very active, or particularly large – then it *will* be stung. Scorpions with slim palps sting, and those with robust palps use their palps to crush.

BELOW The scorpion usually manipulates the prey so that the head is chewed first, which is an effective way of stopping the prey from struggling!

Feeding and digestion

The chelicerae tear the food into minuscule pieces, which are then delivered to the preoral cavity in front of the mouth, a cavity formed by the coxae of the palps and the coxapophyses (hollow protuberances from the coxae of the first two pairs of legs). The process of maceration then begins. Digestive fluids flow from the buccal cavity ('buccal' relates to the cheeks or mouth) onto the food particles, while the coxae compress them, like a juicer working on an orange. The resulting juice is ingested by the scorpion. Dense setae on the chelicerae and in the preoral cavity trap any particles, allowing only liquid to reach the gut. Once the last drops have been squeezed from the food particles, they are discarded. The remains of a scorpion's meal therefore consist only of hard parts such as exoskeleton and wings, which cannot be liquefied.

Remarkably, scorpions are able to go without food for up to a year! This ability was found in certain species living in arid regions – those from humid regions are not able to survive beyond a month. Scorpions can survive starvation for a long time because they have an extremely low metabolic rate, store food very efficiently in the hepatopancreas, and can eat vast amounts of food at one time when it *is* available. For example, *Tityus uniformis* can put on a third of its original weight after one feed! The intersegmental membranes extend out like a concertina, so the mesosoma appears swollen.

Predators and parasites

You might think that a formidable armoury of potent venom and a tough exoskeleton would render an animal virtually untouchable to would-be predators. However, there are a variety of plucky animals willing to take on a scorpion as a potential meal, including birds such as owls, various wild mammals like mongooses, meerkats and baboons, reptiles such as snakes and lizards, amphibians including frogs and toads, and invertebrates like tarantulas, Solifugae, wolf spiders and centipedes. Ants should not be underestimated either, as driver ants are able to kill adult *Pandinus*.

Scorpions are not above eating each other, and cannibalism is fairly common. Ironically, scorpions are often the most significant invertebrate predators of other scorpions. In the species *Smeringurus mesaensis*, for example, more than 25% of the total diet consists of scorpions belonging to the same species! As with spiders, there is occasionally a tendency for sexual cannibalism, particularly where the female is much larger than the male. However, cannibalism doesn't happen very often and a few scorpion species show a tendency towards sociality.

Certain predators have ways of avoiding being stung by scorpions. For example, grasshopper mice, meerkats and baboons break off the metasoma and telson of the scorpion. Other predators, such as the mongoose *Herpestes edwardsi*, and certain tarantulas like *Aphonopelma smithi*, are apparently immune to scorpion stings.

Scorpions are also hosts for nematode and mite parasites. Some scorpions have been found loaded with larval nematodes in their mesosomal and metasomal cavities. Mites from at least seven families have been recorded on scorpions too. Favourite areas for mite infestations seem to be the pectines, and the pleural membranes (those on the sides of the opisthosoma), although mites are also seen scurrying around on heavily sclerotized plates. Parasitized scorpions seem to be unaffected, though, by their uninvited guests. There are no records of scorpions as hosts for insect parasitoids.

Defensive behaviour and aggression

As well as their sting and exoskeleton armour, scorpions have several other defensive adaptations against predators, including predator avoidance, cryptic colouration, stridulation, threat posturing and sociality.

USE OF VENOM Scorpions use their venom in defence. As well as injecting it using their aculeus, some species of *Parabuthus* and *Hadrurus* are apparently able to squirt their venom up to 1 m (40 in) in a forward direction! However, this has never been conclusively demonstrated.

PREDATOR AVOIDANCE The best way to avoid being eaten by a predator is to avoid the predator. Scorpions therefore spend the majority of their time hiding away in places that are difficult for predators to find, such as their burrows. It has been suggested that because of the great pressure of diurnal predation in deserts, scorpions became nocturnal as an avoidance mechanism. However, there are many other potential explanations for scorpions' nocturnal way of life, e.g. the temperature and humidity may be more favourable at night. Diurnal behaviour occurs in quite a few scorpion species, and some may be active both at night and during the day, e.g. *Parabuthus villosus*, from Namibia. Additionally, diurnal activity might occur in normally nocturnal species under certain conditions, such as after the rains. Many scorpions minimize the time they spend foraging out in the open. In this way, although they may not get the maximum amount of food, they minimize their chances of *becoming* food. Small scorpion species may avoid predation by foraging and feeding among vegetation.

CRYPTIC COLOURATION If an animal is the same colour or texture as its surroundings, it will be difficult for predators to locate, and many scorpions are cryptically coloured. Some scorpion species are variable in colour within the species, depending on the habitat in which an individual lives. For example, *Vaejovis hoffmanni* is light-coloured on light soils, and darker-coloured on darker soils. There is clearly a strong selection pressure for this species to match its background, because light-coloured individuals on dark substrate (or vice versa) would stand out and be easy pickings for any passing predator.

STRIDULATION A few genera of scorpions are able to stridulate as a response to a threat stimulus. They make a hissing sound by rubbing together two body parts, which often have cuticular outgrowths or bristles. There are several pairings of body parts that can be used, e.g. pectines and sternites, palp and walking leg, chelicera and prosoma, and telson with abdominal tergites and dorsal surfaces of first three metasomal segments. Although you might imagine that a highly venomous scorpion would sting first, and ask questions later, it is evident from observations of *Parabuthus* that stridulation is the first course of action taken when these animals are disturbed. They move their telson across outgrowths on the exoskeleton, creating a very audible noise. Stridulation is energy efficient, whereas it is clearly costly to use precious venom in defence. However, scorpions often drip venom from the aculeus at the same time as stridulating.

Threat posturing

Many scorpions with large and powerful palps rely on these for defence (and offensive attack), and their venom is rather weak. They use their palps to protect the anterior part of their body (see p.241g), as a medieval knight in combat would have used a shield. However, as a back up, they raise their opisthosoma and use their telson in quick jabs if necessary. Some burrowing scorpions use their chelae to plug the entrance to their burrows. Buthids do not rely on their slim, rather weak palps for protection, but instead they curve their metasoma over so that the telson is positioned over the anterior of the body, ready to strike. Because the metasoma is more developed, a buthid can strike a blow at a predator. If the aculeus is imbedded deeply, the venom becomes effective much more quickly.

RIGHT *Opistophthalmus* sp. in threat posture (Kalahari Desert, Botswana).

SOCIALITY Scorpions tend to be solitary creatures that only come together when the need arises, such as for mating. Many species are far from social, being voracious cannibals. However, there is a small minority of scorpion species (less than 1%) that demonstrate sociality, e.g. burrowing scorpions such as *Pandinus* and *Heterometrus*. The evolution of this strategy is understandable, because burrows are costly to construct in terms of energy, and are vital in defence against predators and adverse environmental factors. It is likely, therefore, to benefit individual conspecifics to stay together in one place.

Fossorial species may also live in colony-type groupings because they have limited chances of dispersal. In a similar way, it has been noted that members of the same family groups in the species *Opisthacanthus cayaporum* congregate in communal chambers in termite mounds. Additionally, constraints in the environment encourage a delay in the dispersal of youngsters, which also encourages sociality.

It seems that the key for sociality is mutual tolerance. Sociality is not to do with how well each individual can defend itself, but how it can live in harmonious co-operation. *Pandinus imperator* is particularly sociable, and this may be due to the breeding of close kin. Cannibalistic females would risk killing their own offspring. Where cannibalism *does* occur, it is found to be of deformed youngsters. Additionally, familiarity between conspecifics, even if they are unrelated, prevents aggression.

Adaptations to extreme physical conditions

ARID CONDITIONS To survive in harsh, arid conditions, with extremes of temperature, an organism must hide away from the sun, and somehow conserve water. Scorpion behaviour and physiology allows them to do exactly these things, and they are amazingly good at it too. 1) Being nocturnal: The majority of scorpions are nocturnal. It is thought that nocturnal activity enables scorpions to avoid extreme climatic conditions, such as the harmful effects of UV radiation from the sun. 2) Making burrows: Scorpions construct burrows for several reasons. One of these is to avoid extreme temperatures. Indeed, many scorpions spend up to 97% of their time in their burrow. Desert scorpions avoid the hostile surface during extremes of temperature and at low humidity levels. Inside the burrow there is very little daily fluctuation in humidity or temperature, and they only need to be a few centimetres below the surface for extreme temperature to be moderated. Burrowing scorpions also avoid being wiped out by fire, which only rages on the surface, leaving them unscathed. 3) Behavioural thermoregulation: Scorpions in very hot conditions, such as the Sonoran Desert, USA in the summer months, go into a dormant state, remaining in their burrows for extended periods until the rains appear. Scorpions retreat into their burrows when it gets too hot outside. As many desert animals do (e.g. lizards), scorpions raise their bodies as far as they can off the scorching substrate in order to reduce heat levels – this is called stilting. 4) Conservation of water: The ability to avoid water loss is obviously of immense importance to desert-dwelling animals. Scorpions are able to conserve limited water supplies in their bodies in several ways. The waxy layer of the exoskeleton is impermeable to water loss. The desert-dwelling scorpion *Hadrurus arizonensis*, from Mexico and USA, has one of the least permeable exoskeletons measured for a terrestrial organism, and its exoskeleton is more than twice as thick as that of *Euscorpius italicus*, which lives in moderately moist areas.

Scorpions also conserve water by excreting very dry, crystalline waste products, such as guanine and uric acid. These compounds are non-toxic so they do not need to be 'diluted' by dissolving in water. Scorpions have narrow spiracles to their book lungs, which are set into the body cavity, keeping water loss by evaporation to a minimum. Some species control the size of the spiracle apertures using muscles, which might also help regulate water loss.

COLD CONDITIONS Scorpions also show adaptations to other extreme physical conditions, such as cold. Certain species have a cold hardiness, enabling them to live at high elevations. Some high-latitude and high-altitude scorpions (e.g. at 4,750 m, or 15,584 ft in the Andes) can 'supercool' – they can withstand temperatures below freezing without body fluids crystallizing and tissues becoming damaged. They can remain at very low temperature for several weeks, and yet return completely to normal on warming! This is because they have a low metabolic rate that helps to conserve energy.

BURROWS Scorpions living on the ground often spend a lot of time in burrows, although they may also live under rocks and in crevices, or under debris on the surface, such as leaf litter. Scorpions usually excavate their burrows themselves, though they may take over abandoned burrows dug by small rodents, lizards or spiders, if available. They use burrows to avoid predators and extreme temperatures, and for moulting, mating, giving birth, protecting young and feeding. Scorpions that live in a burrow are likely to live longer and may have more young. Even scorpions that live under rocks in shallow scrapes dig a deep burrow in which to give birth. Obligate burrowers (those that are restricted to burrows) only come out of their burrows to forage and mate, and for youngsters to disperse.

BURROW CONSTRUCTION Certain species have very specific requirements for the areas in which they dig their burrows. Of top importance is soil type, with the appropriate hardness. Good drainage properties and adequate moisture are also essential. Additionally, several species prefer to construct their burrows among plant roots or under vegetation. Shade, protection from flooding and protection against debris been blown into the burrow are key considerations, too.

Scorpion burrows vary in structure between species. The most basic is a simple shallow scraping under rocks and is constructed by many species. The most common burrow, however, is fairly straight or slightly curved, at a 20–40° angle from the surface. Some burrows are loosely spiralled as they descend. Burrow designs can vary between young and old scorpions even within the same species. Burrows under rocks and other surface objects are rather shallow at less than 10 cm (4 in), because the object provides the necessary barrier against desiccation, while those dug in the open are more spiralled and deeper – they may be as deep as 1 m (40 in), although the average is between 15 and 50 cm (6 and 19¾ in).

Some species improve their homes further by building a discrete feeding chamber next to the entrance, a kind of scorpion dining room. They may also build an enlargement in which to turn around, as it is much better to come out of one's burrow face first. Sometimes burrows are constructed with more than one entrance.

HOMING BEHAVIOUR Burrowing scorpions do not like to stray too far from home. The majority of species only forage up to 1 m (40 in) from their burrow entrance, although some individuals go further. As scorpions age, they tend to get more daring, and expand their foraging range. Individuals of *Smeringurus mesaensis*, for example, demonstrate homing behaviour over quite large distances. On relatively long foraging trips, they have been known to stay over in burrows that they have previously dug. Quite amazingly, nocturnal scorpions are able to navigate back to their burrows by using the stars. They also navigate by landmarks such as the shadows of stones. They use the direction of light to guide their orientation, as well as using their pectines to pick up familiar chemical cues.

Moulting

As with all arachnids, scorpions have to take action when their exoskeleton becomes too small. Before the rather traumatic and somewhat risky process of moulting, a scorpion may become quite antisocial and inactive. In fact, it might even stay rooted to the spot for the last 24 hours before a moult. Once the 12-hour operation gets underway, the exoskeleton ruptures around the front and sides of the carapace and all the limbs and chelicerae are extricated. The large body takes much longer to free, and most of the effort is put in at this point. Once the exhausted creature is finally free of its old self, it briefly expands its soft body by taking in air or by an increase in haemolymph pressure until the exoskeleton begins to harden.

Life history

Reproduction

The adult males of most species go walk-about during the mating season searching for receptive females. Even obligate burrowers leave their burrows and this is one of the few times that they are found in full view on the surface. A male may travel long distances on his quest for a mate, and this improves gene flow through the population.

SPERMATOPHORE STRUCTURE In scorpions, there are two types of spermatophore. Flagelliform spermatophores are only found in the Buthidae, and are so called because they have an elongate flagellum. Lamelliform spermatophores, on the other hand, are found in all other families, and have a lamina. The flagellum and the lamina are important in spreading the female's genital operculum so that the sperm can be inserted into the gonopore. However, spermatophores have not been studied in many species, so there may be many more differences that have yet to come to light.

COURTSHIP Scorpion courtship is a complex process. As with other arachnids, courtship has evolved to enable the female to recognize the male as a potential mate, and to stimulate her to become receptive to the male's advances, so that she ultimately picks up the spermatophore. Among all scorpions, there is some similarity in courtship and mating. Basic behaviours shown by all families are discussed below, along with selected variations. The following headings and information are taken from Polis and Sissom's chapter on life history in *The Biology of Scorpions*.
Initiation: In general, it is the male that initiates courtship. The causes of courtship initiation can be various, and may include substrate vibrations, male recognition of female morphological characters due to physical contact, and male juddering. Male courtship behaviour might also be initiated by female pheromones. For example, pheromone communication is considered to be very important among species whose population densities are low.
Juddering and clubbing: Part of the male's repertoire is bodily juddering. He moves his body back and forth in a series of quivering motions, while keeping his legs stock still to support him. He performs this juddering throughout the initiation process and the dance, as well as during and after spermatophore deposition. It may be due to intense excitation, it may function as a female stimulator, or it may be for species recognition – perhaps all three are important. Clubbing is possibly an inhibited form of aggression – both partners 'club' each other using their metasomas, but no stinging is involved.
Dancing: The 'promenade à deux' is the main part of courtship, which enables the male to find a suitable substrate on which to deposit his spermatophore, while making sure the female will be in attendance. In this dance for two, the male uses his palpal chelae to grip those of the female, and leads her around. Occasionally he gets her in a chelicera-to-chelicera grip. The promenade à deux may take around 30 minutes, depending on the species – the duration is usually determined by how easy it is to find a suitable substrate.
The 'sexual sting': At the start of the promenade, and then occasionally afterwards, the male pierces his partner's palp with his telson. It may stay there for over 20 minutes. It is not certain whether he actually injects any venom, but if he does, it may help to subdue the female's aggression.
Cheliceral massaging: The cheliceral massage, romantically referred to by some authors as 'kissing', occurs during the promenade à deux and during sperm uptake. As with the sexual sting, it is thought to help reduce female aggression. The male kneads the female's chelicerae with his own, while the female is motionless.
Movement of pectines: During the promenade à deux, the male scorpion opens out his pectines, and moves them across the substrate over which he is walking, in an attempt to find an ideal place to deposit his spermatophore.
Sand scraping: On rough substrate, the male scrapes loose particles away with his legs. It is quite possible that he is preparing an area on which he can deposit his spermatophore. He needs a place where the spermatophore will stick and stay in position during transfer of the sperm.

ABOVE Fat-tailed scorpions performing a promenade à deux.

TRANSFER OF THE SPERMATOPHORE When the male is happy that he has found a suitable patch of substrate on which to deposit his spermatophore, he touches his genital aperture to the ground. The spermatophore is then extruded and stuck to the substrate by the sticky basal plate. The male pulls the female over the almost vertical spermatophore. As the female lowers herself onto it, her genital opercula spread and the valves open, allowing the sperm mass in. The species-specific spermatophore catapults the sperm mass into the female's gonopore.

BEHAVIOUR IMMEDIATELY AFTER INSEMINATION Immediately after the couple have disengaged, the female may, in some cases, sway her body backwards and forwards, or from side to side, while it is lifted off the ground. This may encourage the sperm up into the female's reproductive system.

After the sperm is removed by the female, many males and females eat the remaining empty spermatophore. It might seem unappetising, but the spermatophore makes a nutritious meal, which would be wasted if not consumed. Given the opportunity, females consume their smaller partners after being inseminated. It therefore makes perfect sense from the male's point of view for the pair to separate immediately afterwards. This is done rapidly and quite violently. There may be sexual stinging or clubbing by the retreating animal. Mostly it is the male that flees, but surprisingly it may be the female.

POST-MATING Males are capable of producing another spermatophore a week after a mating. This speedy recharging of their sperm stocks means that, theoretically, they are capable of inseminating around 50 females annually – the reality falls vastly short of this, however, as they usually only manage around 1 to 5 matings per year. Cohabitation is known in some species, such as *Euscorpius flavicaudis*, but as the father does not assist in rearing his offspring, his staying with the female must be in order to guard her, and ensuring that he is the last male to mate with her before she gives birth.

PARTHENOGENESIS A few species of scorpions e.g. *Liocheles australasiae* are able to reproduce parthenogenetically. In 1971, a Brazilian scientist called Matthiesen established that the parthenogenetic *Tityus serrulatus* can produce up to 70 offspring in a lifetime without fertilization! This is rare in other arachnids, being found only in a few species of spiders, Acari and Opiliones.

Pregnancy, birth and maternal care

Pregnancy can last 2–18 months (depending on the species), which – at the longer extreme of this range – is around the same gestation period as a white rhino! Its duration can vary within the same species, depending on temperature and food availability. Lower temperature seems to increase the gestation period.

All scorpions are viviparous – they give birth to live young, rather than eggs. This is extremely uncommon in terrestrial arthropods, and is certainly rare in other arachnids, only being found in some mites. Live birth helps to increase the survival of the offspring. The young are also born large, which gives them a good start in life.

The gravid mother retreats to a burrow, or under an object such as a rock, in order to give birth. She then gets into position, with the front part of her body off the substrate and the first and sometimes the second pair of legs curled under her body. This creates the 'birth basket'. The young emerge from the gonopore, either head first or tail first depending on the species, and drop into the basket. Shortly after, they become active and climb up the mother's legs, onto her back. Some species don't produce a birth basket, but drop the young directly onto the substrate. However, they are helped to climb onto the mother's back. The whole birth process may take less than 12 hours, or it may go on for as long as 72 hours. The record is 10 days, which occurs in *Hadogenes*. Females of some scorpion species, e.g. *Centruroides*, may mate again while gravid or carrying recently born first instar young.

CLUTCH AND OFFSPRING SIZE Larger species produce larger offspring and clutches; smaller species produce smaller offspring and clutches. For all scorpions, the litter size is between 1 and 105. Interestingly, this variation is also shown in the Buthidae. Offspring size and clutch size are often greater in larger females within a species. However, there is often a trade off, so that a larger number of offspring results in smaller individuals, or vice versa. Litter size is also affected by food supply and temperature.

Scorpions give birth seasonally or all the year round, depending on the species and the climate in which they live. Some species in the Buthidae have been observed to produce up to five broods in one year, even after just one insemination! These species are iteroparous. Iteroparity is useful in unpredictable habitats, allowing individuals to reproduce rapidly in favourable years, and make up for poorer reproductive success during bad years. Iteroparity may in fact be common among all scorpions.

MATERNAL CARE AFTER BIRTH Very few people are aware of scorpions' good parenting skills. After birth, all scorpion mothers care for their offspring. Maternal care can be somewhat fleeting, lasting only as long as the babies are dependent. Conversely, some mothers invest months in looking after their offspring (e.g. *Smeringurus mesaensis*). It is thought that maternal care has evolved mainly to prevent cannibalism by scorpions of the same species (conspecifics). Mothers are very aggressive to males, but are much more tolerant to females carrying young. This strongly suggests that males are potential cannibals, as are non-pregnant females.

The bond between a mother and her offspring is due to chemoreception and her ability to recognize her young by scent. Normally, youngsters that have fallen off their mother's back are helped back into position, but those that are removed and given a different scent, are eaten or rejected. Such chemical bonding can also create a tolerance of the young. These chemical stimuli are not necessarily species-specific though – they may be across a genus e.g. *Euscorpius*.

During the first instar, juvenile scorpions have to absorb stored nutrients, because they are unable to feed at this stage. Additionally, as they are not highly sclerotized, juvenile scorpions are at great risk of dehydration, so they must obtain water from their mother, while they are clinging to her back.

BELOW A female carries around her young on her back (Skeleton Coast, Namibia) until after their first moult; they may disperse between three days and two weeks after. This protects them from predators, such as other scorpions and ants, and from drowning if their burrow floods in heavy rain.

Usually, the maternal bond weakens with each successive instar. Once the association between a mother and young terminates after dispersal, the female becomes predatory towards her young. However, there are some species in which the maternal bond is greatly extended. For example, the young of some species, such as *Scorpio maurus palmatus*, stay in the burrow with their mother for up to four months before dispersing, and species of *Urophonius* remain for up to six months. *Pandinus* youngsters are the least keen to leave their mother, they may stay for up to two years! Mothers usually limit their foraging time on the surface, often preferring to remain in the relative safety of their burrow, away from predators in a stable microclimate, although some species are known to forage with young on their backs.

Living communally is particularly valuable to young scorpions living in habitats where it is hard to get prey. In the species *Pandinus imperator*, the mother is aggressive and strong and can overpower dangerous or large prey, e.g. millipedes, that would not usually be accessible to juveniles. The mothers and offspring of some species have even been observed to capture food together and drag it into their burrows. Communal foraging also allows juveniles to overpower prey items as a group that they would not be able to cope with on their own. This gives them a better start in life, as they can grow faster. Not all food sharing is necessarily intended, though. Fossorial scorpions tend to eat their prey inside their burrows to avoid predation on the surface. Any leftovers remaining in the burrow can be eaten by the offspring.

Scorpion mothers aren't always as motherly as they first appear. For example, *Pandinus imperator* mothers have been observed in the lab to cannibalize those young that didn't easily climb onto their backs, weeding out those with deformities and weakness. They have also been seen to eat first instar scorpions that drop off their backs. Although this might seem harsh, deformed offspring are unlikely to get to adulthood anyway. In this act, the mother is avoiding investing energy, resources and time in offspring that are not going to reach maturity. She also recoups important energy resources.

Post-embryonic development

Within scorpions as a whole, development is between 1 and 51 days, and is most variable in the Liochelidae, being 10–51 days. The average first instar lasts 10–20 days, but is very fast in buthids, normally lasting around 6 days. Developmental time can vary depending on the amount of food that a scorpion youngster gets over a period of time. Therefore, a well-fed scorpion has a shorter instar duration than an unfed one. A large variation of instar length within a species could therefore be due to variable food supplies. For example, there can be 11 months' difference in the length of an instar in the species *Pandinus gambiensis*. The number of moults that a scorpion has before maturity ranges between 4 and 9, although most species moult 5–6 times. However, in some species, males and females may take a different number of moults before maturity – males may moult fewer times and therefore mature more quickly.

The time taken to reach maturity can be six months, e.g. *Centruroides guanensis* or up to 6 years and 11 months, e.g. *Pandinus gambiensis*. There is a great variation in the age to maturity within families, the average is at least 2–3 years. Buthids seem to mature faster than other families. Scorpions are unusual in the length of time that they spend as juveniles, which is much greater than for many arachnids. Long generation times result in slow population growth.

First instar scorpions have a soft exoskeleton, their sting is non-functional, and most of their sensory organs, such as the trichobothria, are under-developed. However, the pectines are an important exception. These have a few chemosensory peg sensilla, and enable the youngster to recognize their mother. A great deal of differentiation occurs at the moult from the first to the second instar. Second instar scorpions become much hairier, as the setae and trichobothria become apparent. Structures such as the telson and chelicerae harden and darken, as do the tarsal claws, which also appear for the first time. Additionally, the exoskeleton begins to show granulation, and brightly fluoresces under UV light (unlike that of the first instar).

Longevity

Size is not indicative of age, because scorpion growth rates are very variable. Scorpion lifespans vary with species – the average is 2–5 years (e.g. Buthidae). Species in the family Vaejovidae survive for 5–7 years, Scorpionidae for 8 years, and some larger species, such as *Hadrurus arizonensis*, *Pandinus*, *Heterometrus* and *Hadogenes* species, apparently can live beyond 25 years, which is not surprising, given their large size and slow growth rates.

11 Pseudoscorpiones

False scorpions, book scorpions

Pseudoscorpions are fascinating little arachnids. Although it is easy to see how they got their scientific name and subsequently their common name 'false scorpions' – their greatly enlarged claw-like palps and segmented opisthosoma undeniably convey a superficial resemblance to true scorpions. The French name 'faux scorpions' was coined in 1778. This was Latinized as Pseudoscorpiones and then translated into English as 'false scorpions'. However, some pseudoscorpion workers dislike the negative connotations of 'pseudo', feeling that it says more about what the animal isn't than what it is. Those who object to 'pseudoscorpion', because of this superficiality, often use the name Chelonethi instead. In this chapter the name pseudoscorpion is used throughout.

OPPOSITE *Neopseudogarypus scutellatus*, which is endemic to Tasmania, has a highly sclerotized exoskeleton.

BELOW *Chthonius herbaria* is unusual looking with long legs and palps.

BOTTOM The female of *Neobisium sylvaticum* (Liège, Belgium) has an opisthosoma rather like that of a termite.

Unlike their scorpion relatives, pseudoscorpions don't have a 'tail' with a sting, although the majority of species do have venom glands in their palps. They produce silk, like their relatives the spiders, but the silk glands open on their chelicerae, with the glands situated in the prosoma. This silk is not used to trap prey, but to build silken chambers for hibernation, moulting and egg laying. Pseudoscorpions range in size from the miniature *Apocheiridium pelagicum* found in coral polyps in Vietnam, which measures 0.7 mm (3/100 in), to the relative 'giant' from Ascension Island, *Garypus titanius*, which can reach 12 mm (¾ in).

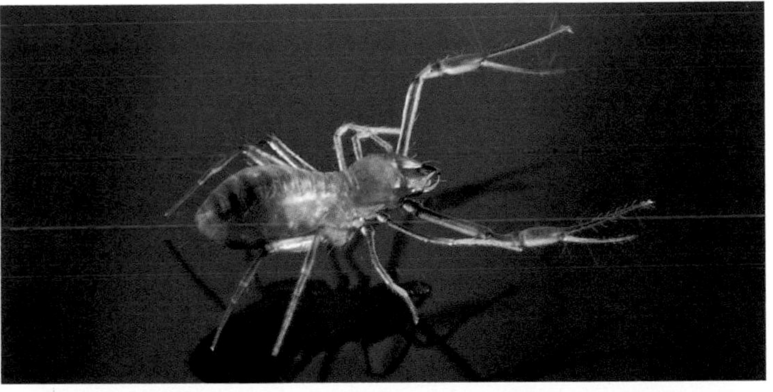

Classification and diversity

The order Pseudoscorpiones is a distinct and homogeneous group, with 3,380 described species. These are classified in 439 genera within 25 families. As an order, pseudoscorpions are generally fairly similar in appearance, although they can be variable in colour, from reddish or golden brown to black, and some species illustrate distinct morphological differences.

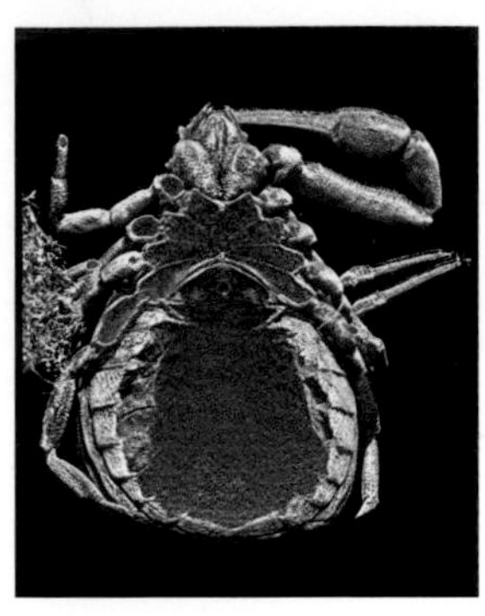

ABOVE If you take a vertical slice through a specimen of *Pseudogarypus banksi* (Duparquet, Canada) you can see that the exoskeleton is extremely thick.

BELOW Dorsal surface (left) and ventral view (right) of a generalized pseudoscorpion.

Anatomy

External anatomy

The surface of the exoskeleton may be smooth, sculptured or granulated, and covered with hairs, which are sensitive to various stimuli. There are two basic types of 'hairs' – setae and trichobothria. Setae are short, bristle-like structures, which are distinctly arranged in groups or rows. They are found on the tergites, sternites, carapace and appendages, and there is an extra concentration of them on the posterior of the opisthosoma. Trichobothria are mainly found on the palps and are used to detect prey. Other sensilla on the exoskeleton include lyriform organs. These are scattered over the body, mainly on the ventral surface, where they act as 'strain gauges' that detect slight deformation of the exoskeleton.

In some species, the exoskeleton is particularly thick. This is an adaptation to dry environments. For example, *Cheiridium museorum* is often found in fairly dry environments – its thick exoskeleton helps to prevent water loss and therefore potentially fatal desiccation. In many species, the exoskeleton is much thinner, and in cavernicolous species it is very pale in colour.

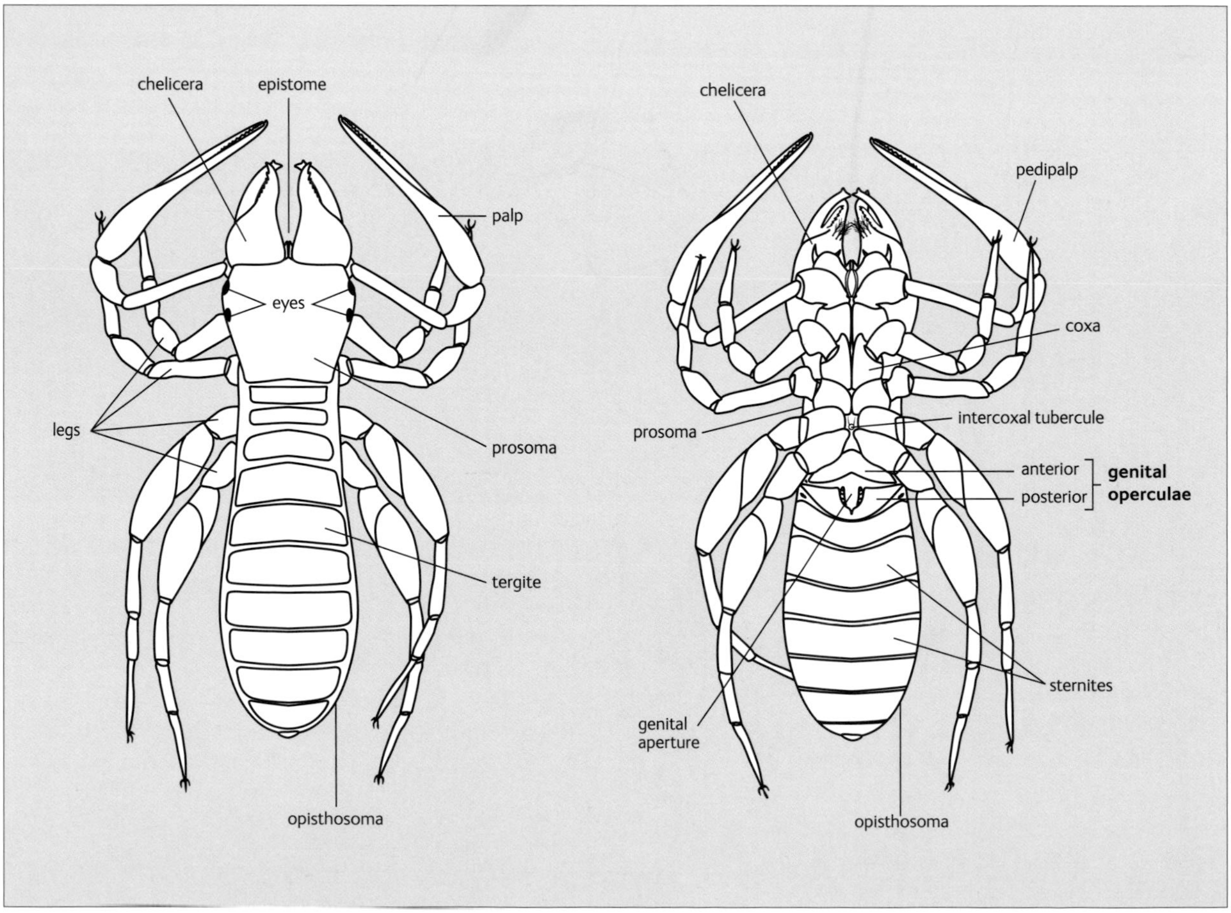

PROSOMA The prosoma is composed of six segments, but these can't be seen as it is covered dorsally by a single large carapace, or scutum, which is variable in shape, from triangular to

square, and may be smooth, have tubercles or bear setae. It is often crossed by one or two transverse furrows. The coxae of the legs cover the ventral side of the prosoma completely in most pseudoscorpions, although they are relatively small. As a result, there are almost no traces of a true sternum, or ventral shield. The coxae are not fused with each other, but they are immovable.

EYES Most pseudoscorpions have two or four eyes, although some cavernicolous species have no eyes at all, such as *Troglochthonius doratodactylus* from southern Europe. The eyes are near the front of the carapace on each lateral margin. The sensitivity of pseudoscorpion eyes to different spectral wavelengths is thought to be comparable to that of the honeybee *Apis mellifera*. They are small, simple and can only tell light from dark, so they do not play much of a sensory role.

CHELICERAE: MULTI-FUNCTIONAL TOOLS The chelicerae are located under the anterior margin of the carapace, and are variable in size depending on the species. They are two-segmented and pincer-like. The first segment is broad and tapers into a curved finger. The second segment is a moveable finger, which articulates against the fixed segment forming a pincer. The chelicerae are multi-functional (a bit like a Swiss army knife), because as well as being used to grasp and macerate prey, they serve as spinning, cleansing and sensory organs. They are also used in nest building to transport bits of detritus.

Among the arachnids, pseudoscorpions are unique in having a galea, or spinneret, that is located at the end of the moveable finger. There are several different forms of galea, which vary from a simple tubercle, to an elongated, slender appendage that may be branched. The galea has an opening to a duct that runs to the silk glands in the prosoma. These can be very large and extend into the opisthosoma, and may even be seen through the intersegmental membranes of living pseudoscorpions. The chelicerae also have several setae that are grouped closely together and form a 'flagellum' (recently renamed 'rallum').

Both chelicerae possess comb-like structures called serrulae that are used as grooming organs for the palps. Each finger of the chelicerae has a 'serrula' – the serrula exterior is on the moveable finger, and the serrula interior is on the fixed finger. The serrula interior is composed of 'blades' in primitive groups, but these are fused in more highly evolved groups.

BELOW LEFT Tree-shaped galea (used for silk production) on the chelicerae of *Geogarypus minor* (Palamos, Spain).

BELOW RIGHT Ventral view of a *Apocheiridium* sp. (from Lesbos, Greece). The chelicera with serrula and galea are clearly visible.

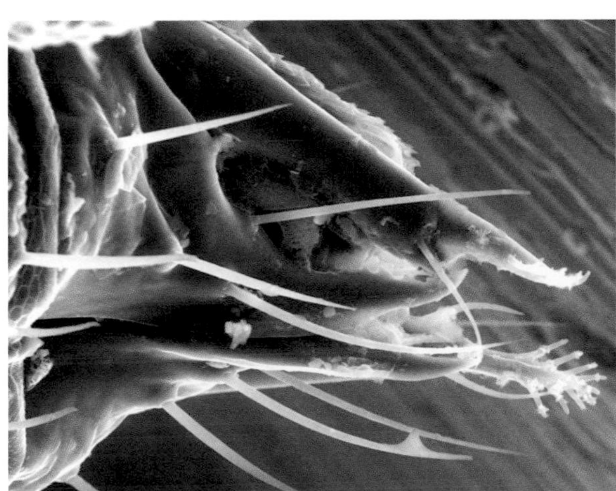

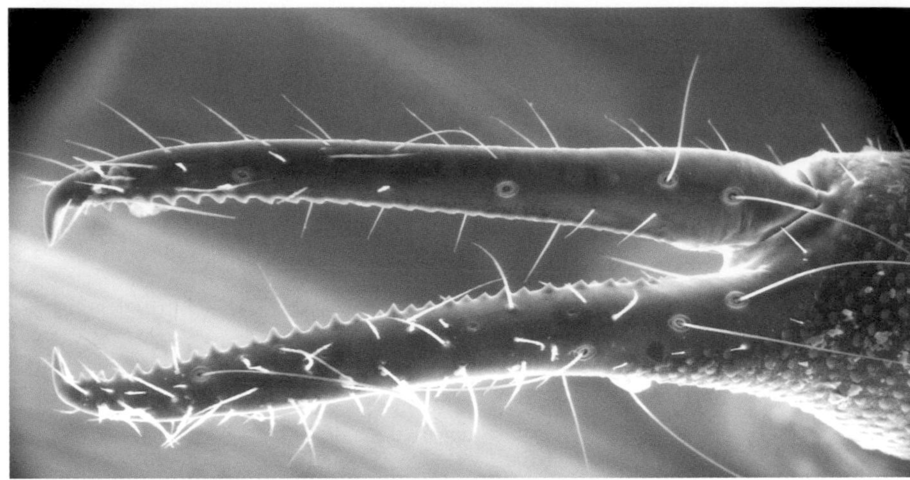

RIGHT Fixed and moveable fingers of the chela of *Geogarypus minor*, Rosas, Spain. The teeth are clearly visible on the inside of the fingers.

BELOW The palps are quite chunky in some species e.g. *Dendochernes cyrneus* (Tenerife, Canary Islands) while they can be long and very slender in others, such as *Acathocreagis granulata* (Catalunya, Spain).

PALPS: FROM ENVENOMATION TO LOCOMOTION In many pseudoscorpion species, the palps strongly resemble those of their much larger relatives, the scorpions, because they have large chelae (pincers). On the inner margins of the fingers, there are usually many teeth.

In the cavernicolous species *Troglochthonius doratodactylus*, there are long spines on the inside of the chelae. These spines may form a catching basket (like those of amblypygids), or may be sensory in function. As with the chelicerae, the chelae of the palps are two-segmented. The structure of the chela is derived from the tibia and tarsus. In other words, the tibia is modified into the base with a fixed finger and the tarsus is modified into a moveable finger, which articulates against the fixed finger to form a pincer. The tips of the fingers curve inwards, and in many species, one or both of them possess the openings of ducts from venom glands, which are located in the finger or in the base of the chela – a feature not seen in any other arachnid order. Certain families, such as the Chthoniidae, don't possess venom glands.

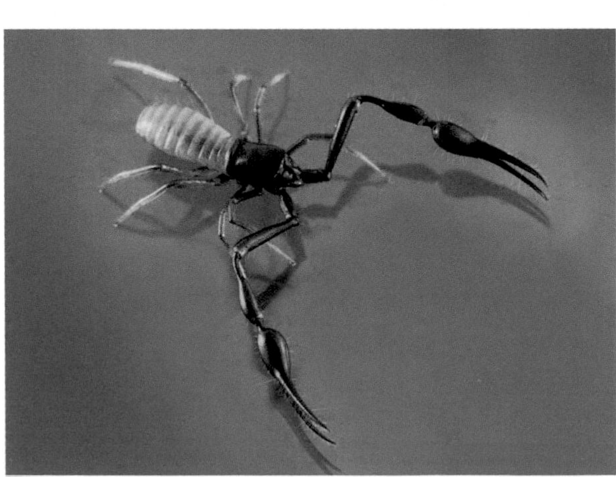

As well as being used to capture and kill prey, the palps are sometimes used in nest building, and are important sensory structures, as they possess a multitude of tactile and chemosensory setae, plus twelve large trichobothria. There are usually eight on the fixed finger and hand, and four on the moveable finger. When they walk, pseudoscorpions hold their palps stretched forward above the ground, with the pedipalpal chelae held open – just like their larger scorpion relatives (see p.271 bottom). The palps can also be used as locomotory structures in cases where the tarsi of the legs are unable to grip the substrate. Pseudoscorpions can therefore pull themselves out of a droplet of water, for example. The palps are also used to turn a pseudoscorpion back up the right way, when it has fallen on its back.

Walking legs with a difference

Usually the first two pairs of legs are different from the third and fourth pairs, in terms of structure, movement and posture. They are relatively small compared to the larger hind legs. Additionally, the femur of the first leg may be slim, whereas the fourth leg is much chunkier. Pseudoscorpions normally walk forwards, but have the ability to run backwards at an astonishing speed. They also have the ability to turn 180° in an instant, to accurately position their palps in response to a stimulus. As already mentioned, the palps aid the legs in locomotion.

BELOW Pseudoscorpions can walk on smooth surfaces, and upside down! They have an arolium (pad) at the tip of the tarsus between the claws of each leg, which enables it to grip and which may work like the scopulae on the tarsi of some spiders.

BELOW The tergites of *Cheiridium museorum*, the house pseudoscorpion, are divided in two down the centre.

OPISTHOSOMA The opisthosoma is broadly joined to the prosoma, so there is usually little mobility between these two parts. However, some pseudoscorpions are much more mobile in this region because they have a special joint, e.g. species in the families Feaellidae and Menthidae. The opisthosoma is noticeably segmented, with 12 segments. There are 11 tergites on the dorsal surface, which may be uniform – as in the Neobisiidae – or divided into two down the centre, as in *Apocheiridium ferum*. There are ten sternites on the ventral surface, the first sternite is missing. The two that follow (belonging to the second and third segments) form the genital opercula. The last segment is reduced, and forms a very small anal cone. This cone can retract into the eleventh segment. There are several long setae on the last few segments, enabling a pseudoscorpion to detect what is creeping up on it from behind. The opisthosoma can greatly expand because of the flexible intersegmental and pleural membranes, which are broad and surround the tergites and sternites.

EXTERNAL GENITALIA In adult pseudoscorpions, the genital atrium is closed to the outside by a large shield called the genital operculum. The opercula in both sexes are formed by sternites 2 and 3 and are named the anterior and posterior genital plates. They have various levels of adaptation depending on the sex. Males that produce rather simple spermatophores have basic opercula, whereas males that produce complex spermatophores have corresponding modifications, such as a deep groove to allow the spermatophore to be released easily.

RESPIRATORY STRUCTURES Pseudoscorpions have tubular tracheae, which open externally at two pairs of slit-like spiracles on the third and fourth ventral abdominal segments. These spiracles are closed by muscles. In some species, the spiracles have their own stigmatic plates, rather than being located on the sternites.

SEXUAL DIMORPHISM: FROM SACS AND SETAE TO SPINES AND SECRETIONS Pseudoscorpions display sexual dimorphism in several ways. As with most arachnids, males are usually smaller than females. In most species in the Cheliferidae, the tarsi of the first pair of legs are modified – often the claws are enlarged in the males. Males of the family Cheliferidae also usually possess coxal sacs, located within the coxae of the fourth pair of legs. The internal walls of the coxal sacs are covered in setae, and a little brush of setae pokes through the opening. Each seta is attached to two glandular canals. They are possibly used in territory marking, as they produce secretions that are rubbed on the ground. Males in the family Withiidae have patches of secretory setae on some of their sternites.

Some species show sexual dimorphism in the proportion of some appendages, such as the palps. In *Dinocheirus tumidus*, the male palps are much more stout and swollen. The palps of several species are variable e.g. the palpal chelae between males and females can be of different size and form. The palpal femur in the male of *Lasiochernes pilosus* is thicker than that of the female, and has brush-like setae. Males of *Levigatocreagris hamatus* from Thailand have a heavy thorn-like spine on the ventral surface of the chelal hand – this structure is unique within the family Neobisiidae. Within species, the galea is more developed in females, which is probably related to the greater amount of silk that the females produce when making maternal silk chambers.

Internal anatomy

CENTRAL NERVOUS SYSTEM Relatively little is actually known about the central nervous system in pseudoscorpions. It is concentrated in the prosoma, as a basic brain – the cerebral ganglion – which surrounds the oesophagus. Unlike spiders, which have two large ganglia (the sub- and supraoesophageal ganglia), pseudoscorpions have a quite simple ganglion consisting of a single combined mass.

CIRCULATORY SYSTEM The heart is elongate and tube-like, and lies dorsally in the first four segments of the body, above the midgut diverticula in the opisthosoma. This very basic pump sends the haemolymph through a short anterior aorta to the anterior of the body. The heart has four pairs of ostia in some species, while others have one or two.

RESPIRATORY SYSTEM The four tracheal trunks leading into the body from the four spiracles are usually wide and short, and possess spiral thickening. The trunks end in a mass of unbranched narrow tracheoles. These penetrate the tissues and deliver oxygen to them. The anterior pair of tracheae serves the prosoma and appendages, while the posterior pair serves the opisthosoma. It is likely that pseudoscorpions breathe passively, as no respiratory movements have ever been noted. Interestingly, it has been recorded that certain species of pseudoscorpions consume oxygen in a rhythmic pattern. For example, the peak consumption of oxygen by *Chthonius ischnocheles* in a 24-hour period is between 1 pm and 3 pm. Not surprisingly, this corresponds to a peak of activity at this time.

DIGESTIVE SYSTEM The coxa of each palp is very large, and has an anterior process called the lamina superior. These, along with the chelicerae on top, the upper lip with its large muscle, plus the lower lip, form an enclosed preoral cavity. This ensures that there is very good suction. The actual mouth is a narrow slit between the lips. The arrangement of the lips and the minute ridges, teeth and folds on the rostrum, maxillae and on the upper lip, ensure that large food particles are filtered out, so only liquid is ingested in this one-way system. In certain pseudoscorpion species, the preoral cavity forms a syringe-like structure. There is likely to be a taste receptor around the preoral cavity. The mouth leads into the pharynx, or gullet, which is a structure with an X-shaped cross-section. The pharynx is attached to a long, cuticle-lined foregut, which is in turn attached to an extensive midgut, which is divided into two areas. The anterior part of the midgut has many diverticula that extend into large areas of the opisthosoma, and may also extend into the prosoma. This area of midgut is very important for digestion, food storage and excretion, because the epithelium contains digestive, excretory and basal cells. The posterior area of the midgut is simply a winding, narrow tube, widening at the end to form a rectal pouch. This is joined to a short, cuticle-lined hindgut, which ends at the anus.

EXCRETORY SYSTEM The key part of the excretory system is the anterior region of the midgut. The extensive diverticula have many excretory cells that store waste material. The midgut diverticula form granules that are excreted, likely to be made up of guanine. From time to time, the waste material or even the entire cell is shed into the midgut. In turn, the waste material is stored in a rectal pouch and voided periodically through the hindgut and out of the anus. This opens through the small anal cone, which is formed from the reduced last segment of the opisthosoma. Pseudoscorpions also have coxal glands, opening at the margin of the coxae of the third pair of legs. Each gland is a long, thin tube ending in an onion-shaped sacculus. These glands are possibly involved with excretion.

INTERNAL GENITALIA In both sexes, the genital atrium (with the associated glands and muscles) varies in structure between families. The female's genitalia basically consist of an ovary, paired oviducts and a genital atrium. The ovary is situated in the opisthosoma. It is hollow and contains many egg follicles that develop in its walls. The oviducts extend from the anterior part of the ovary, and open into the genital atrium, which is closed to the outside by a large shield – the genital operculum. When the eggs are shed from the ovary, they pass through the oviducts, out of the genital atrium, and into the brood pouch. Most pseudoscorpions (except the Chthonioidea) have a pair of erectile little structures called gonopods (also known as gonosacs) attached to the inner side of the genital operculum. These can be elongated through an increase in blood pressure, and protrude from the genital atrium,

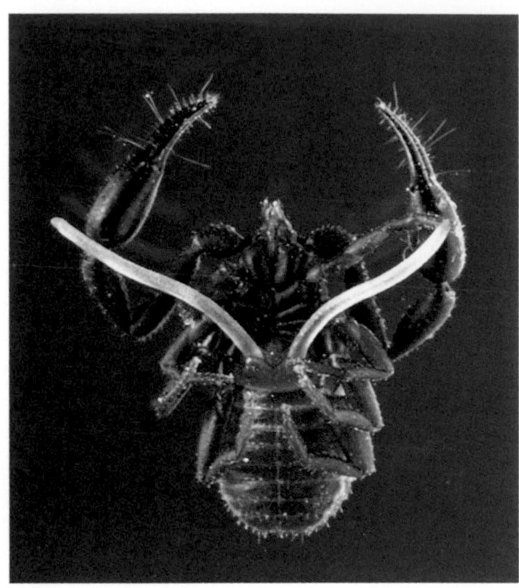

ABOVE Male cheliferids, such as this *Hysterochelifer meridianus* (Crete, Greece), have 'rams horn organs', erectile genital sacs which protrude from the genital atrium. These are in-foldings of the posterior wall of the genital atrium and are elongated through an increase in blood pressure during courtship and mating. These may exude secretions, as females are certainly interested in them, holding their palps very close to the organs during courtship.

where they support the brood pouch while it hardens. The cells of the ovary that don't produce egg follicles produce a glandular secretion, the accumulation of which makes the opisthosoma swell. The secretion is passed into the brood pouch to nourish the developing embryos.

Species in several families, e.g. Chernetidae and Cheliferidae, also have one or two spermathecae, which are cuticular tubes that extend from the atrium. Spermathecae store sperm that will be used to fertilize eggs over a period of months, or even years. The ability of pseudoscorpions with spermathecae to store sperm for long periods has, in part, enabled such species to spread into unstable habitats like animal nests and dead trees.

The internal genitalia of male pseudoscorpions are very complex in structure, because they produce and store sperm, and form and deposit spermatophores. However, only the basics are discussed here. The testis is located below the midgut in the opisthosoma and is lobed in various ways. It is a hollow structure like the female's ovary, and contains follicles where the spermatids develop. A pair of vasa deferentia (ducts which convey sperm from the testes to the exterior) extend from the anterior part of the testis and are comparable to the female's oviducts. The vasa deferentia enlarge in diameter to form seminal receptacles, which empty into the genital atrium. The atrium has variable folds to mould the spermatophore head, and many accessory glands associated with spermatophore formation.

Distribution

Pseudoscorpions occur in the tropics and subtropics all over the world. They are also found in temperate zones, as well as colder regions, such as Scandinavia. *Wyochernes asiaticus* is found in the Yukon Territory of arctic Canada (at 69°10'N). The highest altitude record is held by *Bisetocreagris kaznakovi*, found in the Himalayas at 4,810 m (15,780 ft) above sea level! Most species have a restricted distribution.

RIGHT Some species, such as *Chelifer cancroides*, are almost cosmopolitan, as they have probably been distributed by human activity.

LEFT *Microbisium brevifemoratum* is a rare species that lives exclusively in *Sphagnum* moors and swamps (Belgium).

A multitude of habitats

Because of their small size, pseudoscorpions are often overlooked. However, these engaging little animals can be abundant in most habitats. Some species have a wide habitat preference e.g. *Chthonius ischnocheles* can be found in habitats as diverse as birds' nests, rotten wood, salt marshes and under rocks on mountainsides. Others have a much narrower habitat preference. What is essential in all these habitats is the availability of small crevices in which the animals can hide away. High humidity is preferred by most species, although some, e.g. certain Olpiidae and Garypidae, can tolerate very dry environments such as deserts.

BELOW Pseudoscorpions live in a wide range of habitats but what is essential is the availability of small crevices in which to hide.

Forests and woodlands provide a great variety of suitable habitats, and large numbers of pseudoscorpions are found there. Some species live in forests that are annually flooded for several months every year, and have to adapt to this harsh environment. For example, *Brazilatemnus browni* is found in the seasonally inundated forests of the Rio Negro valley in Brazil, among other habitats. It survives by living in the soil during dry periods, and climbing up trees when the floods arrive. Some species, such as *Beierius walliskewi* (from South Africa), are found in caves, where they live in guano. In fact, it's not this unappetising waste product that interests them as a food source, but the other organisms that live there too, such as mites.

Some pseudoscorpion species, such as *Chthonius kewi* (from Britain), are found in shingle and salt marsh habitats. Other species live their lives on sand dunes and the seashore. *Dactylochelifer latreillii*, for example, can be found hiding in

ABOVE *Neobisium maritimum* from Europe, which lives in the intertidal zone, is able to survive being submerged by the tide twice a day because its finely constructed nest retains an air bubble when under water. Oxygen diffuses into the nest, and carbon dioxide dissolves into the surrounding water. It has even been found to survive freezing at −4°C (25°F) in seawater! Some pseudoscorpions are able to survive immersion in seawater for weeks, e.g. *Pselaphochernes litoralis*.

maritime plants on sand dunes. Pseudoscorpions can also be found under stones, driftwood and other beach debris. *Halobisium occidentale* (from North America) lives in the spray zone just above the high tide line.

Pseudoscorpions can be found in nests of many different types. For example, the common species *Chthonius ischnocheles* lives in bird nests, squirrel drays, bat roosts, and even in mole nests. Some species are even found in termite nests. The European *Chelifer cancroides* is often found in beehives. South African *Ellingsenius sculpturatus* and *E. fulleri* appear to be restricted to the nests of wild and domesticated honeybees. *E. fulleri* was previously known from South Africa, but has now been discovered in the Middle East, Cyprus and Spain. It may be that human activity introduced this species to Europe. These pseudoscorpions live alongside the bees in their hives, and are sometimes found clinging on to the adult worker bees. Although the honeybees appear to be totally unaffected by these miniature invaders, a researcher called Vachon discovered in 1954 that *Ellingsenius hendrickxi* killed and fed upon bees in captivity. However, the role of *Ellingsenius* in hives is still under discussion.

Very rarely, pseudoscorpions can be found living on mammals. Some species of *Megachernes* live on rats and *Lasiochernes* on moles. It is easy to see how this probably evolved from living in the nests of these creatures. However, until *Chiridiochernes platypedipalpus* was discovered living on a rat in Sulawesi, no pseudoscorpion with specific modifications to this way of life had been discovered. *C. platypedipalpus* has modified long legs enabling easy movement among the coarse hairs of the rat's body. The flattened palps allow easy access around the hairs to the good food source of mites and lice that live on the rat. Even its body is heavily sclerotized with tergites and sternites that fit together tightly. This arrangement is likely to prevent damage by the rat when it scratches its fur, and additionally it would prevent desiccation.

Many species are associated with man-made habitats, and can be found living in houses, within crevices in furniture, for example. The so-called Aristotle's book scorpion, *Cheiridium*

RIGHT The myrmecophilous *Chernes vicinus* lives exclusively in the nests of *Lasius* ants. The pseudoscorpion here has built its white silk chamber in one of the tunnels of the ant nest (Belgium).

museorum (from Europe), seems to favour more academic surroundings, being found occasionally feeding upon psocids (book lice) that munch on the microscopic moulds and mildews binding glue of old books. However, although the species name suggests it is to be found in museums, these are now an unlikely habitat because of the high levels of environmental control in most modern museums. Pseudoscorpions are also found in other types of buildings such as greenhouses, barns, chicken sheds and grain stores, as well as in gardens.

Dispersal – by hitchhiking

When you are only a few millimetres long, with short legs and no wings, dispersal can be somewhat problematic. So how *do* you disperse? Many pseudoscorpions hitchhike a ride on flies and beetles. It is particularly effective with these insects, because they fly, and can carry their tiny passengers some distance. The pseudoscorpion attaches itself by one of its palpal chelae to a leg and can't be easily dislodged. This association is called phoresy.

The European pseudoscorpion species *Lamprochernes nodosus* and *Pselaphochernes scorpioides*, mainly found in dung and compost heaps, often attach themselves to flies that have also visited these odoriferous havens. Wood-boring longhorn beetles are good dispersal agents for pseudoscorpions that live in and under the bark of dead wood, such as the European *Dendrochernes cyrneus*.

Some pseudoscorpions actually benefit their invertebrate carriers. When *Cordylochernes scorpioides*, hitches a ride on a large beetle (such as Cerambicidae) for example, it also helps to rid it of parasitic mites by eating them! Crane flies are also popular as free taxis, a single individual has been observed carrying eight passengers.

Pseudoscorpions don't appear to limit their choices though, because a single species can be found attached to a variety of insects. *Neocheiridium triangulare* uses Lepidoptera to get around. Arachnids are used for transportation too – *Pselaphochernes dubius* regularly uses harvestmen. *Lasiochernes pilosus*, which lives in mole nests, actually uses moles instead of invertebrates for transportation! Interestingly, pseudoscorpions themselves occasionally become 'phoretic hosts' to rhabditoid nematodes. The nematode will attach itself to the pseudoscorpion by its head and be transported along to a new microhabitat.

LEFT Many species of pseudoscorpions disperse by hitchhiking rides from other larger and much more mobile animals, such as this chernetid pseudoscorpion on a small tineid moth (Scaravand, Iran).

General biology and behaviour

Silk

Unlike spiders, pseudoscorpions don't use their silk to capture prey – they usually only use it to produce cocoon-like silken chambers. Silken chambers are important to pseudoscorpions in a number of ways. They are used in moulting, hibernation, giving birth and protection of eggs, and less commonly as retreats.

When a chamber is used as a retreat, the creature often sits with its palps protruding through the opening, so that it can grab potential prey items passing by. Moulting nests are smaller than brood nests, because they are constructed to fit the size of the nymph.

Pseudoscorpions are good little housekeepers, keeping their nests clean by jettisoning any corpses of dead prey organisms. However, they are rather cautious as they don't leave the nest to do this. Pseudoscorpions appear not to have a sense of direction because they normally can't find their way back to their own nest if they leave it to search for a mate or for food! This doesn't bother them though, because they either take over an empty nest (from another pseudoscorpion that clearly needed a compass) or they construct a new one.

Brood nests are made out of the same silk and detritus particles as those built for moulting and hibernation, and are lined with silk. Variations on a theme include the flattened lens-shaped nests of the Cheiridiidae, e.g. *Apocheiridium*, and the Sternophoridae, e.g. *Idiogaryops*, which are constructed without the use of detritus particles.

RIGHT A typical chamber is built in the following way. The pseudoscorpion first finds a good site. It then spins a mat of silken threads that are attached to the ground. In some cases, the beast uses its palps and chelicerae to incorporate small particles of detritus – wood, sand grains or bits of discarded insect remains, for example – into the silk to give support to the walls and to camouflage the chamber. The final result is an igloo-shaped or disc-shaped silken chamber. The extremely small species *Apocheiridium ferum* sits on a disc-shaped silken chamber under bark (Llansa, Spain).

SILK FOR SEX There are examples, however, of silk being used in a much more unusual way. For instance, adult males in the species *Serianus carolinensis* will produce a special kind of silk in their rectal pockets, which is used to lay signal threads marking the way to the spermatophores. Another example, which is even more amazing, is found in the species *Cordylochernes scorpioides*. Individuals crawl under the elytra of the large harlequin beetle *Acrocinus longimanus* in order to disperse from tree to tree. Large adult pseudoscorpion males are very canny, because they use the opportunity to mate with dispersing females while they are still tucked under the beetle's elytra! Males tend to hang around for a couple of weeks on their beetle transport. Such longevity *in situ* throws up the problem of how to stay attached for this prolonged period of time. A male can't hang on by his chelal claw for such a long time, and he also needs both palps for mating. So instead he constructs himself a safety harness out of silk, which attaches him to the beetle's opisthosoma. He can then mate in safety without fear of a mid-flight disembarkation! Pseudoscorpions have also been seen to hang from harlequin beetles by silk threads – it is possible that they may actually use the threads to re-board the beetles.

OVER-WINTERING IN SILK Pseudoscorpions that live in leaf litter avoid the worst of the cold temperatures by migrating further down into the litter when temperatures drop, and coming back to the surface when the weather warms up. By such vertical migration through leaf litter and soil, pseudoscorpions are able to avoid freezing, which would be lethal to them. When they have migrated downwards, nymphs and adults produce silk chambers to help protect against the cold. They sometimes share their shelters with one or two other individuals. They remain in these shelters for a long time when the temperature is low. A female *Chthonius ischnocheles*, for example, can stay in her chamber for up to 250 days. Growth grinds to a halt over winter, and nymphs hibernate as deutonymphs. These beasts are more sensitive to the cold than proto- or tritonymphs, so they tend to stay in their protective silk chambers for longer. Overwintering only occurs in cold regions and is not known for tropical species. It's likely that pseudoscorpions recycle their chambers in the spring, when they use them for moulting.

Food and feeding

FOOD PREFERENCES: FROM LARVAE TO LICE Because of their modest size, pseudoscorpions are often limited to correspondingly small-sized prey. Favourites include small flies, collembolans, mites, ants, beetle larvae, small worms and psocids. Pseudoscorpions with syringe-like mouthparts are able to feast on larger prey because it is not necessary for it to be chewed or carried around. There are some extremely rare records of pseudoscorpions being found on children's heads, and it was thought that they might feed on the eggs (nits) and juveniles of head lice! However, it seems unlikely that they were feeding there.

It has been discovered that pseudoscorpions can tell between animal, plant and sugar extracts. It seems that they use chemoreceptors in the oral cavity rather than surface or contact chemoreceptors. If they grasp unpalatable prey in their chelicerae, they can determine that it is not edible and discard it.

PREY CAPTURE As is the case for most arachnids, pseudoscorpions are active predators. Species in the family Chthoniidae, for example, have a preference for active insects, which they stalk much like a cat with a mouse. Other pseudoscorpions, e.g. species in the Chernetidae,

prefer less active prey, for which they lie in wait, and grab when it is near. Pseudoscorpions can detect the presence of moving prey up to 15 mm (¾ in) away, by picking up the tiniest of air movements using trichobothria on their palps. The prey is then grasped by the palps. Those pseudoscorpions with venom glands paralyze or kill the prey in this way, hold it until it is still, and then pass it to the chelicerae. Those pseudoscorpions without venom glands (the Chthoniidae) pass the prey very quickly to the chelicerae.

FEEDING: MASTICATION AND IMBIBATION There are two different ways that pseudoscorpions feed. Certain families, such as the Neobisiidae, masticate their prey by crushing and pulling the tissues with their chelicerae, whereas more 'advanced' families, such as the Garypidae, rip a small hole, or several holes, in the prey's body wall with their chelicerae. The syringe-like mouthparts are then pressed against the hole, or inserted into it.

In all pseudoscorpions, as with nearly all arachnids, there is extra-intestinal digestion. The salivary glands (called Wassmann's glands) are located in the anterior-ventral area of the prosoma, and open on the upper lip; they secrete fluid into the preoral cavity, along with secretions from the midgut. These fluids are poured onto or into the prey, and the liquefied soup is sucked up into the mouth by the contractions of the large muscle on the upper lip. They are then sucked through the foregut to the midgut, by the action of the pharyngeal pump. Only a ball of undigested parts of the prey remains at the end of the meal. The pseudoscorpion doesn't let its guard down while feeding – it keeps its palps free in case it needs to ward off any potential food thieves.

Lipids and glycogen are stored in the cells of the diverticular midgut and surrounding epithelium. As a result, a pseudoscorpion's opisthosoma can be rather inflated. The animal, so stocked, is then able to go without further food for several weeks – even some months. However, given the opportunity, a food-inflated individual will start to feed again only a few days later, if food is available.

Predators and parasites

There are no specific predators of pseudoscorpions. However, pseudoscorpions occasionally fall prey to creatures such as spiders, mites, ants and ground beetles that live under tree bark or in leaf litter. Mermithid nematodes occasionally parasitize pseudoscorpions. Their effects on the unfortunate victim are fatal, as they ultimately completely occupy the body cavity of the host, and cause all internal tissues (including the gonads) to atrophy (wither away). Fortunately for pseudoscorpions, the infection rate is low. Wasps and mites also occasionally parasitize pseudoscorpions.

Defensive behaviour, aggression and co-operation

Many pseudoscorpions possess an aggressiveness that belies their diminutive size. Those species with well-armoured palps wave them at each other in a conflict situation, and then use them to grasp their opponent. They will usually fight until one gives way, but in some cases the fight may end in the death of one of the opponents! To add insult to injury (literally), cannibalism of the loser might also occur. Not surprisingly, those with slimmer palps are less aggressive, avoiding fights by running backwards and pulling their palps into their bodies.

These solitary beasts sometimes maintain territories through chemoreception. However, in some species, such as the particularly easy-going *Paratemnus elongatus*, there is a high level of

social organization. Co-operative foraging among colony members results in this pseudoscorpion species being able to predate unusually large prey, which could not be captured by an individual. This high level of social organization is possibly linked to interrelatedness among colony members.

Ecological significance

Pseudoscorpions are small, but they play a significant ecological role. Certain species perform a very useful service by being predators of, and therefore regulating, potentially harmful insects and mites. For example, it has been discovered that pseudoscorpions living in beehives feast upon small and sluggish microarthropod fauna that live there, including different species of moth larvae, beetle larvae and mites, such as *Varroa*. Since *Varroa* mites predate honeybees, and are therefore unwanted houseguests in hives, pseudoscorpions are being considered for use in the bee industry as a form of biological control. Pseudoscorpions also play a part in regulating numbers of small soil animals.

Grooming

As for many arachnids, grooming appears to be an important operation. The palpal fingers are in need of regular cleaning, because they have many sensory structures. Pseudoscorpions will often stop what they are doing to clean these important structures particularly after they have touched something with their palps or after they have eaten. To clean its palps, a pseudoscorpion will pull them through the chelicerae, where they are cleaned by the serrulae (see p.273 right). A few genera are more fastidious than the rest, as they clean their whole bodies. For example, *Chthonius* uses a fluid excreted from its preoral cavity, possibly produced by the salivary or coxal glands. The fluid is then wiped over the whole body by the legs and palps.

Coxal spines are found in the superfamilies Feaelloidea and Chthonioidea, and in certain species these spines are used to clean the legs. Such cleaning is assisted by washing fluid, which collects around the spines. The tarsi are pulled against inward-facing bristles that are scattered on the coxal joints. Several legs are cleaned at the same time in various combinations, so that three legs are always on the ground to hold up the creature. After washing is over, the cleaning fluid is ingested.

Moulting

There are two moults during embryonic development and three post-embryonic moults once the nymph has hatched from its egg, before adulthood is reached. When pseudoscorpions moult, they undertake this major task in the relative safety of a silk nest. The moulting chamber may be constructed in a crevice, so it will end up roughly cylindrical. If it is constructed in the open, it is not constrained by enclosing walls, in which case it will end up domed like an igloo. Pseudoscorpions always become torpid (very inactive) before a moult. The torpid individual has an inflated body, outstretched legs and palps, and can't move. Even if disturbed, the animal can't defend itself with its palps, or even walk away. This inactive state may last from a few days up to a week. During this time, the new exoskeleton is forming under the old one. At last the old exoskeleton ruptures at the front of the prosoma, and the animal can escape. The freshly clothed beast is white and soft. During this time, the animal inflates itself to make room for later internal growth, until the exoskeleton hardens and becomes a reddish or yellowish brown. This sclerotization takes several days, after which the animal is able to leave the confines of its moulting chamber.

Life history

Reproduction

SPERMATOPHORE STRUCTURE Apart from a few species (for example, those in the genus *Microbisium*) where parthenogenesis occurs, pseudoscorpions reproduce sexually and transfer their sperm indirectly using spermatophores. Pseudoscorpion spermatophore structure can be divided into four different categories: 1) spermatophores with a simple packet and simple stalk (Neobisiidae, Cheiridiidae); 2) spermatophores with a simple packet and complex stalk (Chthoniidae); 3) spermatophores with a complex packet and a simple stalk (Chernetidae); 4) spermatophores with a complex packet and a complex stalk (Cheliferidae).

Spermatophores that have highly complex structures may be species specific. The complexity of the genital atrium and other reproductive structures is related to the complexity of the spermatophore, as are the different mating behaviours. The simplest spermatophores are found among species that do not mate.

There are various ways in which the spermatophore is transferred from the male to the female, and not all involve courtship mating 'dances'. These methods were described in detail by Peter Weygoldt in his *Biology of Pseudoscorpions* and subsequently further discussed by Gerald Legg in *Pseudoscorpions*, among others.

SPERMATOPHORE TRANSFER WITH NO MATING There are many species of pseudoscorpions that do not indulge in 'foreplay', as they don't have a mating dance, e.g. *Chthonius tetrachelatus* and *Cheiridium museorum*. Not only that but the adult male doesn't even require the presence of a female to encourage him to produce a spermatophore. In a somewhat haphazard and rather wasteful fashion, the male will deposit many spermatophores in an area. The large number increases the likelihood that sperm will eventually be transferred to a female. This method of transfer is only effective in humid environments, because the sperm package dries out fast. The male presses his genital opening on the ground, in order to attach the spermatophore to the substrate. Once his work is done, he wanders off. In some species, the male makes sure that his spermatophores are always fresh by destroying old spermatophores and producing new ones.

Chemoreception plays a great part in pseudoscorpion reproduction. It is particularly important in pseudoscorpions where there is no mating. Instead of being directed to the spermatophore by the male, the female has to find it herself, and this she does by chemoreception. In order to pick up the spermatophore, she steps over it on stretched legs. To rid herself of the empty spermatophore, she rubs her underside on the ground.

NO MATING BUT SPERMATOPHORE FORMATION ONLY IN THE PRESENCE OF FEMALES This method of sperm transfer is slightly less wasteful, as the male only produces spermatophores in the presence of a female. However, it is still not ideal, because he will go into spermatophore production mode even if the female is not receptive. Males of *Serianus carolinensis* (from North Carolina, USA) will help females to locate their spermatophores by producing silk threads that form a path towards them. Males of this species are unique within the order in having abdominal silk glands.

SPERM TRANSFER WITH MATING DANCES Mating between male and female, whether it involves firm body contact or not, is the most efficient way of transferring sperm. Mating

ensures not only that there is a female present, but also that she is receptive and eager to pick up the spermatophore. Thus, there is no wastage of sperm material. The involvement of mating dances is considered to indicate much more highly evolved reproductive behaviour than the simple modes of spermatophore transfer without mating first. Mating dances are only performed by species in the superfamily Cheliferoidea (Atemnidae, Cheliferidae and Chernetidae).

There is a basic pattern running through most mating dances with contact throughout. The male grasps one or both of the female's palpal hands and pulls her backwards and forwards several times. The male gives a display with his palps, forelegs or chelicerae (or with all of them together) and then transfers the spermatophore. The whole phase may take from 15 minutes up to more than an hour.

Of course, there are variations depending on the species involved. In *Lasiochernes pilosus*, the male has a brush on his palpal femur, which the female touches with her palps. It is thought that this has a great stimulation effect, especially as the male is trembling at the time. In *Parachernes litoralis*, the male gently strokes the female's palpal fingers with the serrula exterior on one of his chelicerae during the mating dance. In some species, the male pushes the female backwards and forwards after she has taken up the sperm. In *Dinocheirus tumidus*, the male produces several spermatophores during courtship of an individual female.

In the family Cheliferidae, the male performs mating dances usually without touching the female. However, once he has produced his spermatophore, he firmly assists her in taking it up. An example of mating without firm body contact throughout is demonstrated by *Dactylochelifer latreillii*. The male often initiates his dance by grabbing the female's palpal hands. Then he displays his 'ram's horn' organs (see p.278 top) and starts to vibrate his body. He releases the female at this early stage, but continues his display. Encouraged by his display, the female approaches the male slowly. When she is near to his 'ram's horn' organs, he stops vibrating and steps forward, with palpal hands held high. This encourages her to step backwards and he retreats too. These forward and backward movements will be performed several times. Once the male produces his spermatophore, which takes a swift two minutes, he steps back to allow the female to approach. When she steps over it, he dashes forward to assist her in the uptake. This he does by holding her palps and opening her genital atrium with his forelegs, which have modified tarsi for this very task. During this process, he presses her against the sperm package. Once the uptake is complete, the male and female go their separate ways. The entire process, from the start of the dance until uptake, may take up to 30 minutes.

Egg laying and brood pouches

Pseudoscorpions lay between 3 and 40 eggs, which are fairly large and typically ovoid, or globular, depending on the species. There isn't always a breeding season with pseudoscorpions as the females of some families are receptive throughout the year. However, females in the Chernetidae don't need to be inseminated later in the summer, as they have stored sperm. In certain species, one generation is produced each year, while others are capable of producing two generations.

Most species carry the eggs around in a brood pouch while the eggs fully develop, until the babies hatch and disperse. The brood pouch is attached to the genitalia, and it is produced in a similar way to that of Amblypygida – it is formed from a liquid secretion from

ABOVE A female of the family Chernetidae carrying larvae in a brood sac.

the large accessory glands of the genital atrium, which hardens into shape. The gonopods hold the brood pouch in place while it hardens. The eggs are then laid into the interior of the pouch. Females of *Neobisium muscorum* produce such large brood pouches that they need to lift their opisthosomas in order to carry them! If the mother is disturbed, she will abandon her precious load. Some species, such as *Chthonius tetrachelatus*, don't produce a brood pouch. Instead the eggs are first suspended in fluid at the female's genital atrium and then glued together in a mass. This gluey mass forms a sac-like structure, in which the eggs are embedded.

Pseudoscorpion eggs tend to have limited yolk reserves, so as the embryos develop in the brood pouch, they are further nourished by a milk-like fluid formed by the ovary of the mother and secreted from the genital atrium. In order to absorb nutrients essential for development, the embryos develop a pumping organ, unique to pseudoscorpions. In later moults, this forms the mouth region. In some species, such as *C. tetrachelatus* and *Chelifer cancroides*, the protonymphs attach themselves to the brood pouch as soon as they have hatched and continue to 'suckle' on the nutritive fluid. The mother, in turn, continues to supply them with very large quantities of nutritive fluid. After around four days and a massive increase in size, the protonymphs finally leave the brood nest. It has been observed in some species that once the protonymphs hatch, they sit on their mother's back for a few days until they disperse.

BROOD NESTS In many species of pseudoscorpion, females build little igloos of silk to protect themselves and the developing eggs in their brood pouch. Construction starts a few days before the eggs are laid. A good nest-building site is a small crevice or a depression in the substratum. Usually, each female constructs her own nest. However, in the common European species *Neobisium carcinoides*, females are particularly sociable (or perhaps it's a case of safety in numbers), as a pair may build a nest between them, which they share, sitting side by side. The mother stays in the brood nest with her egg sac without feeding until the protonymphs hatch. Unfortunately, most females discard their sac if their brood nest is opened, but some hardier ones may repair it and continue brooding.

Brooding mothers usually leave their brood nests after the protonymphs have hardened after their first moult, or around the time of the development of the protonymphs into deutonymphs, such as in *Chthonius tetrachelatus*. The youngsters will disperse soon afterwards. To help their protonymphs escape the confines of the silken brood nest, mothers of the species *Cheiridium museorum* rip open the nest allowing the youngsters to emerge. The thrifty mother may eat the brood pouch before she leaves. Some mothers leave the brood pouch in the nest while the embryos develop. Others drop the brood pouch, but still stay with it until the embryos hatch and disperse. A female usually drops the brood pouch after the protonymphs have left.

It has recently been discovered that the species *Paratemnoides nidificator*, which lives in Neotropical savannah, sometimes indulges in matriphagy. In times of food shortage, a mother will leave the brood nest and allow her protonymphs to consume her bodily tissues, without moving. It is possible that this act reduces cannibalism among the offspring.

Post-embryonic development

Pseudoscorpions have a total of six instars – pre-larva, larva, protonymph, deutonymph, tritonymph and adult. In some species, such as *Chthonius tetrachelatus*, it takes a mere ten days from fertilization of the eggs to the hatching of the protonymph, whereas in other species such as *Neobisium carcinoides*, it takes up to five weeks. However, although this is a very short developmental time, once the protonymphs have hatched, they might stay at this nymphal stage for some time before moulting into a deutonymph. Protonymphs may be free-living or remain in the brood nest. By the time pseudoscorpion youngsters reach the deutonymphal stage, they are all free living. It can take from three months to over a year to develop into adults, depending on the species. Interestingly, developmental time varies considerably, depending on the temperature at which the nymphs are kept. For example, the generation time of *Chthonius ischnocheles* at 25°C (77°F) is 90 days, whereas at 15°C (59 °F), it increases to 250 days.

Nymphs look like a generalized version of their parents, but obviously there are certain differences apart from undeveloped genitalia, such as the pale colour of their palps due to a lack of sclerotization. The developmental stages of most species can be determined by the number of trichobothria on the moveable finger of the palp. Additionally, there is a difference in the proportions of particular structures compared to those of the adults, as a result of allometry. Morphological abnormalities are not uncommon in pseudoscorpions. They can be both internal and external. Such aberrations may occur in all developmental stages. The most obvious external deformities include abnormalities of the tergites and sternites (whereby they fuse or split, or are even absent), and alteration in the segmentation of appendages. Other peculiarities are changes in the distribution of setae and trichobothria, and variations in the surface of the exoskeleton.

Longevity

Unfortunately, the longevity of pseudoscorpions is a bit of an unknown quantity. It is safe to say that longevity is probably largely affected by climatic conditions, or – in the case of laboratory animals – by how they are kept in culture. It has been discovered that some species can live up to an impressive five years, cultured *Chelifer cancroides* up to four years, and *Pselaphochernes scorpioides* up to three years. Ageing pseudoscorpions become less agile. They lose their ability to climb, and can't get back up again if they fall on their backs, so they tend to stay at home in their crevices and only venture out for food. However, these hungry little beasts still attempt to capture food even when dying. Old males may fail in holding a territory, and geriatric pseudoscorpions also suffer from a breakdown in bodily immunity, so they are more likely to be attacked by parasites and fungi.

12 Solifugae

Camel spiders, wind spiders, wind scorpions, sun spiders

Solifugids are solitary and often highly aggressive creatures, with huge jaws and an attitude to match! They are unique in possessing fan-like structures, called racquet organs or malleoli, on the ventral surface of their fourth pair of legs, as well as suctorial organs on their palps. Males are unique in having flagella on their chelicerae. They are archaic creatures, which represent an ancient lineage. Solifugids are mainly nocturnal or crepuscular, resulting in their scientific name – 'fugae' means avoidance in Latin and 'soli' refers to the sun. This is rather misleading, however, because there are some diurnal species that are very active under the hot sun – hence their common name 'sun spider'. Solifugids have many other names, such as solpugids, camel spiders, wind spiders, wind scorpions, jerrymanders, jerrymunglums and haircutters.

Solifugids have had a bad reputation among desert-dwelling people for centuries. Stories abound of such creatures invading homes at night and inflicting venomous bites. These are unfounded, although certainly solifugid aggression is a fact – indeed, soldiers stationed in Egypt and the Middle East during the First World War would entertain themselves by staging fights between solifugids, and between solifugids and scorpions. Often, the winner would devour the loser.

OPPOSITE Solifugids are usually very hairy; *Galeodes granti* has a mane of silky hair on its legs.

LEFT The smallest solifugid is *Eusimonia orthoplax*, measuring 80 cm (3¼ in). At the other end of the scale, *Galeodes arabs* and possibly *G. caspius* (shown here, from North Africa, Middle East and Kazakhastan) have been recorded measuring nearly 15 cm (6 in) in leg span.

Classification and diversity

There are currently 1,100 species of solifugids within 141 genera, in 12 families, though this is considered to be a gross underestimate. Solifugids are rather primitive and fairly homogeneous in appearance, across the order. However, there are variations.

At first glance, some species look much like mice, because they are short-legged, with dark hairy bodies and can run fast. There are also differences in size and shape too. For example, species in the Solpugidae are fairly 'elegant' with long legs that have lengthy setae, while species in the Hexisopodidae are short-legged and stocky, with particularly vast jaws. Species in the Ammotrechidae tend to be much smaller.

RIGHT Nocturnal species, such as this *Gluvia dorsalis*, are more muted in colour than diurnal species (Algarve, Portugal).

BELOW Diurnal species of solifugids are sometimes brightly coloured, or variegated, such as *Paragaleodes pallidus* (shown here from Kazakhstan), and *Mummucia variegata* from South America. Such patterning helps to break up the outline of the animal, providing camouflage that protects it against potential predators.

Anatomy

External anatomy

The solifugid exoskeleton is well developed. Not surprisingly, it is thicker in the prosomal region and thinner in the opisthosomal region. Pigment in the exoskeleton is associated with the prosoma and the palps. A solifugid can be covered in numerous spines and sensilla, depending on the species. They also have slit sensilla. Although they do not possess normal trichobothria, they do have long hairs that are sensitive to air currents and this helps them capture prey. The exoskeleton can be covered in very long setae (see p.296).

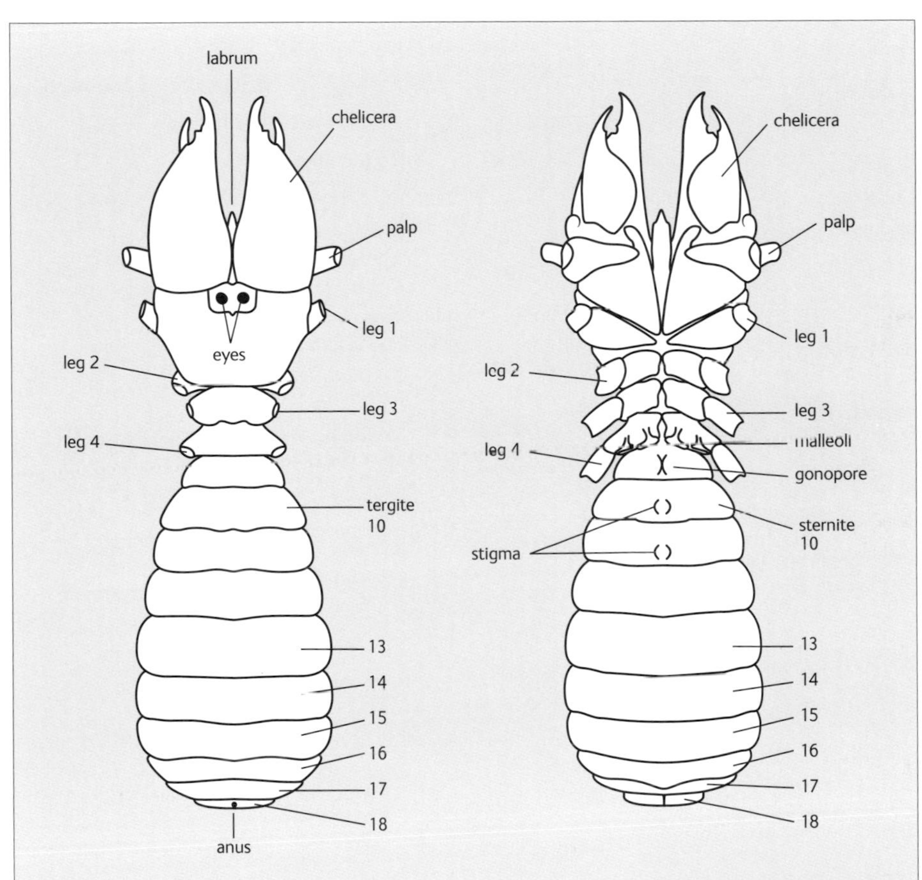

LEFT Generalized dorsal (left) and ventral (right) surfaces of an adult solifugid.

BELOW The prosoma of this male *Chelypus lennoxae* from South Africa has no sternites or sternum, as the immovable leg coxae meet in the centre.

PROSOMA Solifugids are often called camel spiders because many species have a prominent arched structure on the peltidium, which is reminiscent of a camel's hump. The peltidium is shaped in this way to house the muscles that operate the enormous jaws. It is divided into at least three sclerites, called the propeltidium (the hump), the mesopeltidium and the metapeltidium. This arrangement allows flexibility between segments 4–7. This is a primitive feature, as in most other arachnids these prosomal sclerites have fused together, as in scorpions and pseudoscorpions.

Giant jaws

Solifugids are the great white sharks of the arachnid world, as they have enormous jaws. As well as for prey capture and to tear apart and macerate tissues, the chelicerae are used to dig burrows, as well as in threat posturing, fighting and mating. The chelicerae project in front of the prosoma between the palps and are formed of two parts, which taper to a point. The movable lower finger articulates against the fixed upper finger in the vertical plane. Both fingers usually have teeth on their inner edge – the number and location varies depending on the species. It is possible to estimate the age of a solifugid by looking at the teeth on the chelicerae. They can be worn down due to long-term use, and so appear blunted in older individuals. The teeth are often less numerous, smaller and blunter in males. This may be a benefit for both sexes during courtship, as the male is less likely to burst the female's opisthosoma as he massages it with his chelicerae! The inner surfaces of the chelicerae may have stridulatory ridges, depending on the species. When rubbed together, these produce sounds.

BELOW Solifugid chelicerae are substantially larger than those found in any other arachnid orders. Here, *Chelypus lennoxae* from South Africa.

TUBERCULATED EYES Solifugids have a large pair of median eyes and a small vestigial pair on the edges, which are inserted into pits on the peltidium, on the front edge. Like those of Opiliones, solifugid eyes (the median pair) are located on a tubercle. The vestigial eyes have an atrophied lens, but in some species they are capable of detecting movement and differences in light intensity. Solifugids appear to have sensitivity to objects that are moving overhead, an adaptation that may have arisen in response to predation by birds.

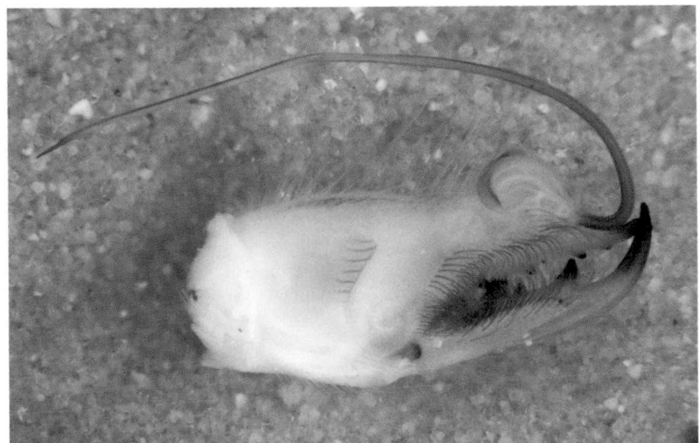

MALE FLAGELLA During the last moult before sexual maturity in males, a flagellum forms on each chelicera, except in the Eremobatidae, which have bristles instead. Flagella can vary in structure considerably between species. In some species, the flagellum sits in a socket-like groove and can be elevated slightly through several positions, while in other species it is immovable as it is fixed at its point of attachment. The function of this structure has yet to be agreed – it has been suggested that it may be used in territorial displays during the mating season, it may temporarily store and release glandular secretions such as pheromones, or it may be used during mating for inserting into the female's gonopore. However, males that have had their flagella removed are still able to mate, and males in the Eremobatidae don't have flagella to begin with, so their function is far from clear.

ABOVE LEFT The flagella forms on the chelicerae and can be extremely long, as in males of *Solpugista bicolor* from Namibia.

ABOVE RIGHT Flagellum of a male *Blossia longipalpis* from Namibia.

PALPS Solifugid palps are leg-like (as in adult female spiders) and elongate. The palps are primarily tactile organs and have numerous sensory hairs on the tips. There are also several hundred receptor hairs on each palp, which are thermoreceptors, olfactoreceptors (for 'smelling') or hygroreceptors (for detecting moisture). As a solifugid moves around, it holds its palps off the ground and extended forward. The palps tap the substrate to locate prey and potential mates. The openings of the coxal glands can be found near the base of each palp.

Solifugid palps are unlike any other arachnid palps because they have an adhesive sucker, called the suctorial organ, on the tip of the tarsus of each one. Each sucker is formed from a cuticular pad that has a complex structure. Basically, there is a layer of very fine, epicuticle overlying numerous branching, stalked structures. This arrangement creates a surface that has many furrows and ridges, providing a large surface area for adhesion. When not in use, it is tightly folded inside the palp and covered by two valves made from cuticle. It can be erected quickly when required for use – contraction of muscles around the organ and an increase in haemolymph pressure help to fan out the organ, ready for action. The name 'suctorial organ' is something of a misnomer in fact, because it is possible that adhesion is achieved through intermolecular electrostatic attractions (van der Waal's forces), rather than by suction, as with scopulae in spiders. It is thought that suctorial organs evolved for prey capture, although they also enable the animal to climb slippery surfaces. Additionally, solifugids sometimes use their suctorial palps to gather up water droplets when drinking.

BELOW In some species, palps can be as long as the entire body, but in others they are much shorter.

ANTENNIFORM LEGS Solifugids do not walk on their first pair of legs, these slender appendages tend to be reduced in size, but can vary in length between species. They are used as sensory structures, and are equivalent to antenniform legs in amblypygids and uropygids. As the animal moves around, the antenniform legs feel the substrate. There may be vestigial claws at the tips in a handful of species.

WALKING LEGS The second, third and fourth pairs of legs are used for locomotion. Each leg has a fixed coxa, complex articulation and lacks a patella, which are all adaptations enabling the animal to run swiftly. Solifugids are classed as cursorial, which means that they are adapted to or specialized for running. The common names, wind spiders and wind scorpions, are very appropriate, as anyone who has ever tried to catch one will testify – they certainly run like the wind! Indeed, many solifugids can run at over 50 cm (6 in) per second, although they cannot keep up this exhausting pace for long. A longer length of leg also greatly helps the animal to attain speed, although in some families (Hexisopodidae) they are comparatively short and stout with strong spines, as they are adapted to rapid burrowing. The legs possess tarsal claws that enable the animal to grip the substrate on which it is moving.

BELOW Some species of solifugid that live on sandy substrates such as *Metasolpuga picta*, have spatulate setae on the second pair of walking legs. These setae are adaptations for digging in sand. The walking legs also function as sensory organs as they have a variety of sensory hairs on the lower leg segments.

Racquet organs

Racquet organs, or malleoli, are unique to solifugids. There are generally ten malleoli in total, located on the ventral side of the fourth pair of legs of both sexes. The length of the stalk varies between species and is shorter on shorter-legged creatures.

The malleoli are chemosensory, and they probe the substrate beneath as the solifugid walks along, to detect prey and potential mates. Interestingly, there is some flexibility in the malleoli to allow this probing movement, because there are joints at the base of the fan, and at the base of the stalk. Each malleolus is 'wired up' to a solifugid's sensory system, by a network of nerves. The nerves of each malleolus are bunched into a sensory ganglion, just above the point where the stalk attaches to the leg. Nerves run down the length of the stalk and through the fan, terminating in a small opening within a groove on the ventral edge of the fan. The opening exposes the nerve tips to the elements, much in the same way as the chemosensory hairs of a spider. Malleoli are considered to be equivalent to pectines on scorpions.

BELOW Malleoli are fan-shaped structures on stalks, which stand out vertically from the exoskeleton, like flattened miniature mushrooms.

RESPIRATORY ORGANS Solifugids have an extensive tracheal system, which opens to the outside of the body through three pairs of spiracles – one pair is located behind the coxae of the second pair of legs (stigmata on the prosoma are unique to the order), one pair is on the second opisthosomal segment, and the third pair is on the third opisthosomal segment. There may also be an additional spiracle on the posterior part of the opisthosoma. The spiracles possess muscles that can open and close them.

In general, solifugids have a similar aerobic capacity and an almost identical gas exchange system to that of insects. This means that they are able to have sustained aerobic activity while searching around for food, and are capable of living in hypoxic (low oxygen) environments. In fact, their movements are characterized by great bursts of speed, followed by sessions of sustained walking, in contrast to other arachnids, which tire more quickly. This suggests that solifugids have a more advanced respiratory system. It has been suggested that the gas exchange control system of solifugids is highly convergent with that of insects.

OPISTHOSOMA The opisthosoma is broadly joined to the prosoma. It has ten segments, which may decrease in size towards the anus (see p.295 bottom). Each segment consists of a dorsal tergite and a ventral sternite; each segment is separated by a wide region of soft intersegmental membranes. These membranes allow the opisthosoma to have great flexibility – it can swell after feeding and during egg production. The membranes also enable the animal to move easily. The anus is usually at the end of the opisthosoma, although in some species it is located on the ventral surface.

EXTERNAL GENITALIA A solifugid's genitalia open to the outside world through a longitudinal slit called the genital aperture, on the ventral side of the first segment of the opisthosoma (see p.293 right). The genital aperture is covered by a pair of genital opercula that are very similar in both sexes.

SEXUAL DIMORPHISM Sexual dimorphism in solifugids can be quite obvious or rather subtle, depending on the characters involved: 1) as in nearly all arachnids, the males are smaller in body length than females; 2) males have longer legs and a more slender opisthosoma, and usually narrower chelicerae and fewer teeth; 3) males of many species have flagella on their chelicerae, which females never possess, and they vary in form between species (see p.295 top left and right); 4) the racquet organs are broader and more prominent in males.

Internal anatomy

CENTRAL NERVOUS SYSTEM There is very little known about the solifugid nervous system. However, it has been established that there is a 'brain' (the dorsal cerebral ganglion) that is fused to the underlying suboesophageal ganglion, located in the prosoma. Various nerves run off the ganglionic mass. The dorsal cerebral ganglion innervates the chelicerae, the mouthparts and the eyes. The suboesophageal ganglion innervates the palps, walking legs, racquet organs and the genitalia. An opisthosomal nerve leads to the opisthosomal ganglion. Various other minor nerves connect to other structures, such as the sensory hairs. Solifugids have shown a form of association learning. For example, they stop their cursorial foraging in order to catch prey, at artificial lighting. They apparently associate these lights with the presence of insect prey. In captivity, it takes only a week for them to learn to feed from forceps.

CIRCULATORY SYSTEM There is a lack of information on cardiac function and circulation in solifugids. However, it is known that they have a tubular, elongated heart with eight pairs of ostia. Unlike most arachnids, the heart is located in both the opisthosoma and prosoma, and runs from segments 4 to 13. There are two pairs of ostia in the prosoma and six pairs in the opisthosoma.

RESPIRATORY SYSTEM A solifugid's extensive tracheal system resembles the interconnecting pipes of a plumbing system. Solifugids have a greater elaboration of tracheae than any other arachnid, and the tubes are also wider than in other orders. From each of the spiracles in the prosoma, a series of tubes branches out. These tubes run to the anterior as well as the posterior, thereby providing a ready supply of oxygen to all the extremities. The tracheae in the prosoma also connect up with those in the opisthosoma, which are extensively branched. In general, there are two main tracheae that run through the opisthosoma from front to back – these connect with the pericardial trachea (tube leading from the vicinity of the heart).

DIGESTIVE SYSTEM Food is macerated and liquefied by the chelicerae and mouth, and then sucked into the long tubular oesophagus, which runs the length of the beak. A section of the oesophagus located in front of the ganglionic mass has muscles that contract and distend the tube, thereby acting as a pump. The length of the oesophagus has shortened, as in some other arachnids, such as spiders, so the sucking chamber is located behind the brain. The oesophagus runs through the ganglionic mass, bends in an S-shape and then connects to the midgut. The midgut has several diverticula branching off into the prosoma, which store ingested nutrients – in *Galeodes* there are four pairs. The opisthosomal diverticula are much more developed. These extensive 'larders' allow a solifugid to survive without food for long periods of time. The midgut runs in a straight line to the hindgut, which has a specialized enlarged posterior section called the stercoral pocket. The final absorption of water from the fecal matter takes place within the complex folded walls of this pocket, allowing the solifugid to conserve valuable moisture.

EXCRETORY SYSTEM The solifugid excretory system is a modification of the basic design found in other arachnids, with the usual coxal glands and Malpighian tubules. However, there are a few differences. For example, there is a mucus segment connecting the sacculus (a derivative of the coelom) with a long tubule that receives nitrogenous waste. Another difference is that the excretory pore of the coxal gland is located near the coxae of the palps, whereas it is found close to other coxae in spiders, scorpions and pseudoscorpions. In solifugids, the Malpighian tubules are attached to the stercoral pocket, while in other arachnids they are usually attached between the midgut and hindgut. The anus is found on the terminal segment. Waste matter can be ejected forcibly, like water from a high-pressure hose.

INTERNAL GENITALIA The internal genitalia of both male and female are similar to those of other arachnids. A female solifugid has a pair of elongate ovaries, lying longitudinally in the opisthosoma. The shape and length of the ovaries vary between species. For example, in some species they fold back on themselves and are rather truncated – ending in the third segment.

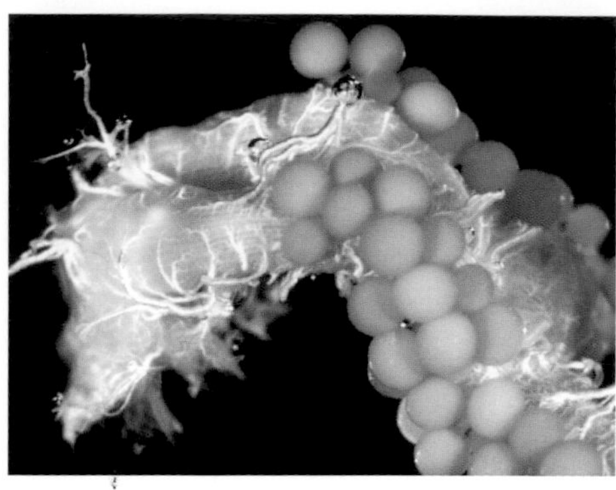

ABOVE It appears that the eggs develop on the outer walls of the ovary like grapes attached to a vine, but the eggs are actually covered by the delicate wall of the uterus, as in this *Galeodes* sp.

In other species, however, they have minimal coiling, are quite elongate and end in the seventh segment. In all species, the anterior ends of the ovaries form oviducts, which merge into the genital chamber (atrium). The genital chamber exits the body through the genital aperture.

A male solifugid's testes consist of long seminiferous tubules that are elaborately intertwined like a plate of spaghetti. They surround the Malpighian tubules and the diverticula of the midgut. The seminiferous tubules merge into the genital chamber, which merges into the vas deferens. The whole system terminates in a common duct that exits through the genital aperture.

Distribution and habitats

Most solifugids are found in tropical and subtropical desert regions in the Americas, Europe, Asia and Africa. Three families – the Eremobatidae, Mummuciidae and Ammotrechidae – are only found in the New World (the western hemisphere including North and South America), while eight families are restricted to the Old World (the eastern hemisphere including Europe, Asia and Africa). The Daesiidae is the only family to be found in both New and Old Worlds, with a range in Asia, Africa, Chile and Argentina. The Galeodidae and Hexisopodidae are found in Africa, the Rhagodidae in India, the Ammotrechidae in the West Indies, and the Eremobatidae in North America. Curiously, solifugids are not found in Australia, even though there are deserts there. They are also not found in Madagascar.

Solifugids are usually restricted to arid habitats. As with scorpions, their diversity is greatest in deserts. In fact, because solifugids are one of the dominant arthropod predators in arid ecosystems, they are considered to be indicator organisms for desert environments. They live in sand dunes and desert scrub, rocky hillsides and gravel plains, and many species can be found under decaying vegetation, beneath rocks, in crevices, or in burrows. Substrate seems to be the main factor in determining habitat preferences. Certain species are restricted to particular microhabitats, e.g. *Hemiblossia* species are found under stones on slopes with

RIGHT Solifugids (here from Namibia) are most diverse in southern Africa and southwest USA. Because most species are found in Africa, is has been suggested that they may have originated there.

bushes but no grasses, while *Blossiola* species are restricted to grassy, rocky slopes. Most solifugids avoid open areas with no vegetation cover, as vegetation provides protection from possible predation, as well as sites for prey capture. A few species inhabit surprising places for creatures generally at home in the desert, e.g. *Dinorhax rostrumpsittaci* (from Southeast Asia) is found on the edge of the rainforest, and *Gylippus rickmersi* is found at 3000 m (10,000 ft) in mountains north of the Hindu Kush, in Central Asia.

General biology and behaviour

Venom: are solifugids venomous?

For many years, solifugids have been reported in several countries as being venomous (such as a species of *Eremobates* in Mexico). There have also been unusual remedies for 'venomous' bites, such as one used in Azerbaijan, where people boiled solifugids in oil and then used the oil to rub on a solifugid bite to neutralize the so-called venom. In fact, although they can be aggressive, solifugids do not appear to produce venom. The symptoms of some people's reactions to their bites, such as localized pain and swelling, are much more likely to be caused by tissue damage and bacterial infection of the wound inflicted by the chelicerae.

Food preferences

Solifugids are unarguably carnivorous. Some species often eat a wide range of prey e.g. *Eremorhax striatus*. However, across the order as a whole, the usual choice of food items depends on the family (or genus) in question. In experiments with specimens of *Eremobates mormonus*, it was discovered that these solifugids had a much higher capture rate of small prey items and tended to avoid larger prey. Also, some solifugids may reject prey items if they produce defensive chemicals, or if they are too highly sclerotized or too large.

Although usually generalist, a few species have restricted preferences. For example, a species of *Eremobates* (from Colorado) feeds on bees, while *Chelypus hirsti* (from Africa) and *Eremobates durangonus* (from North America) specialize on termites. Other species are associated with ants. The species *Eremobates pallipes* (from Colorado) has a taste in food that is potentially useful to humans – it has been reported to feed extensively on the bed bug *Cimex lectularius*. Solifugids are insatiable feeders – adult females quite regularly stuff themselves so full of food that they find it difficult to get around. Such greed is not limited to adults – a miniscule (5 mm, or ¼ in, long) immature specimen of *Galeodes* was seen to cram down 100 flies in just 24 hours!

BELOW Solifugids are opportunistic predators that normally prey on insects. However, some feed on anything they can overpower, including earthworms, centipedes, spiders, scorpions, each other, snakes, small lizards, and even rodents. Some are skilled enough to capture birds. This *Galeodes* sp. is feeding on newborn rodent prey (North Africa).

Prey capture

FAST AND FURIOUS FORAGING Solifugids are usually nocturnal or crepuscular hunters, but some are diurnal. They are extremely active hunters with some species covering large distances in search of prey. Their normal foraging methodology is very distinctive. They run vigorously in a wide zigzag pattern that is apparently random. They stop regularly, possibly to detect prey underground. Solifugids extend their palps in front of their body (in the same way as scorpions do), and tap them regularly on the ground to locate prey. They search thoroughly among detritus, clumps of grass, burrows and stones for food – they also dig if food is detected below ground.

There are some exceptions to this usual foraging strategy. Some species stalk their prey, e.g. *Galeodes arabs* stalks flies, and *Eremorhax magnus* stalks beetles. Some forage beneath the ground, such as the Hexisopodidae (see pp.293 bottom and 294), and some are more sedentary, such as termite feeding solifugids. Solifugids usually move over the ground in search of prey, but some species are extremely good climbers. They are able to climb on vegetation such as bushes and trees, and also scale the walls of buildings. In fact, they can be quite athletic in their quest for food.

Some workers consider that solifugids are not as good as other wandering arachnids (i.e. spiders and scorpions) at capturing prey, as their method of prey capture is usually very hit-or-miss. However, some solifugids congregate at lights, where flying insects are attracted, and wait for the prey to come to them. This could be seen as being similar to a wandering spider's sit-and-wait strategy, which appears to be the more efficient hunting method.

LOCATION OF PREY Solifugid use various combinations of mechanoreceptive, visual and chemical senses in prey location, depending on the species. Species such as *Eremobates palisetulosus* are able to locate live prey beneath the ground by detecting substrate vibrations with the mechanoreceptors on their legs and palps, as well as chemoreception by the malleoli. They dig out the prey with their chelicerae. The South African *Solpugyla globicornis* uses mechanoreception too. Solifugids respond quickly on contact of the prey with the legs and palps. Some solifugids orientate towards substrate vibrations, but only attack on contact with the prey. Some seem to locate prey using only tactile cues e.g. *Eremobates durangonus*. *Galeodes granti* uses both tactile and visual stimuli to capture prey and has often been seen capturing prey near artificial lights.

When a solifugid has located its prey, it moves very fast indeed to capture it. The more hungry the animal is, the higher the capture rate. Not surprisingly, solifugids tend to ingest their prey more quickly when they are more hungry. Depending on the relative size and type of the prey, the solifugid may either first grip the prey with the suctorial organs on the palps and then pull it to the chelicerae, or grab the prey using the chelicerae straightaway. Sometimes, the chelicerae stab the prey when it is first caught. Once prey has been seized, a solifugid tends to stop running. This may be because it needs to prepare the prey item and to concentrate on feeding.

Feeding and digestion

The enormous chelicerae are used to macerate the food. They move backwards and forwards alternately, pulling at the prey's tissues. The tissues are ground to a pulp between the inner surfaces of the chelicerae, as the teeth grind together. This process has been called the 'cheliceral

mill'. A feeding solifugid can be heard crunching loudly on the hapless prey's exoskeleton! If the prey is elongate (such as millipede), it is held crossways in the jaws, and fed through the cheliceral mill from one end to the other, like laundry through a mangle. Prey objects with shorter and stouter bodies tend to be rotated through the cheliceral mill. Large organisms may be chewed apart at the intersegmental membranes. Smaller solifugids use their chelicerae to scoop out the soft parts from the hard exoskeleton. Larger solifugids, however, often crush the prey totally.

Solifugids may be voracious predators, but they do not necessarily consume every part of their prey. Depending on the prey type, they may sever and discard body parts that have a high level of chitin, such as the head, wings and legs. This is called prey preparation and allows the animal to target those parts of the prey containing the most nutrients. In acridid grasshoppers, even the intestinal tract may be discarded. In fact, solifugids seem to have preferences for specific body parts and tissues. The more meaty hind legs of grasshoppers are usually severed and then put through the cheliceral mill several times, until all of the tissues have been ingested.

LEFT A solifugid using its chelicerae to tear at a beetle (Namibia).

A solifugid has a beak made up of the rigid dorsal labrum and ventral labium fused together along their edges. The mouth is at the tip of the beak and is elongated and compressed. The beak projects forward between the bases of the palps and is ventral to the chelicerae. The action of the chelicerae enables the beak to penetrate the tissues. As the solifugid grinds its food, there is a rhythmic pumping of the pharynx, which sucks the liquefied food into the gut. The dorsal labrum has rows of setae that filter the partly liquefied food.

Predators and parasites

These ferocious beasts are not without their own predators, however. In the arid ecosystems where they flourish, they are an important source of food for other animals that live there. A range of animals feast upon solifugids including insects, spiders, scorpions, other solifugids, reptiles (e.g. geckos), birds (e.g. owls and raptors) and small mammals (e.g. bat-eared fox and black-backed jackal). Solifugid eggs and nymphs are eaten by small insects, such as ants and silverfish. Large solifugids, e.g. *Solpuga hostilis* from South Africa, satisfy birds such as buzzards, shrikes, owls and roadrunners, as well as mongoose, meerkats, and even foxes. Parasitoid wasps also use solifugids as hosts for their young.

Defensive behaviour and aggression

A typical threat display involves the solifugid holding open and moving its jaws, while lifting its palps and antenniform legs aloft. Solifugids also sway on their legs, which is especially apparent in long-legged species. They have also been known to raise their opisthosoma to an almost vertical position – it is suggested that this helps to protect the opisthosoma, and it may also be a form of scorpion mimicry, though this has not been proved. Old World solifugids also have the ability to stridulate when provoked, which may deter attackers. If disturbed, a solifugid often demonstrates displacement activity, such as furious digging. A threat display is usually followed by flight or, more frequently, fight.

Anyone who has ever encountered solifugids will attest to their aggressive nature! People have reported being chased by the ferocious invertebrates, and having had to fend them off. In an attack, a solifugid moves forward and bites with its huge jaws. Such hostility is clearly reflected in the scientific name of the South African species *Solpuga hostilis*. Solifugid aggression is not related to size – even small species can have real attitude. However, the level of aggression depends on the species, e.g. North American species are less aggressive and much more likely to run away. In captivity, solifugids become tame quickly and lose their aggression, unless startled.

There are many reports of solifugids fighting scorpions. In cases where they win, solifugids have been known to sever or crush the scorpion's 'tail', thus immobilizing the sting. However, it is usually the scorpion that wins. Certainly, fights between solifugids are much more evenly matched. Some conflicts are over before they begin, because one animal runs away; if a fight does takes place, the winner is the one to grab the other's opisthosoma. The winner may eat the loser, or carry its body around in its jaws, before discarding it.

Such aggression is not present in the first and second instars – they are in fact gregarious. Indeed, second instar siblings have shown communal feeding. Additionally, siblings may gather together in a loose huddle. If stimulated mildly, the group may move away together, but a stronger stimulus will result in the group breaking up, each individual going its own way.

Adaptations to extreme physical conditions

Solifugids, like scorpions, are highly adapted to life in arid habitats, even though they are subjected to very high daytime temperatures, low temperatures at night and low relative humidity. Their adaptations include being nocturnal, making burrows, being tolerant of extremes of temperature, behavioural thermoregulation and conservation of water. Being nocturnal is a very good way of avoiding extreme daytime temperatures, and the majority of solifugids have adapted to this way of life. Many larger species are nocturnal and spend the day deep inside their burrows or under rocks to avoid high temperatures and maintain a higher level of humidity.

Several smaller species are diurnal, and are active in the hotter parts of the day – hence their common name, sun spiders. For example, *Galeodes fatalis* runs around in hot sunshine in the middle of the day. Such diurnal behaviour is rare in other arachnids, apart from the Araneae. There are several diurnal species that live in Africa. These have an amazing capacity to withstand great differences in temperature, and many are surprisingly tolerant of drought and heat. For example, *Galeodes granti* can survive for at least 24 hours at 50°C (122°F) at low humidity (below 10%). Others are active up to a scorching 61°C (142°F) at ground level. At the other extreme, *Eremobates palpisetulosus* can survive temperatures as low as 3.3°C (37.9°F). This capacity to survive a large drop in temperature is an adaptation to cold desert nights. Optimal temperatures are in the range 20–35°C (68–95°F), with low-elevation animals preferring 24–27°C, and those living at higher elevations preferring 21–24°C (70–75°F). Solifugids are also tolerant of submersion in water, which allows them to survive flash floods and very heavy rain.

Diurnal species shelter in the shade when it is available. They have been observed to raise their opisthosoma off the substrate, which is probably a way of reducing heat gain. Like other arthropods that live in arid regions, solifugids may possess morphological adaptations to their hot habitats, such as a thicker exoskeleton. However, experiments on *Eremobates palpisetulosus* have shown that this species does not have a good cuticular barrier to evaporative water loss, which *is* found in scorpions from the same desert habitat. This is probably because *Eremobates palpisetulosus* is nocturnal and is thus active when the humidity is higher. Water is normally obtained from prey, but if a solifugid is particularly thirsty it may drink from free water. They also have a physiological adaptation – the stercoral pocket – that removes water from fecal material, so that water is conserved, and there is a concentration of nitrogenous waste.

Burrowing

Having a burrow is very important to a solifugid. A burrow has many functions, including protection from predators, extreme temperatures and low humidity, and provision of a refuge while moulting, digesting large meals, egg laying and hibernating.

DEPRESSIONS, CREVICES AND HOLES Many species construct burrows, and might spend nine months of the year inside. However, not all solifugids dig themselves a burrow. Some species hide away in small depressions under rocks, inside rock crevices, or under decaying vegetation. Some species may take over abandoned burrows. In the winter months in desert regions such as the Sudan, juvenile solifugids probably hide themselves away in such places as termite nests, where they can obtain plenty of food at a time when the desert itself is fairly devoid of life.

BURROW CONSTRUCTION Burrows are effectively of two different types. Those constructed beneath rocks and other forms of cover are called depression burrows, and are bowl-shaped and shallow. These tend to be used as overnight resting places. They are dug rather rapidly, taking only around seven minutes to construct. The other type is a tube burrow, which is cylindrical and between 4 cm and 30 cm (1½ in and 11¾ in) deep. Depth and angle vary depending on the substrate, and the age and size of the animal digging the burrow, as well as its sex. Egg-laying burrows may be deeper than those used for shelter, and these may take up to an hour to construct. A species may construct either type of burrow, depending on the circumstances. Burrows may extend for several metres, such as those constructed by *Galeodes granti*. The advantage of such as long burrow is that the solifugid can escape the heat by going into the depths.

In most solifugid species, the chelicerae are used to bite at the substrate to loosen it. Like a dog burying a bone, the solifugid kicks the soil out from under its opisthosoma as it digs, and uses the chelicerae to 'plough' the dug-out substrate from the burrow. The palps can be used to assist the chelicerae by flattening the removed substrate. Solifugids may also remove the substrate away from the excavation site. For example, an individual of *M. picta* was observed carrying large pebbles up to five times its body weight, away from its burrow entrance! In many cases, solifugids plug the burrow entrance with excavated sand, or dead leaves, while this is less common in old burrows that have been taken over. Not surprisingly, burrows used for egg laying are much more thoroughly plugged than those used just for shelter. Some individuals may dig 40 or more burrows in their short lifetime.

Life history

Unfortunately, the life history of most solifugid species is unknown. However, studies have been undertaken on a few species from three families, showing that there are distinctive differences between those species in the New World and in the Old World.

Reproduction

SEARCHING FOR A MATE Adult males spend most of their above-ground activity searching for potential mates. They may travel very long distances during their quest for a female. In *Metasolpuga picta*, the male searches for the female in straight-lines, not in zigzag runs as would be associated with food searching. The male stops at intervals to carry out intensive searches. The female attracts the male (possibly by emitting pheromones from her burrow), and the male then alters his searching strategy by moving in a circular pattern in order to home-in on the female. The male locates the female's burrow and starts to dig, at which point the female comes up to meet the male.

COURTSHIP There are variations in mating behaviour between different species, even those within the same genus. Courtship in *Galeodes caspius*, *G. sulfuripes* (Old World), and *Eremobates* species (New World) all in the family Galeodidae, are compared below.
Galeodes caspius: The male lulls the female into a torpor by stroking her with his palps – this prevents attack by the female. Next, the male grabs her rapidly with his chelicerae, carries her along and then puts her on the ground. This grabbing also induces a submissive state. The male massages the female's opisthosoma by chewing with his chelicerae.

Galeodes sulfuripes: The male displays to the female by raising and lowering his palps, and the female goes into a threat display in response. However, the female needs only to touch the male's palps with her own to go into a torpor. As the female falls to the ground, her opisthosoma is twisted towards her prosoma. As with *G. caspius*, the male massages the female's opisthosoma by chewing with his chelicerae.

Eremobates spp.: In the initial phase, the male, female or both demonstrate a threat display, which goes further than that in the galeodids, because they rock backwards and forwards on their legs. The male strokes the female with his palps. The receptive female drops to the ground in a submissive position, with prosoma bent back over the opisthosoma, closed chelicerae and relaxed legs. The male grasps the female behind the propeltidium and turns her on her back (in *E. duranonus*), or on her side (in *E. palpisetulosus*). He inserts the fixed finger of his chelicerae into the female's genital opening. The female is then put back in walking position, with opisthosoma still bent backwards. The male massages the female's opisthosoma by chewing with his chelicerae. If the female tries to escape at this stage, the male strokes her with his first pair of legs and palps until she calms down.

SPERM TRANSFER Sperm transfer in solifugids is a little different to that of pseudoscorpions and scorpions, because a solifugid's spermatophore is not fixed to the ground with a stalk. In the Solpugidae and the Galeodidae (Old World), the male massages the female with his chelicerae while he is releasing a spherical spermatophore from his genital opening. It may be deposited either onto the substrate, or onto the female's dorsal surface (as in *Metasolpuga picta*). It is thought that placing it onto the female's opisthosoma instead of the ground prevents it drying out. He stops massaging her, clutches the spermatophore in his chelicerae and places it into the female's genital opening.

In *G. sulfuripes* (Galeodidae) the female lies on her side while he places his chelicerae right into the genital opening, whereas in *G. caspius* she lies on her back. He does this while producing sperm. He catches the sperm with his chelicerae as it emerges from his genital opening, by twisting his body. He then places the spermatophore into the female's genital aperture, and then massages the area briefly – this is thought to ensure that the sperm is released from the spermatophore whilst inside the female. The male makes a quick getaway before the female recovers.

In the Eremobatidae (New World), the male transfers a droplet of seminal fluid onto the female's genital opening directly from his own. This appears to be pushed in by chewing movements of the chelicerae. If the female is trying to escape at this stage, she is calmed down by stroking. When the male has finished, the female might recover very quickly and run away, they may fight, or they may both run away.

Where females are unreceptive to males, death can occur for either sex, although it is much more likely to be the male that suffers. However, some males are able to escape a fatal attack by autotomising a leg.

Egg laying and maternal care

GRAVID AND GREEDY A female carrying fertilized eggs is even more ravenous than usual, and increases her food intake. However, she stops feeding altogether for up to five days before egg laying, while she busies herself with excavating a burrow, finding an empty burrow, or setting up a nest in rock crevices.

RIGHT Shortly before oviposition, the eggs can be seen easily through the exoskeleton of the distended opisthosoma of *Galeodes granti*.

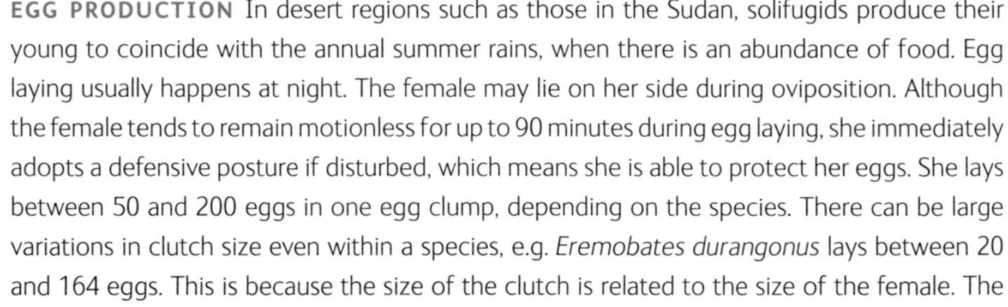

EGG PRODUCTION In desert regions such as those in the Sudan, solifugids produce their young to coincide with the annual summer rains, when there is an abundance of food. Egg laying usually happens at night. The female may lie on her side during oviposition. Although the female tends to remain motionless for up to 90 minutes during egg laying, she immediately adopts a defensive posture if disturbed, which means she is able to protect her eggs. She lays between 50 and 200 eggs in one egg clump, depending on the species. There can be large variations in clutch size even within a species, e.g. *Eremobates durangonus* lays between 20 and 164 eggs. This is because the size of the clutch is related to the size of the female. The

longest eggs in the largest clutch size are laid by one of the largest species, *Solpuga caffra* (from Africa).

Some solifugids (Eremobatidae) are able to lay more than one batch in a season, and a few individuals of *Eremobates durangonus* have been observed to lay five batches! Interestingly, the number of eggs per batch increased slightly each time, though none of the eggs actually hatched from the fourth and fifth batches. Being able to produce more than one batch of eggs should increase reproductive success, though this may not be the case in this example. However, these observations were made on laboratory animals, so things may be different under natural conditions.

BELOW Solifugid eggs are usually oval in shape, cream, yellow or light brown in colour, with a leathery surface that is pitted or covered in truncated papillae (small nipple-like projections).

GUARDING THE BROOD The degree of protection of the young varies among solifugid species. Some genera, such as the North American *Eremobates* and *Ammotrechella*, lay their eggs at the base of a deep burrow. They then plug and camouflage the burrow entrance and disappear, never to be seen again. However, the burrow provides the eggs with a relatively consistent microclimate, and the plugging protects them from predation. Some species show developed maternal care, with the female remaining with her eggs until they hatch and moult from the post-embryo into the first instar nymphs. Female solpugid and galeodid species guard the young in the burrow, valiantly attacking any would-be invaders. Some *Eremobates* species, such as *E. marathoni*, also guard their offspring. Such maternal care results in a significant increase in nymphal survival rates.

Post-embryonic development

The various stages in a solifugid's life-cycle are egg, post-embryo (larva), up to 8–10 nymphal instars, and adult. They are chiefly univoltine – that is, one generation reaches maturity each year. The eggs take around three to four weeks to hatch. Post-embryos vary in appearance between species. They may have poorly developed appendages, or the palps and legs may be well developed but have no sensory hairs at this stage. However, all are pale in colour, and essentially immobile; they are only able to wriggle slightly. The chelicerae are not fully developed, and the malleoli are not yet visible. It takes just over one week before they moult into the first instar nymphs, which are non-feeding but much more like the adults in appearance. They are quite active at this stage. Another week later they moult into second instar nymphs capable of feeding and burrowing.

Before moulting, a solifugid fasts for about a month, then gets into the distinctive moulting position, with legs and palps facing upwards and to the rear. It takes around one week for the exoskeleton to harden, during which time the solifugid does not eat. Post-embryos and first instar nymphs are gregarious, enabling them to stay within the safety of the oviposition chamber without eating each other. After moulting into the second instar stage, however, their social nature changes considerably, and they become cannibalistic. This when they disperse from the nest. From this point onwards, they lead a relatively solitary existence. Males possibly have one fewer instars than females, and reach adulthood slightly earlier than females.

Longevity

Males live a shorter life than females, have a relatively short mating period, and die shortly after mating. This might be because they eat a lot less and use a lot more energy while they are looking for a mate. Females live slightly longer than males, perhaps because after mating the eggs need time to develop. Solifugids are rather short-lived animals – one year is usually quoted as the maximum lifespan. However, some individuals kept in captivity have survived for up to two years.

References

References relating to specific orders are listed under that chapter reference list.

CHAPTER 1

Cohen, A.C., 1995, Extra-oral digestion in predaceous terrestrial arthropods, *Ann R Entom*, 40: 85–103.

Dunlop, J.A., 1996, Systematics of the fossil arachnids. *Rev Suisse Zool*, vol. hors série, 173–184.

Giribet, G. & Kury, A.B., 2007, Phylogeny and biogeography, in Pinto-da-Rocha, R., Machado, G. & Giribet, G. (eds), *Harvestmen: The Biology of Opiliones*, Harvard University Press, 597 pp.

Giribet, G., Edgecombe, G.D., Wheeler, W.C. & Babbitt, C., 2002, Phylogeny and Systematic Position of Opiliones: A combined analysis of chelicerate relationships using morphological and molecular data, *CLADEC*, 18: 5–70.

Grzimek, B., Schlager, N., Olendorf, D. & McDade, M.C., 2003, *Grzimek's Animal Life Encyclopedia*, vol. 2, Gale Publishing.

Harvey, M.S., 2002, The neglected cousins: what do we know about the smaller arachnid orders? *J Arachnol*, 30: 357–372.

Proctor, H.C., 1998, Indirect sperm transfer in arthropods: Behavioural and Evolutionary trends, *Ann R Entom*, 43: 153–174.

Punzo, F., 1998, *The Biology of Camel-Spiders (Arachnida, Solifugae)*, Kluwer Academic Publishing, 301 pp.

Savory, T.H., 1964, *Arachnida*, Academic Press, London and New York, 291 pp.

Scharff, N. & Enghoff, H., 2005, *Arachnida*, Zoological Museum, University of Copenhagen (unpublished).

Selden, P.A., 1993, Fossil arachnids: recent advances and future prospects, *Qld Mus Mem*, 33 (2): 389–400.

Shultz, J.W., 2000, Skeletomuscular anatomy of the harvestman Leiobunum aldrichi (Weed, 1893) (Arachnida: Opiliones) and its evolutionary significance, *Zool J Linn Soc*, 128: 401–438.

Shultz, J.W., 1990, Evolutionary morphology and phylogeny of Arachnida, *CLADEC*, 6: 1–38.

van der Hammen, L., 1989, *An Introduction to Comparative Arachnology*, SPB Academic Publishing, 576 pp.

van der Hammen, L., 1985, Comparative studies in Chelicerata III, Opilionida, *Zool Verh*, 220: 1–60.

Walter, D. & Proctor, H., 1999, *Mites: Ecology, Evolution and Behaviour*. CABI Publishing, 322 pp.

Weygoldt, P. & Paulus, H.F., 1979, Untersuchungen zur Morphologie, Taxonomie und Phylogenie der Chelicerata, *Z Zool Syst Evol*, 17: 85–116, 177–200.

Wheeler, W.C., & Hayashi, C.Y., 1998, The phylogeny of the extant chelicerate orders, *CLADEC*, 14: 173–192.

WEBSITES

Scorpion Systematics Research Group
http://scorpion.amnh.org

The Arachnids: Systematics
http://www.members.tripod.com/Spinnenman/ArachnEigensch.htm

The Scorpion Files
http://www.ub.ntnu.no/scorpion-files

CHAPTER 2 – ARANEAE

Aitchison, C.W., 1978, Spiders active under snow in southern Canada, *Symp Zool Soc Lond*, 42: 139–148.

Biere, J.M. & Uetz, G.W., 1981, Web orientation in the spider *Micrathena gracilis* (Araneidae), *ECOL*, 62: 236–244.

Binford, G.J., 2001, Differences in venom composition between orb-weaving and wandering Hawaiian *Tetragnatha* (Araneae), *Biol J Linn Soc*, 74: 581–595.

Bleckmann, H., 1985, Discrimination between prey and non-prey wave signals in the fishing spider *Dolomedes trition* (Pisauridae), in Kalmring & Elsner (eds), *Acoustic & Vibrational Communication in Insects*, Paul Parney, Berlin, 215–222.

Bristowe, W.S., 1958, *The World of Spiders*, Collins, London.

Brunet, B.S., 1994, *The Silken Web: A Natural History of Australian Spiders*, Reed Books, 208 pp.

Cangialosi, K.R., 1990, Social spider defense against kleptoparasitism, *Behav Ecol Sociobiol*, 27(1): 49–54.

Cloudsley-Thompson, J.L., 1983, Desert adaptations in spiders, *J Arid Environ* 6: 307–317.

Davy, G.C.L., 1994, The 'Disgusting' Spider: The role of disease and illness in the perpetuation of fear of spiders, *Soc Anim*, 2 (1): 17–25.

Devito, J. & Formanowicz, D.R., 2003, The effects of size, sex and reproductive condition on thermal and desiccation stress in a riparian spider (*Pirata sedentarius*, Araneae, Lycosidae), *J Arachnol*, 31: 278–284.

Diaz, J.H., 2004, The global epidemiology, syndromic classification, management and prevention of spider bites, *Am J Trop M*, 71 (2): 239–250.

Eberhard, W.G., 2001, Under the influence: Webs and building behaviour of *Plesiometa argyra* (Araneae, Tetragnathidae) when parasitized by *Hymenoepimecis argyraphaga* (Hymenoptera, Ichneumonidae), *J Arachnol*, 29: 354–366.

Eberhard, W.G., Barrantes, G. & Weng, J-L, 2006, The mystery of how spiders extract food without masticating prey, *Bull Br Arachnol Soc*, 13 (9): 372–376.

Eberhard, W.G. & Levi, H.W., 2006, Camouflage and chemical defense in a bolas spider, *Mastophora caesariata* sp.n. (Araneae, Araneidae), *Bull Br Arachnol Soc*, 13 (9): 337–340.

Edwards, R.L., Edwards, E.H. & Edwards, A.D., 2003, Observations of *Theotima minutissimus* (Araneae, Ochyroceratidae), a parthogenetic spider, *J Arachnol*, 31: 274–277.

Elgar, M.A. & Allan, R.A., 2004, Predatory spider mimics aquire colony-specific cuticular hydrocarbons from their ant model prey, *Naturwissen*, 91(3): 143–147.

Foelix, R.F., 1996, *Biology of Spiders* (2nd edn), Oxford University Press, 330 pp.

Fromhage, L. & Scheider, J.M., 2006, Emasculation to plug up females: the significance of pedipalp damage in *Nephila fenestrata*, *Behav Ecol*, 17 (3): 353–357.

Gemeno, C., Yeargan, K.V. & Haynes, K., 2000, Aggressive chemical mimicry by the Bolas Spider *Mastophora hutchinsoni*: identification and quantification of a major prey's sex pheromone components in the spider's volatile emissions, *J Chem Ecol*, vol. 26, no. 5: 1235–1243.

Gibbs, A.G. *et al.*, 1998, Effects of temperature on cuticular lipids and water balance in a desert *Drosophila*: is thermal acclimation beneficial? *J Exp Biol*, 201: 71–80.

Gorb, S.N. *et al.*, 2006, Brief communications: Silk-like secretion from tarantula feet, *NATRA*, vol. 443: 407.

Heidger, C., 1988, Ecology of spiders inhabiting abandoned mammal burrows in South African savanna, *Oecologia*, 76(2): 303–306.

Hillyard, P.D., 1997, *Collins Gem Spiders Photoguide*, Harper Collins Publishers, 254 pp.

Hoefler, C.D., Chen, A. & Jakob, E.M., 2006, The potential of a jumping spider, *Phidippus clarus*, as a biocontrol agent, *J Econ Entomol*, 99(2): 432–436.

Holland, C., Terry, A.E., Porter, D. & Vollrath, F., 2006, Comparing the rheology of native spider and silkworm spinning dope, *Nat Mater*, 1–4.

Husby, J.A. & Zachariassen, K.E., 1980, Antifreeze agents in the body fluid of winter active insects and spiders, *Cell Mol Life Sci*, vol. 36, no. 8: 963–964.

Isbister, G.K., White, J., Currie, B.J., Bush, S.P., Vetter, R.S. & Warrell, D.A., 2005, Spider bites: Addressing mythology and poor evidence, *Am J Trop M*, 72(4): 361–367.

Jackson, R.R. *et al.*, 2001, Jumping spiders (Araneae: Salticidae) that feed on nectar, *J Zool*, 255: 25–29.

Kirchner, W., 1987) Behavioural and physiological adaptations to cold in Nentwig, N. (ed.), *Ecophysiology of Spiders*, Springer, Berlin, 66–77.

Kuhn-Nentwig, L., Schaller, J. & Nentwig, W., 2004, Biochemistry, toxicology and ecology of the venom of the spider, *Cupiennius salei* (Ctenidae), *TOXIA*, 43 (5): 543–553.

Land, M.F., 1985, The morphology and optics of spider eyes, in Barth, F.G. (ed), *Neurobiology of Arachnids*, Springer Verlag, Berlin.

Low, A.M., 1983, Untersuchungen zur cuticularen und pulmonren transpiration der vogelspinne *Eurypelma californicum*, Dupl. Thesis, University of München.

Marshall, S.D. & Uetz, G.W., 1990, Incorporation of urticating hairs into silk: a novel defense mechanism in two neotropical tarantulas (Araneae, Theraphosidae), *J Arachnol*, 18: 143–149.

Nelson, X.J. & Jackson, R.R (2006, A predator from East Africa that chooses malaria vectors as preferred prey, *PLoS ONE*, 1(1):e132.

Oxford, G.S. & Gillespie, R.G., 1998, Evolution and ecology of spider colouration, *Ann R Entomol* 43: 619–43.

Pollard, S.D., Beck, M.W. & Dodson, G.N., 1995, Why do male crab spiders drink nectar? *Anim Behav*, 49(6): 1443–1448.

Punzo, F., 2007, *Spiders: Biology, Ecology, Natural History and Behaviour*, Brill, Leiden, 428 pp.

Roberts, M., 1996, *Spiders of Britain and Northern Europe*, Collins Field Guide, Harper Collins, 383 pp.

Scharff, N. & Enghoff, H., 2005, *Arachnida*, Zoological Museum, University of Copenhagen.

Schneider, J.M., 2002, Reproductive state and care giving in *Stegodyphus* (Araneae: Eresidae) and the implications for the evolution of sociality, *Anim Behav*, 63 (4): 649–658.

Schneider, J.M. & Elgar, M.A., 2001, Sexual cannibalism and sperm competition in the golden orb-web spider *Nephila plumipes* (Araneoidea): female and male perspectives, *Behav Ecol*, vol. 12, no. 5: 547–552.

Schneider, J.M. & Lubin, Y., 1998, Intersexual conflict in spiders, *Oikos*, 83: 496–506.

Seibt, U. & Wickler, W., 1988, Interspecific tolerance in social *Stegodyphus* spiders (Eresidae, Araneae), *J Arachnol*, 16: 35–39

Stalhandske, P., 2001, Nuptial gift in the spider *Pisaura mirabilis* maintained by sexual selection, *Behavioral Ecology*, 12(6): 691–697.

Suter, R.B., Stratton, G.E. & Miller, P.R., 2004, Taxonomic variation among spiders in the ability to repel water: surface adhesion and hair density, *J Arachnol*, 32: 11–21.

Toyama, M., 2001, Adaptive advantage of matriphagy in the foliage spider *Chiracanthium japonicum* (Araneae: Clubionidae), *J Ethol*, vol. 19, 2: 69–74.

Vachon, M., 1957, Contribution à l'étude

du développement postembryonnarie des araignées. Première note. Généralités et nomenclature des stades, *Bull Soc Zool France*, 82: 337.

Vogelei, A. & Greissl, R., 1989, Survival strategies of the crab spider *Thomisus onustus* Walckenaer 1806 (Chelicerata, Arachnida, Thomisidae), *Oecologia*, 80(4): 513–515.

Wheeler, G.S., Mccaffrey, J.P. & Johnson, J.B., 1990, Developmental biology of *Dictyna* spp. (Araneae: Dictynidae) in the laboratory and field, *Am Midl Nat*, 123: 124–134.

Williams, J.L., Moya-Laran, J. & Wise, D.H., 2006, Burrow decorations as anti-predatory devices, *Behav Ecol*, 17 (4): 586–590.

Wolska, H., 1957, Preliminary investigations on the thermic perferndum of some insects and spiders encountered in snow, *Folia boil, Krakow*, 5: 195–208, in Aitchison, C.W., 1978, Spiders active under snow in southern Canada, *Symp Zool Soc Lond*, 42: 139–148.

WEBSITES

Arachnid developmental stages
http://www.atshq.org/

Sociality in spiders
http://www.entomology.cornell.edu/public/IthacaCampus/ProgramPages/Rayor.html

Spider elimination
http://www.pestproducts.com/spider.htm

Spider silk programme using transgenic goat technology
http://www.nexiablotech.com

The Baboon Spiders of South Africa
http://www.scienceinafrica.co.za/2002/november/baboon.htm

The World Catalogue of Spiders
http://www.research.amnh.org/entomology/spiders/catalog

CHAPTER 3 – AMBLYPYGI

Cohen, A.C., 1995, Extra-oral digestion in predaceous terrestrial arthropoda, *Ann R Entom*, 40: 85–103.

Harvey, M.S., 2002, *Catalogue of the Smaller Arachnid Orders of the World*, CSIRO Publishing, 385 pp.

Hebets, E.A. & Chapman, R.F., 2000, Electrophysiological studies of olfaction in the whip spider *Phrynus parvulus* (Arachnida, Amblypygi), *J Insect Physiol*, 44: 1441–1448.

Ladle, R.J. & Velander, K., 2003, Fishing behaviour in a giant whip spider, *J Arachnol*, 31: 154–156.

Pinto-Da-Rocha, R., MacHado, G. & Weygoldt, P., 2002, Two new species of *Charinus* Simon, 1892 from Brazil with biological notes (Arachnida, Amblypygi, Charinidae), *J Nat Hist*, 36 (1): 107–118.

Prendini, L., Weygoldt, P. & Wheeler, W.C., 2005, Systematics of the *Damon variegates* group of African whip spiders (Chelicerata: Amblypygi): Evidence from behaviour, morphology and DNA, *Organisms, Diversity & Evolution*, 5, 203–236.

Weygoldt, P., 1995, The development of the phrynichid 'hand': notes on allometric growth and introduction of the new generic name *Euphrynichus* (Arachnida, Amblypygi), *Zool Anz*, 234: 75–84.

Weygoldt, P., 1997/98, Mating and spermatophore morphology in Whip Spiders, *Zool Anz*, 236: 259–276.

Weygoldt, P., 2000, *Whip Spiders (Chelicerata: Amblypygi) Their Biology, Morphology and Systematics*, Apollo Books, Stenstrup, 163 pp.

CHAPTER 4 – UROPYGI

Ahearn, G.A., 1970, Water balance in the whipscorpion *Mastigoproctus giganteus* (Lucas) (Arachnida: Uropygi), *Comp Biochem Physiol*, 35: 339–353.

Crawford, C.S. & Cloudsley-Thompson, J.L., 1971, Water relations and desiccation-avoiding behaviour in the vinegaroon *Mastigoproctus giganteus* (Arachnida: Uropygi), *Ent Exp App*, 14: 99–106.

Dunlop, J., 1994, Whip scorpions, *Journal of the British Tarantula Society*, 9: 9–11.

Eisner, T., Meinwald, J., Monro, A. & Ghent, R., 1961, Defense mechanisms of Arthropods: I, The composition and function of the spray of the whipscorpion *Mastigoproctus giganteus* (Lucas) (Arachnida, Pedipalpida), *J Insect Physiol*, 6: 272–298.

Geethabali & Moro, S.D., 1988, The general behavioural patterns of the Indian whip scorpion *Thelyphonus indicus*, *Rev Arachnol*, 7 (5): 189–196.

Harvey, M.S., 2007, The smaller arachnid orders: diversity, descriptions and distributions from Linnaeus to the present (1758 to 2007), in Zhang, Z.-Q. & Shear, W.A. (eds), Linnaeus Tercentenary: Progress in Invertebrate Taxonomy, *Zootaxa*, 1668: 363–380.

Harvey, M.S., 2003, *Catalogue of The Smaller Arachnid Orders of the World*, CSIRO Publishing, 385pp.

Harvey, M.S., 2002, The Neglected Cousins: What do we know about the smaller arachnid orders? *J Arachnol*, 30: 357–372.

Haupt, J. & Mueller, F., 2004, New products of defense secretion in South East Asian whip scorpions (Arachnida: Uropygi: Thelyphonida), *Z Naturforsch*, 59(7–8): 579–581

Haupt, J., 2000, Biology of whipscorpions (Uropygi, Thelyphonida), *Mem Soc Entomol Ital*, 78 (2): 305–319.

Haupt, J., 1996, Fine structure of the trichobothria and their regeneration during moulting in the whip scorpion *Typopeltis crucifer* Pocock, 1894, *Acta Zool (Stockh)*, 77 (2): 123–136.

Itokawa, H., Kano, R. & Nakajima, T., 1985, Chemical investigations of the spray of the Asian whipscorpion *Typopeltis stimpsoni* (Wood, 1862), *Jpn J Sanit Zool* , 36 (1): 65–66.

Moro, S.D. & Bali, G., 1988, Prey catching behaviour of the Indian whip scorpion *Thelyphonus indicus*, *Rev Arachnol*, 7 (5): 197–203.

Moro, S.D. & Bali, G., 1986, The topography of slit sense organs in the whip scorpion *Thelyphonus indicus* (Arachnida, Uropygida), *Verhandlungen des naturwissenschaftlichen*, 28: 91–105, in Haupt, J., 2000) Biology of whipscorpions (Uropygi, Thelyphonida), *Mem Soc Entomol Ital*, 78 (2): 305–319.

Punzo, F., 2006, Responses of the whipscorpion *Mastigoproctus liochirus* (Arachida, Uropygi) to environmental humidity, *J Environ Biol*, 27(4): 619–622.

Punzo, F., 2005, Habituation, avoidance learning, and spatial learning in the giant whipscorpion *Mastigoproctus giganteus* (Lucas) (Arachnida, Uropygi), *Bull Br Arachnol Soc*, 13(4): 138–144.

Punzo, F., 2001, Substrate preferences and the relationship between soil characteristics and the distribution of the giant whip scorpion, *Mastigoproctus giganteus* (Lucas) (Arachnida, Uropygi) in Big Bend National Park, *Bull Br Arachnol Soc*, 12 (1): 35–41.

Punzo, F., 2000, Diel activity patterns and diet of the giant whip scorpion *Mastigoproctus giganteus* (Lucas) (Arachnida, Uropygi) in Big Bend National Park (Chihuahuan Desert), *Bull Br Arachnol Soc*, 11 (9): 385–387.

Punzo, F. & Reeves, 2001, Geographical variation in male courtship behaviour of the giant whip scorpion *Mastigoproctus giganteus* (Lucas) (Arachnida, Uropygi), *Bull Br Arachnol Soc*, 12 (2): 93–96.

Rajashekhar, K.P. & Bali, G., 1987, 'Giant' fibres I: the ventral nerve cord of the whip scorpion *Thelyphonus indicus* Stoliczka, *Curretn Science*, 56 (24): 1300–1301.

Ramachandra, Y.L. & Bali, G., 1990, Brood care in the whip scorpion *Thelyphonus indicus*, *11th International Congress: IUSSI*, India.

Rowland, J.M. & Adis, J., 2002, Uropygi (Thelyphonida), in Adis, J. (ed.), *Amazonian Arachnida and Myriapoda*, 449–456.

Scharff, N. & Enghoff, H., 2005, *Arachnida*, Zoological Museum, University of Copenhagen (unpublished).

Schmidt, J.O., Dani, F.R., Jones, G.R. & Morgan, E.D., 2000, Chemistry, ontogeny and role of pygidial gland secretions of the vinegaroon *Mastigoproctus giganteus* (Arachnida: Uropygi), *J Insect Physiol*, 46(4): 443–450.

Shultz, J.W., 1993, Muscular anatomy of the giant whipscorpion *Mastigoproctus giganteus* (Lucas) (Arachnida: Uropygi) and its evolutionary significance, *Zool J Linn Soc*, 108: 335–365.

Thomas, R.H. & Zeh, D.W., 1984, Sperm transfer and utilization strategies in arachnids: ecological and morphological constraints, in Smith, R.L. (ed.), *Sperm competition and the evolution of animal mating systems*, Orlando, Florida Academic Press, 179–221.

Tetlie, O.E. & Dunlop, J.A., 2008, *Geralinura carbonaria* (Arachnida, Uropygi) from Mazon Creek, Illinois, USA, and the origin of subchelate palps in whip scorpions, *J Paleontol*, 82: 299–312.

Van der Hammen, L., 1989, *An introduction to comparative arachnology*, SPB Academic Publishing, 576 pp.

Weygoldt, P., 1988, Sperm transfer and spermatophore morphology in the whip scorpion *Thelyphonus linganus* (Arachnida: Uropygi: Thelyphonidae), *J Zool (Lond)*, 215: 189–196.

Weygoldt, P., 1971, Notes on the life history and reproductive biology of the giant whip scorpion *Mastigoproctus giganteus* (Lucas) (Uropygi, Thelyphonidae) from Florida, *J Zool (Lond)*, 164: 137–147.

Weygoldt, P., 1970, Courtship behaviour and sperm transfer in the giant whip scorpion *Mastigoproctus giganteus* (Uropygi, Thelyphonidae), *Behav*, 36 (1): 1–8.

Yogi, S. & Haupt, J., 1977, Analyse des Wehrsekretes be idem Geibelscorpion *Typopeltis crucifer* Pocock, *Acta Arachnol*, 27: 53–56.

WEBSITES

Arachnid developmental stages
http://www.atshq.org/

Invertebrate Anatomy Online
http://webs.lander.edu/rsfox/invertebrates/

CHAPTER 5 – SCHIZOMIDA

Adis, J., 2001, Abundance and phenology of Schizomida (Arachnida) from a secondary upland forest in Central Amazonia, *Rev Suisse Zool*, 108(4): 879–889.

Alberti, G. & Palacios-Vargas, J.G., 1987, Fine structure of spermatozoa and spermatogenesis of *Schizomus palaciosi*, Reddell and Cokendolpher, 1986 (Arachnida: Uropygi, Schizomida), *PROTA*, 137: 1–14.

Armas, L.F., 1989, Depredación de *Schizomus portoricensis* (Arachnida: Schizomida) por *Phrynus marginemaculatus* (Arachnida: Amblypygi), *Miscelanea Zoologica*, Instituto de Ecologia y Sistematica, Academia de Ciencias de Cuba, 46: 3.

Brach, V., 1976, Development of the whipscorpion *Schizomus floridanus*, with notes on behaviour and laboratory culture, *Bull Sth Calif Acad Sci*, 74 (3): 97–103.

Briggs, T.S. & Hom, K., 1966, A new Schizomid Whip-Scorpion from California with Notes on the Others, *Pan-Pac Entomol*, 42 (4): 270–274.

Cokendolpher, J.C., 1988, Review of the Schizomidae (Arachnida, Schizomida) of Japan and Taiwan, *Bull Natl Sci Mus Ser A (Zool)*, 14 (4): 159–171.

Cokendolpher, J.C. & Sites, R.W., 1988, A new species of eyed *Schizomus* (Schizomida: Schizomidae) from Java, *Acta Arachnol*, 36: 79–85.

Firstman, B., 1973, The relationship of the chelicerate arterial system to the evolution of the endosternite, *J Arachnol*, 1: 1–54.

Gravely, F.H., 1915, Notes on the habits of Indian insects, myriapods and arachnids, *Rec Indian Mus*, 11: 482–539.

Harvey, M.S., 2007, The smaller arachnid orders: diversity, descriptions and distributions from Linnaeus to the present (1758 to 2007), in Zhang, Z.-Q. & Shear, W.A. (eds), Linnaeus Tercentenary: Progress in Invertebrate Taxonomy, *Zootaxa*, 1668: 363–380.

Harvey, M.S., 2002, Short-range endemism among the Australian fauna: Some examples from non-marine environments, *Invertebrate Systematics*, 16: 555–570.

Harvey, M.S., 2001, New cave-dwelling schizomids (Schizomida: Hubbardiidae) from Australia, *Rec West Aust Mus*, Supplement no. 64: 171–185.

Harvey, M.S., 2000, A review of the Australian schizomid genus *Notozomus* (Hubbardiidae), *Qld Mus Mem*, 46 (1): 161–174.

Harvey, M.S. & Humphreys, W.F., 1995, Notes on the genus *Draculoides* Harvey (Schizomida: Hubbardiidae), with the description of a new troglobitic species, *Rec West Aust Mus*, Supplement no. 52: 183–189.

Harvey, M.S., 1992, The Schizomida (Chelicerata) of Australia, *Invertebr Taxon*, 6: 77–129.

Hilton, W.A., 1916, The central nervous system and simple reactions of a rare whip scorpion, *J Entomol Zool*, 8 (2): 73–79.

Humphreys, W.F., 1990, The biology of a troglobitic schizomid (Chelicerata: Arachnida) from caves in the semi-arid Cape Range, Western Australia, *Acta Zool Fenn*, 1990: 181–186.

Humphreys, W.F., 1989, The biology of *Schizomus vinei* (Chelicerata : Schizomida) in the caves of Cape Range, Western Australia, *J Zool (Lond)*, 217: 177–201.

Lawrence, R.F., 1958, Whipscorpions Uropygi) from Angola, the Belgian Congo and Mossambique, *Publ Cult Cia Diamantes Angola*, 40: 71–79.

Miyazaki, K., Ueshima, R. & Makioka, T., 2001, Structure of the female reproductive system and oogenetic mode in a schizomid, *Schizomus sawadai* (Arachnida, Schizomida), *Invertebr Reprod Dev*, 40 (1): 1–7.

Reddell, J.R. & Cokendolpher, J.C., 2002, Schizomida, 387–398, in Adis, J. (ed.), *Amazonian Arachnida and Myriapoda*, Pensoft Publishers, 590 pp.

Reddell, J.R. & Cokendolpher, J.C., 1995, *Catalogue, Bibliography and Generic Revision of the Order Schizomida (Arachnida)*, Speleological Monographs, no.4, Texas Memorial Museum.

Rowland, J.M., 1973, A new genus and several new species of Mexican schizomids (Schizomida: Arachnida), *Occasional Papers*, The Museum, Texas Tech University, 11, 23 pp.

Rowland, J.M., 1972, The brooding habits and early development of *Trithyreus pentapeltis* (Cook), (Arachnida: Schizomida), *Entomol News*, 83: 69–74.

Sturm, H., 1973, Zur Ethologie von *Trithyreus sturmi* Kraus (Arachnida, Pedipalpi, Schizopeltidia), *Z Tierpsychol*, 33: 113–140.

Van der Hammen, L., 1989, *An Introduction to Comparative Arachnology*, SPB Academic Publishing, 576pp.

Vine, B., Knott, B. & Humphreys, W.F., 1988, Observations on the environment and biology of *Schizomus vinei* (Chelicerata: Schizomida) from Cape Range, Western Australia, *Rec West Aust Mus*, 14 (1): 21–34.

WEBSITES

Australasian Arachnological Society (AAS) http://www.australasian-arachnology.org/arachnology/schizomida

CHAPTER 6 – PALPIGRADI

Adis, J., Scheller, U., de Morais, J.W., Condé, B. & Rodrigues, J.M.G., 1997, On the abundance and phenology of Palpigradi (Arachnida) from central Amazonian upland forests, *J Arachnol*, 25: 326–332.

Condé, B. & Adis, J., 2002, Palpigradi, in Adis, J. (ed.), *Amazonian Arachnida and Myriapoda*, Pensoft Publishers, Sofia, Moscow, 590 pp.

Condé, B., 1996, Les Palpigrades, 1885–1995: acquisitions et lacunes, *Rev Suisse Zool*, hors serie 1: 87–106.

Condé, B., 1988, 46, Palpigradida, in Higgins, R.P. & Thiel, H. (eds), *Introduction to the Study of Meiofauna*, Smithsonian Instituion Press, Washington D.C. 488 pp.

Condé, B., 1984, Les Palpigrades: quelques aspects morpho-biologiques, *Rev Arachnol*, 5 (4): 133–143.

Erhard, C., 1998, *Eukoenenia austriaca* from the catacombs of St Stephen's Cathedral in the centre of Vienna and the distribution of palpigrades in Austria (Arachnida: Palpigradida: Eukoeneniidae), *Senken Biol*, 77 (2): 241–245.

Firstman, B., 1973, The relationship of the chelicerate arterial system to the evolution of the endosternite, *J Arachnol*, 1: 1–54.

Harvey, M.S., 2002, The Neglected Cousins: What do we know about the smaller arachnid orders? *J Arachnol*, 30: 357–372.

Harvey, M.S., 2003, *Catalogue of The Smaller Arachnid Orders of the World*, CSIRO Publishing, 385pp.

Harvey, M.S., 2007, The smaller arachnid orders: diversity, descriptions and distributions from Linnaeus to the present (1758 to 2007), in Zhang, Z.-Q. & Shear, W.A. (eds), Linnaeus Tercentenary: Progress in Invertebrate Taxonomy, *Zootaxa*, 1668: 363–380.

Harvey, M.S., táhlavsk, F. & Theron, P.D., 2006, The distribution of *Eukoenenia mirabilis* (Palpigradi: Eukoeneniidae): a widespread tramp, *Rec West Aust Mus*, 23: 199–203.

Ludwig, M. & Alberti, G., 1991, Fine structure of the midgut of *Prokoenenia wheeleri* (Arachnida, Palpigradi), *Zoologische Beitraege*, 34 (1): 127–134.

Monniot, F., 1966, Un palpigrade interstitial: *Leptokoenenia scurra*, n.sp., *Rev Ecol Biol Sol*, 3: 41–64.

Pointer, M., 1992, A Guide to prehistoric Chelicerates, Part 5: The Schizomida and Palpigradida, *Journal of the British Tarantula Society*, vol. 7, 4: 10–11.

Rowland, J.M. & Sissom, W.D., 1980, Report on a fossil Palpigrade from the Tertiary of Arizona, and a review of the morphology and systematics of the order (Arachnida: Palpigradida), *J Arachnol*, 8: 69–86.

Van der Hammen, L., 1969, Notes of mouthparts of *Eukoenenia mirabilis* (Grassi) (Arachnidea: Palpigradida), *Zool Meded (Leiden)*, 44: 41–45.

Van der Hammen, L., 1982, Comparative studies in Chelicerata II, Epimerata (Palpigradi and Actinotrichida), *Zool Verh*, 196: 1–70.

Van der Hammen, L., 1989, *An Introduction to Comparative Arachnology*, SPB Academic Publishing, 576 pp.

Zagmajster, M. & Kovác, L., 2006, Distribution of palpigrades (Arachnida, Palpigradi) in Slovenia with a new record of *Eukoenenia austriaca* (Hansen, 1926), *Natura Sloveniae*, 8 (1): 23–31.

CHAPTER 7 – RICINULEI

Adis, J, Messner, B. & Platnick, N., 1999, Morphological structures and vertical distribution in the soil indicate facultative plastron respiration in *Cryptocellus adisi* (Arachnida, Ricunulei) from Central Amazonia, *Studies Neotrop Fauna Environ*, 34(1): 1–9.

Adis, J.U., Platnick, N.I., De Morais, J.W., & Gomes Rodrigues, J.M., 1989, On the abundance and ecology of Ricinulei (Arachnida) from Central Amazonia, Brazil, *J N Y Ent Soc*, 97(2): 133–140.

Cooke, J.A.L., 1967, Observations on the biology of Ricinulei (Arachnida) with descriptions of two new species of *Cryptocellus*, *J Zool (Lond)* 151: 31–42.

Dunlop, J., 1994, Ricinuleids, *Journal of the British Tarantula Society*, 10, 2: 55–58.

Harvey, M.S., 2003, *Catalogue of The Smaller Arachnid Orders of the World*, CSIRO Publishing, 385 pp.

Klompen, H., 2000, Prelarva and larva of *Opilioacarus (Neocarus) texanus* (Chamberlin and Mulaik) (Acari: Opilioacarida) with notes on the patterns of setae and lyrifissures, *J Nat Hist*, 34: 1977–1992.

Legg, G., 1977, Sperm transfer and mating in *Ricinoides hansenii* (Ricinulei: Arachnida), *J Zool (Lond)*, 182: 51–61.

Ludwig, M., Palacios-Vargas, J.G. & Albertif, G., 2005, Cellular details of the midgut of *Cryptocellus boneti* (Arachnida: Ricinulei), *J Morphol*, 220 (3): 263–270.

Pittard, K & Mitchell, R.W., 1972, Comparative Morphology of the Life Stages of *Cryptocellus pelaezi* (Arachnida, Ricinulei), *Graduate Studies*, Texas Tech University, 1: 1–77.

Platnick, N.I. & Pass, G., 1982, On a new Guatemalan *Pseudocellus* (Arachnida, Ricinulei), *Am Mus Novit*, no. 2733, 1–6, American Museum of Natural History.

Talarico, G., Palacios-Vargas, J.G. & Alberti, G., 2008, The pedipalp of *Pseudocellus pearsei* (Ricinulei, Arachnida): ultrastructure of a multifunctional organ, *Arthropod Struct Dev*, 37 (6): 511–521.

Talarico, G, Garcia Hernandez, L.F. & Michalik, P., 2008, The male genital system of the New World Ricinulei (Arachnida): Ultrastructure of spermatozoa and spermiogenesis with special emphasis on its phylogenetic implications, *Arthropod Struct Dev*, 37: 396–409.

CHAPTER 8 – ACARI

Andrews, J.H.R., 1998, A progress report on scabies in humans, in Halliday, R.B., Walter, D.E., Proctor, H.C., Norton, R.A. & Colloff, M.J. (eds), *Acarology: Proceedings of the 10th International Congress*, CSIRO Publishing, 657 pp.

Baker, A.S., 1999, *Mites and ticks of domestic animals: an identification guide and information source*, The Natural History Museum, London, 240 pp.

Baker, E.W., Evans, T.M., Gould, D.J., Hull, W.B. & Keegan, H.L., 1956, *A manual of parasitic mites of medical or economic importance*, National Pest Control Association Inc., New York.

Briese, D.T, & Cullen, J.M., 1998, The use and usefulness of mites in biological control of weeds, in Halliday, R.B., Walter, D.E., Proctor, H.C., Norton, R.A. & Colloff, M.J. (eds), *Acarology: Proceedings of the 10th International Congress*, CSIRO Publishing, 657pp.

Brodie, T.A., Holmes, P.H. & Urquhart, G.M., 1986, Some aspects of tick-borne fever of sheep, *Veterinary Records*, 118: 415–418, in Ogden, N.H., Casey, A.N.J., French, N.P. & Woldehiwet, Z., 2002, A review of studies on the transmission of *Anaplasma phagocytophilum* from sheep: implications for the force of infection in endemic cycles, *Exp Appl Acarol*, 28: 195–202.

Bukva, V., 1993, Sexual dimorphism in the hair follicle mites (Acari: Demodecidae): scanning electron microscopy of *Soricidex dimorphus*, *Folia Parasitologica* 40(1): 71–9.

Burns, E.C. & Melancon, D.G., 1977, Effect of imported fire ant (Hymenoptera: Formicidae) invasion on lone star tick (Acari: Ixodida) populations, *J Med Entomol*, 14: 247–9.

Carswell, F., 1988, State of the art: mites and human allergy, *Immun*, 65: 497–500.

Chagas, C.M. *et al.*, 1998, *Brevipalpus* mites (Acari: Tenuipalpidae) as vectors of plant viruses, in Halliday, R.B., Walter, D.E., Proctor, H.C., Norton, R.A. & Colloff, M.J. (eds), *Acarology: Proceedings of the 10th International Congress*, CSIRO Publishing, 657 pp.

Clift, A. & Terras, M.A., 1998, A quantitative study of phoresy in *Microdispus lambi* (Acari: Microdispidae) in eastern Australia, in Halliday, R.B., Walter, D.E., Proctor, H.C., Norton, R.A. & Colloff, M.J. (eds), *Acarology: Proceedings of the 10th International Congress*, CSIRO Publishing, 657 pp.

Colloff, M.J., 1987, Mite fauna from passenger trains in Glasgow, *Epidem Inf Bull*, 98: 127–130.

Domrow R., 1991, Acari Prostigmata (excluding Trombiculidae) parasitic on Australian vertebrates: an annotated checklist, keys and bibliography, *Invertebrate Taxonomy*, 4: 1238–1376.

Ebermann, E., 1988, *Imparipes (Imparipes) pselaphidorum* n.sp., a new scutacarid species phoretic upon African beetles (Acari, Scutacaridae; Coleoptera, Pselaphidae), *Acarol*, 29 (1): 35–42.

Ebermann, E. & Palacios-Vargas, J.G., 1988, *Imparipes (Imparipes) tocatlphilus* n.sp. (Acari, Tarsonemina, Scutacaridae) from Mexico and Brazil: first record of ricinuleids as phoresy hosts for scutacarid mites, *Acarol*, 29 (4): 347–354.

Filimonova, S.A., 2001, Internal Anatomy of Females of *Myobia murismusculi* (Schrank) (Acari: Trombidiformes, Myobiidae), *Entomol R* c/c of *Entomol Obozrenie*, vol. 81, no. 9: 1049–1058.

Gerson, U., 1998, Trends in research on acarine biocontrol agents, in Halliday, R.B., Walter, D.E., Proctor, H.C., Norton, R.A. & Colloff, M.J. (eds), *Acarology*,

Proceedings of the 10th International Congress, CSIRO Publishing, 657 pp.

Gerson U., Smiley R.L. & Ochoa R., 2003, *Mites (Acari) for Pest Control*, Blackwell, Oxford, 539 pp.

Hillyard, P.D., 1996, *Ticks of North-West Europe*, Synopsis of the British Fauna (New Series), no. 52. Field Studies Council, 178 pp.

Hughes, A.M., 1976, *The mites of stored food and houses*, Technical Bulletin 9, Ministry of Agriculture, Fisheries and Food, HMSO, 400 pp.

James, D.G., 1998, History and perspectives of biological mite control in Australian horticulture using exotic and native phytoseiids, in Halliday, R.B., Walter, D.E., Proctor, H.C., Norton, R.A. & Colloff, M.J. (eds), *Acarology: Proceedings of the 10th International Congress*, CSIRO Publishing, 657 pp.

Jovani, R. & Serrano, D., 2001, Feather mites (Astigmata) avoid moulting wing feathers of passerine birds, *Anim Behav*, 62 (4): 723–727.

Krips, O., 2000, *Plant effects on biological control of spider mites in the ornamental crop Gerbera*, Wageningen University, 113 pp.

Lindquist, E.E., Sabelis, M.W. & Bruin, J. (eds), 1996, *Eriophyoid mites: their biology, natural enemies and control*, vol. 6, Elsevier, 790 pp.

Lwande, W., Ndakala, A.J., Hassanali, A., Moreka, L., Nyandat, E., Ndungu, M., Amiani, H., Gitu, P.M., Malonza, M.M., & Punyua, D.K., 1999, *Gynandropsis gynandra* essential oil and its constituents as tick (*Rhipicephalus appendiculatus*) repellents, *Phytochemistry*, 50 (3): 401–405.

McFarlane, J.A., 1989, *Guidelines for pest management research to reduce stored food losses caused by insects and mites*, Bull. no. 22, Overseas Development Natural Resources Institute, Odnri, 62 pp.

Mbati, P.A., Hlatshwayo, M., Mtshali, M.S., Mogaswane, K.S., de Waal, T.D. & Dipeolu, O.O., 2002, Ticks and tick-borne diseases of livestock belonging to resource-poor farmers in the eastern Free State of South Africa, *Exp Appl Acarol*, 28: 217–224.

Morris, M.J.A. & Rimmer, J., 1998, Allergenicity of the predator dust mite *Cheyletus tenuipilis* (Acari: Cheyletidae): a preliminary study, in Halliday, R.B., Walter, D.E., Proctor, H.C., Norton, R.A. & Colloff, M.J. (eds), *Acarology: Proceedings of the 10th International Congress*, CSIRO Publishing, 657 pp.

Newell, I.M., 1945, *Hydrozetes* Berlese (Acari, Oribatoidea): the occurrence of the genus in North America and the phenomenon of levitation, *Trans Conn Acad Arts Sci*, 36: 253–275.

Newell, I.M., 1967, Abyssal Halacaridae (Acari) from the Southeast Pacific, *Pacific Insects*, 9 (4): 693–708.

Norton, R.A., Kethley, J.B., Johnston, D.E. & O'Connor, B.M., 1993, Phylogenetic perspectives on genetic systems and reproductive modes of mites, 8–99, in Wrensch, D.L. & Ebbert, M.A. (eds), *Evolution and diversity of sex ratio in insects and mites*, Chapman & Hall, New York.

Nuzzaci, G. & Alberti, G., 1996, Internal anatomy and physiology, in Lindquist, E.E., Sabelis, M.W. & Bruin, J. (eds), 1996, *Eriophyoid mites: their biology, natural enemies and control*, vol. 6, Elsevier, 790 pp.

Oberem, P.T. & Malan, F.S., 1984, A new cause of cattle mange in South Africa: *Psorergates bos* Johnston, *J South African Vet Med Ass*, 55 (3): 121–122.

Ogden, N.H., Casey, A.N.J., French, N.P. & Woldehiwet, Z., 2002, A review of studies on the transmission of *Anaplasma phagocytophilum* from sheep: implications for the force of infection in endemic cycles, *Exp Appl Acarol*, 28: 195–202.

Ostoja-Starzewski, J.C., Eyre, D., Cannon, R.J. & Bartlett, P., 2007, Update on Fuchsia gall mite *Aculops fuchsiae* Keifer, *Plant Pest Notice Leaflet no. 51*, Central Science Laboratory, York.

Parker, G.A., 1970, Sperm competition and its evolutionary consequences in the insects, *Biol R*, 45: 525–567.

Platts-Mills T.A.E., 1982, Reduction of bronchial hyper-reactivity during prolonged allergen avoidance, *LANCA*, 2: 675.

Proctor, H.C., 2003, Feather mites (Acari: Astigmata): Ecology, behavior, and evolution, *Ann R Entom*, 48: 185–209.

Rott, A.S. & Ponsonby, D.J., 1998, Control of two-spotted spider mite *Tetranychus urticae* Koch (Acari: Tetranychidae) on edible crops in glasshouses using two interacting species of predatory mite, in Halliday, R.B., Walter, D.E., Proctor, H.C., Norton, R.A. & Colloff, M.J. (eds), *Acarology: Proceedings of the 10th International Congress*, CSIRO Publishing, 657 pp.

Rounsevell, D.E. & Greenslade, P., 1988, Cuticle structure and habitat in the Nanorchestidae (Acari: Prostigllnata), *Hydrobiol*, 165 (1): 209–212.

Sinclair, A.N., 1990, The epidermal location and possible feeding site of *Psorergates ovis*, the sheep itch mite, *Aust Vet J*, 67 (2): 59–62.

Sonenshine, D.E., 1991, *Biology of Ticks* (vol. 1), Oxford University Press, 447 pp.

Sonenshine, D.E., 1993, *Biology of Ticks* (vol. 2), Oxford University Press, 465 pp.

Steiner, M.Y. & Goodwin, S., 1998, Phytoseiids with potential for commercial exploitation in Australia, in Halliday, R.B., Walter, D.E., Proctor, H.C., Norton, R.A. & Colloff, M.J. (eds), *Acarology: Proceedings of the 10th International Congress*, CSIRO Publishing, 657 pp.

Strachan, D.P., 1998, House dust mite allergen avoidance in asthma, *BMJ*, 317: 1096–1097.

Stuen, S., Bergstrom, K. & Palmer, E., 2002, Reduced weight gain due to subclinical *Anaplasma phagocytophilum* (formerley *Ehrlicha phagocytophila*) infection, *Exp Appl Acarol*, 28: 209–215.

Uspensky, I., 2002, Preliminary observations on specific adaptations of exophilic ixodid ticks to forest or open country habitats, *Exp Appl Acarol*, 28: 147–154.

Vanderhoof-Forschner, K., 1997, *Everything you need to know about Lyme disease and other tick-borne disorders*, John Wiley & Sons, 237 pp.

Walter, D. & Proctor, H., 1999, *Mites: Ecology, Evolution and Behaviour*, CABI Publishing, 322 pp.

Yoder, J.A., 1995, Allomonal defence secretions of the American dog tick *Dermacentor variabilis* (Acari: Ixodidae) promote clustering, *Exp Appl Acarol*, 19 (12): 695–705.

WEBSITES

Acariasis fact sheet
http://www.cfsph.iastate.edu/Factsheets/pdfs/acariasis.pdf

Australasian Arachnological Society
http://www.australasian-arachnology.org/arachnology/acari/mesostigmata

Chiggers fact sheet
http://ohioline.osu.edu/hyg-fact/2000/2100.html

Details and treatments for tick paralysis
http://balgownievet.com.au/html/pet_illnesses/paralysis_ticks.html

Details for tick paralysis
http://www.avma.org/reference/zoonosis/znticpar.asp

Encyclopedic Reference of Parasitology: Mites
http://parasitology.informatik.uni-wuerzburg.de/login/frame.php

House Dust Mites and the Built Environment: A Literature Review
http://www.ucl.ac.uk/bartlett-housedustmites/Publications/Publications/review10Oct02.pdf

Hydracarina, water mites
http://www.tolweb.org/Hydracarina/2606/1998.08.09

Information sheet on scrub typhus
http://www.health.nt.gov.au/library/scripts/objectifyMedia.aspx?file=pdf/9/55.pdf&siteID=1&str_title=Scrub%20Typhus.pdf

Mites infesting stored food fact sheet
http://ohioline.osu.edu/hyg-fact/2000/2152.html

Plant pests, animal parasites, food mites and dust mites
http://kendall-bioresearch.co.uk/mite.htm

Prostigmata
http://www.australasian-arachnology.org/arachnology/acari/prostigmata/

Scrub typhus
http://emedicine.medscape.com/article/1055424-overview

Tickbase http://www.icttd.nl

Varroa control
http://www.beeworks.com/informationcentre/essentialoil.html

Varroa control
http://www.bijenziekten.wur.nl/UK/varroa/

Varroa Mite
http://creatures.ifas.ufl.edu/misc/bees/varroa_mite

Veterinary website
http://www.peteducation.com

CHAPTER 9 – OPILIONES

Acosta, L.E. & Machado, G., 2007, Diet and foraging, in Pinto-da-Rocha, R., Machado, G. & Giribet, G. (eds), *Harvestmen: The Biology of Opiliones*, Harvard University Press, 597 pp.

Cokendolpher, J.C. & Mitov, P.G., 2007, Natural enemies, in Pinto-da-Rocha, R., Machado, G. & Giribet, G. (eds), *Harvestmen: The Biology of Opiliones*, Harvard University Press, 597 pp.

Cokendolpher, J.C., 1993, Pathogens and parasites of Opiliones (Arthropoda: Arachnida), *J Arachnol*, 21 (2): 120–146.

Curtis, D.J. & Machado, G., 2007, Ecology, in Pinto-da-Rocha, R., Machado, G. & Giribet, G. (eds), *Harvestmen: The Biology of Opiliones*, Harvard University Press, 597 pp.

Edgar, A.L., 1971, Studies on the Biology and Ecology of Michigan *Phalangida* (Opiliones), *Misc Pub Mus Zool*, University of Michigan, 144: 1–64.

Giribet, G. & Kury, A.B., 2007, Phylogeny and biogeography, in Pinto-da-Rocha, R., Machado, G. & Giribet, G. (eds), *Harvestmen: The Biology of Opiliones*, Harvard University Press, 597 pp.

Gnaspini, P., 2007, Development, in Pinto-da-Rocha, R., Machado, G. & Giribet, G. (eds), *Harvestmen: The Biology of Opiliones*, Harvard University Press, 597 pp.

Gnaspini, P. & Hara, M.R., 2007, Defence mechanisms, in Pinto-da-Rocha, R., Machado, G. & Giribet, G. (eds), *Harvestmen: The Biology of Opiliones*, Harvard University Press, 597 pp.

Guffey, C., 1999, Costs associated with leg autotomy in the harvestmen *Leiobunum nigripes* and *Leiobunum vittatum* (Arachnida: Opiliones), *Can J Zool*, 77(5): 824–830.

Hara, M.R., Gnaspini, P., & Machado, G., 2003, Male egg guarding behaviour in the Neotropical harvestman *Ampheres leucophcus* (Mello-Leitão 1922) (Opiliones, Gonyleptidae), *J Arachnol*, 31: 441–444.

Holmberg, R.G., Angerilli, N.P.D., & LaCassè, L.J., 1984, Overwintering aggregations of *Leiobunum paeseleri* in caves and mines (Arachnida, Opiliones), *J. Arachnol*, 12: 195–204.

Machado, G. & Macias-Ordóñez, R., 2007, Reproduction, in Pinto-da-Rocha, R., Machado, G. & Giribet, G. (eds), *Harvestmen: The Biology of Opiliones*, Harvard University Press, 597 pp.

Machado, G. & Macias-Ordóñez, R., 2007, Social behaviour, in Pinto-da-Rocha, R., Machado, G. & Giribet, G. (eds), *Harvestmen: The Biology of Opiliones*, Harvard University Press, 597 pp.

Machado, G., Pinto-da-Rocha, R. & Giribet, G., 2007, What are harvestmen? in Pinto-da-Rocha, R., Machado, G. & Giribet, G. (eds), *Harvestmen: The Biology of Opiliones*, Harvard University Press, 597 pp.

Machado, G. & Oliveira, P.S., 2002, Maternal care in the Neotropical Harvestman, *Bourguyia albiornata* (Arachnida: Opiliones): oviposition site selection and egg protection, *Behav*, 139 (11–12): 1509–1524.

Mora, G., 1990, Paternal care in a

neotropical harvestman, *Zygopachylus albomarginis* (Arachnida, Opiliones : Gonyleptidae), *Anim Behav*, vol. 39 (3): 582–593.
Morse, D.H., 2001, Harvestmen as commensals of crab spiders, *J Arachnol*, 29: 273–275.
Newton, B.L. & Yeargan, K.V., 2001, Predation of *Helicoverpa zea* (Lepidoptera: Noctuidae) eggs and first instars by *Phalangium opilio* (Opiliones: Phalangiidae), *J Kans Entomol Soc*, 74: 199–204.
Pinto-da-Rocha, R., & Giribet, G., 2007, *Harvestmen: The Biology of Opiliones*, Harvard University Press, 597 pp.
Ramirez, E.N. & Antonio, A., 1994, Maternal care in a Neotropical harvestman *Acutisoma proximum* (Opiliones, Gonyleptidae), *J Arachnol*, 22: 179–180
Sankey, J.H.P. & Hillyard, P.D., 2005, *Harvestmen*, The Linnean Society of London, Synopses of the British Fauna, no. 4, 3rd Edition.
Santos, F.H., 2007, Ecophysiology, in Pinto-da-Rocha, R., Machado, G. & Giribet, G. (eds), *Harvestmen: The Biology of Opiliones*, Harvard University Press, 597 pp.
Santos, F.H. & Gnaspini, P., 2002, Notes on the foraging behaviour of the Brazilian cave harvestman *Goniosoma spelaeum* (Opiliones, Gonyleptidae), *J Arachnol*, 30: 177–180.
Savory, T.H., 1964, *Arachnida*, Academic Press, London and New York.
Schmitz, A., 2005, Metabolic rates in harvestmen (Arachnida, Opiliones): the influence of running activity, *Physiological Entomology*, 30: 75–81.
Shultz, J.W. & Pinto-da-Rocha R., 2007, Morphology and functional anatomy, in Pinto-da-Rocha, R., Machado, G. & Giribet, G. (eds), *Harvestmen: The Biology of Opiliones*, Harvard University Press, 597 pp.
Shultz, J.W., 1990, Evolutionary morphology and phylogeny of Arachnida, *CLADEC*, 6: 1–38.
Tourinho, Ana Lúcia M. & Kury, A.B., 2001, A review of *Holcobunus* (Arachnida, Opiliones, Sclerosomatidae), *Bol Mus Nac Zool Rio*, 461: 1–22.
van der Hammen, L., 1986, Acarological and arachnological notes, *Zool Meded (Leiden)*, 60: 217–230.
Weygoldt, P. & Paulus, H.F., 1979, Untersuchungen zur Morphologie, Taxonomie und Phylogenie der Chelicerata, *Z Zool Syst Evol*, 17: 85–116, 177–200.
Willemart, R.H., 2001, Egg covering behaviour of the Neotropical harvestman *Promitobates ornatus* (Opiliones, Gonyleptidae), *J Arachnol*, 29: 249–252.

WEBSITES

Invertebrate Anatomy Online
http://webs.lander.edu/rsfox/invertebrates/
http://www.mcz.harvard.edu/Departments/InvertZoo/gonzalo.htm
http://www.museunacional.ufrj.br/mndi/Aracnologia/opiliones.html

CHAPTER 10 – SCORPIONES

Braunwalder, M.E., 2001, Scorpions of Switzerland: summary of a faunistic survey, in Fet, V. & Selden, P.A. (eds), *Scorpions 2001: In Memoriam Gary A. Polis*, British Arachnological Society, Burnham Beeches, Bucks, xi & 404 pp.
Brown, C.A., 2001, Allometry of offspring and number in scorpions, in Fet, V. & Selden, P.A. (eds), *Scorpions 2001: In Memoriam Gary A. Polis*, British Arachnological Society, Burnham Beeches, Bucks, xi & 404 pp.
Cloudsley-Thompson, J.L., 1990, Scorpions in mythology, folklore and history, in Polis, G.A. (ed.), *The Biology of Scorpions*, Stanford University Press, 587 pp.
Cloudsley-Thompson, J.L., 1959, *Spiders, scorpions, centipedes and mites*, Oxford, Pergamon, 278 pp.
Cloudsley-Thompson, J.L., 2001, Scorpions and spiders in mythology and folklore, in Fet, V. & Selden, P.A. (eds), *Scorpions 2001: In Memoriam Gary A. Polis*, British Arachnological Society, Burnham Beeches, Bucks. xi & 404 pp.
Due, A.D., 2001, Historical biogeography, in Brownell, P. & Polis, G. (eds), *Scorpion Biology & Research*, Oxford University Press, 431 pp.
Dumortier, B., 1963, Morphology of sound emission apparatus in Arthropoda, in Busnel, R.G. (ed.), *Acoustic behaviour of animals*, Amsterdam, Elsevier.
Farley, R., 2001, in Brownell, P. & Polis, G. (eds), *Scorpion Biology & Research*, Oxford University Press, 431 pp.
Fet, V., Sissom, W.D., Lowe, G. & Braunwalder, M.E., 2000, *Catalog of the Scorpions of the World* (1758–1998), The New York Entomological Society.
Fleissner, G. & Fleissner, G., 2001, Night vision in desert scorpions, in Fet, V. & Selden, P.A. (eds), *Scorpions 2001: In Memoriam Gary A. Polis*, British Arachnological Society, Burnham Beeches, Bucks. xi & 404 pp.
Frost, L.M, Butler, D.R, O'Dell, B. & Fet, V., 2001, A coumarin as a fluorescent compound in scorpion cuticle, in Fet, V. & Selden, P.A. (eds), *Scorpions 2001: In Memoriam Gary A. Polis*, British Arachnological Society, Burnham Beeches, Bucks, xi & 404 pp.
Gaffin, D.D., 2001, Electrophysiological properties of sensory neurons in peg sensilla of *Centruroides vittatus* (Say, 1821) (Scorpiones: Buthidae), in Fet, V. & Selden, P.A. (eds), *Scorpions 2001: In Memoriam Gary A. Polis*, British Arachnological Society, Burnham Beeches, Bucks, xi & 404 pp.
Hadley, N.F., 1990, Environmental physiology, in Polis, G.A., (ed.), *The Biology of Scorpions*, Stanford University Press, 587 pp.
Hjelle, J.T., 1990, Anatomy and morphology, in Polis, G.A. (ed.), *The Biology of Scorpions*, Stanford University Press, 587 pp.
Keegan, H.L., 1998, *Scorpions of medical importance*, Fitzgerald Publishing, 140 pp.
Locket, A., 2001, Eyes and vision, in Brownell, P. & Polis, G. (eds), *Scorpion Biology & Research*, Oxford University Press, 431 pp.
Loret, E. & Hammock, B., 2001, Structure and neurotoxicity of venoms, in Brownell, P. & Polis, G. (eds), *Scorpion Biology & Research*, Oxford University Press, 431 pp.
Mahsberg, D., 2001, Brood care and social behaviour, in Brownell, P. & Polis, G. (eds), *Scorpion Biology & Research*, Oxford University Press, 431 pp.
Matthiesen, F.A., 1971, The breeding of *Tityus serrulatus* Lutz and Mello, 1927, in captivity, *Revista brasileira de pessquisas medicas e biologicas* 4 (4–5): 299–300.
McCormick, S.J. & Polis, G.A., 1990, Prey, predators and parasites, in Polis, G.A. (ed.), *The Biology of Scorpions*, Stanford University Press, 587 pp.
Polis, G.A., 1990, *The Biology of Scorpions*, Stanford University Press, 587 pp.
Prendini, L., 2006, New South African flat rock scorpions (Liochelidae: *Hadogenes*), *Am Mus Novit*, 3502, 32 pp.
Prendini, L., 2001, Substratum specialization and speciation in southern African scorpions: the Effect Hypothseis revisited, in Fet, V. & Selden, P.A. (eds), *Scorpions 2001: In Memoriam Gary A. Polis*, British Arachnological Society, Burnham Beeches, Bucks, xi & 404 pp.
Prendini, L., & Wheeler, W.C., 2005, Scorpion higher phylogeny and classification, taxonomic anarchy, and standards for peer review in online publishing, *CLADEC*, 21: 446–494.
Root, T.M., 1990, Neurobiology, in Polis, G.A. (ed.), *The Biology of Scorpions*, Stanford University Press, 587 pp.
Schofield, R.M.S., 2001, Metals in cuticular tsructures, in Brownell, P. & Polis, G. (eds), *Scorpion Biology & Research*, Oxford University Press, 431 pp.
Shachak, M & Brand, S., 1983, The relationship between sit-and-wait foraging stategy and dispersal in the desert scorpion *Scorpio maurus palmatus*, *Oeco Planta*, 60: 371–77.
Simard, J.M. & Watt, D.D., 1990, Venoms and toxins, in Polis, G.A. (ed.), *The Biology of Scorpions*, Stanford University Press, 587 pp.
Warburg, M.R., 2001, Scorpion reproductive strategies, potential and longevity: an ecomorphologist's interpretation, in Fet, V. & Selden, P.A. (eds), *Scorpions 2001: In Memoriam Gary A. Polis*, British Arachnological Society, Burnham Beeches, Bucks, xi & 404 pp.
Warburg, M.R. & Polis, G.A., 1990, Behavioural responses, rhythms and activity patterns, in Polis, G.A. (ed.), *The Biology of Scorpions*, Stanford University Press, 587 pp.

WEBSITES

Scorpion Systematics Research Group
http://scorpion.amnh.org
The Scorpion Files
http://www.ub.ntnu.no/scorpion-files

CHAPTER 11 – PSEUDOSCORPIONES

Adis, J. *et al*, 1988 Adaptation of an Amazonian pseudoscorpion (Arachnida) from dryland forests to inundation forests, *ECOL*, 69 (1): 287–291.
André, H.M. & Jocqué, R., 1986, Definition of stases in spiders and other arachnids, *Mem Soc R Belge Ent*, 33: 1–14.
Chamberlin, J.C., 1931, The arachnid order Chelonethida, *Stanford Univ Publ Univ Ser Biol Sci*, 7 (1): 1–284.
Gabbutt, P.D., 1969, Life-histories of some British pseudoscorpions inhabiting leaf litter, *Systematics Association Publication no.8: The Soil Ecosystem*, J.G. Sheals (ed.), 229–235.
Harvey, M.S., 2007, The smaller arachnid orders: diversity, descriptions and distributions from Linnaeus to the present (1758 to 2007), in Zhang, Z.-Q. & Shear, W.A. (eds), Linnaeus Tercentenary: Progress in Invertebrate Taxonomy, *Zootaxa*, 1668: 363–380.
Harvey, M.S., 1991, *Catalogue of the Pseudoscorpionida*, V. Mahnert (ed.), Manchester University Press, 726 pp.
Judson, M.L.I., 1993, The gonosacs ('gonopods') of female pseudoscorpions (Arachnida,Chelonethi), *Bull Soc Neuchatel Sci Nat*, 116 (1): 117–124.
Judson, M.L.I., 1990, Observations on the form and function of the coxal spines of some chthonioid pseudoscorpions from Cameroon (Arachnida, Chelonethida), Proceedings XI International Congress of Arachnology, *Acta Zool Fenn*, 190: 195–198.
Judson, M.L.I., 1990, The remarkable protonymph of *Pseudochthonius* (Chelonethi, Chthoniidae), *C. R. XIIème Coll Europ Arachnol, Bull Soc Euro Arachnol*, (hors série) 1: 159–165.
Judson, M.L.I., 1990, Redescription of the bee-associate *Ellinsenius fulleri* (Hewitt & Godfrey) (Arachnida, Chelonethi, Cheliferidae) with new records from Africa, Europe and the Middle East, *J Nat Hist*, 24: 1303–1310.
Leclerc, P. & Mahnert, V., 1988, A new species of the genus *Levigatocreagris* Curcic (Pseudoscorpiones: Neobisiidae) from Thailand, with remarkable sexual dimorphism, *Bull Br Arachnol Soc*, 7 (9): 273–277.
Legg, G., 2003, *Galea*: Newsletter of the Pseudoscorpion Group, no. 6, October.
Legg, G., 2002, *Galea*: Newsletter of the Pseudoscorpion Group, no. 5, Nov..
Legg, G., 1998, *Galea*: Newsletter of the Pseudoscorpion Group, no. 1, Nov..
Legg, G. & Jones, R.E., 1988, *Pseudoscorpions*, Synopses of the British Fauna, no. 40, The Bath Press, Avon, 159 pp.
Mahnert, V. & Adis, J., 2002, Pseudoscorpiones, in Adis, J. (ed.), *Amazonian Arachnida and Myriapoda*, Pensoft Publishers, pp.367–380
Muchmore, W.B., 1990, A pseudoscorpion from arctic Canada (Pseudoscorpionida, Chernetide), *Can J Zool* vol. 68: 389–390.
Muchmore, W.B., 1990, Pseudoscorpionida, in *Soil Biology Guide*, D.L. Dindal (ed.), John Wiley & Sons, Inc., 503–527.
Muchmore, W.B., 1972, A remarkable pseudoscorpion from the hair of a rat (Pseudoscorpionida, Chernetidae), *Proc Biol Soc Wash*, vol. 85, no. 37: 427–432.
Poinar, G.O. & Curcic, B.P.M., 1992,

Parasitism of Pseudoscorpions (Arachnida) by Mermithidae (Nematoda), *J Arachnol*, 20: 64–66.
Tizo-Pedroso, E. & Del-Claro, K., 2005, Matriphagy in the neotropical pseudoscorpion *Paratemnoides nidificator* (Balzan 1888)(Atemnidae), *J Arachnol*, 33, 3: 873–877.
Weygoldt, P., 1969, *The Biology of Pseudoscorpions*, Harvard, 145 pp.
Zeh, D.W. & Zeh, J.A., 1991, Novel use of silk by the Harlequin beetle-riding pseudoscorpion *Cordylochernes scorpioides* (Pseudoscorpionida, Chernetidae), *J Arachnol*, 19: 153–154.
Zeh, J.A. & Zeh, D.W., 1990, Cooperative foraging for large prey by *Paratemnus elongatus* (Pseudoscorpionida, Atemnidae), *J Arachnol*, 18: 307–311.
WEBSITES
ARC Plant Protection Research Institute South Africa
http://www.arc.agric.za/institutes/ppri/main/divisions/biosysdiv/pseudoscorpion.htm

Use of the name pseudoscorpion
http://www.european-arachnology.org/society/esa-bulletin14.shtml

CHAPTER 12 – SOLIFUGAE

Brownell, P.H. & Farley, R.D., 1974, The organization of the malleolar sensory system in the solpugid *Chanbria* sp., *Tissue & Cell*, 6: 471–485.
Cloudsley-Thompson, J.L., 1977, Adaptational biology of Solifugae (Solpugida), *Bull Br Arachnol Soc*, 4 (2): 61–71.
Cloudsley-Thompson, J.L., 1961, Some aspects of the physiology and behaviour of *Galeodes arabs*, *Ent Exp App* 4: 257–263.
Cushing, P.E., Brookhard, J.O., Kleebe, H-J., Zito, G. & Payne, P., 2005, The suctorial organ of the Solifugae (Arachnida, Solifugae), *Arthropod Struct Dev*, 34: 397–406.
Dean, W. R. J. & Milton, S.J., 1991, Prey capture by *Solpuga chelicornis* Lichtenstein (Solifugae, Solpugidae), *J Entomol Soc South Afr*, 54: 266–267.
Grasshoff, M., 1978, A model of the evolution of the main chelicerate groups, *Symp Zool Soc Lond*, 42: 273–284.
Harvey, M.S., 2007, The smaller arachnid orders: diversity, descriptions and distributions from Linnaeus to the present (1758 to 2007), in Zhang, Z.-Q. & Shear, W.A. (eds), Linnaeus Tercentenary: Progress in Invertebrate Taxonomy, *Zootaxa*, 1668: 363–380.
Harvey, M.S., 2003, *Catalogue of the Smaller Arachnid Orders of the World*, CSIRO Publishing, 385 pp.
Haupt, J., 1982, Hair regeneration in a solpugid chemotactile sensillum during moulting (Arachnida, Solifugae), *Wilhelm Roux' Arch Dev Biol*, 191 (2): 137–142.
Klann, A.E., Gromov, A.V., Cushing, P.E., Peretti, A.V. & Alberti, G., 2008, The anatomy and ultrastructure of the suctorial organ of Solifugae (Arachnida), *Arthropod Struct Dev*, 37 (1): 3–12.
Lamoral, B., 1975, The structure and possible function of the flagellum in four species of male solifuges of the family Solpugidae, *Proc 6th Int Cong Arachnol*, 136–141.
Lighton, J.R.B. & Fielden, L.J., 1996, Gas exchange in wind spiders (Arachnida, Solifugidae): independent evolution of convergent control strategies in solifugids and insects, *J Insect Physiol*, 42: 347–357.
Muma, M.H., 1982, Solpugida, 102–4, in Parker, S.P. (ed) *Synopsis and classification of living organisms, vol. 2*, McGraw-Hill, New York.
Muma, M.H., 1967, Basic behaviour of North American Solpugida, *Fla Entomol*, 50: 115–23.
Muma, M.H., 1966, Burrowing habits of North American Solpugida (Arachnida), *PSYSA* 73: 251–60.
Muma, M.H., 1966, The life cycle of *Eremobates durangonus* (Arachnida: Solpugida), *Fla Entomol*, 49: 233–242.
Muma, M.H., 1966, Feeding behaviour of North American Solpugida (Arachnida), *Fla Entomol*, 49: 199–216.
Muma, M.H., 1966, Egg deposition and incubation for *Eremobates durangonus* with notes on the eggs of other species of Eremobatidae (Arachnida: Solpugida), *Fla Entomol*, 49: 23–31.
Punzo, F., 1998, *The Biology of Camel-Spiders* (Arachnida, Solifugae), Kluwer Academic Publishers, 301 pp.
Punzo, F., 1997, Dispersion, temporal patterns of activity, and the phenology of feeding and mating behaviour, in *Eremobates pedipalpisetulosus* (Solifugae, Eremobatidae), *Bull Br Arachnol Soc*, 10: 303–307.
Punzo, F., 1995, Feeding and prey preparation in the solpugid *Eremorphax magnus* Hancock (Solpugida: Eremobatidae), *Pan-Pac Entomol*, 71 (1): 13–17.
Punzo, F., 1994, Intraspecific variation in response to temperature and moisture in *Eremobates pedipalpisetulosus* Fichter (Solpugida, Eremobatidae) along an altitudinal gradient, *Bull Br Arachnol Soc*, 9 (8): 256–262.
Punzo, F., 1994, An analysis of feeding and optimal behaviour in the solpugid *Eremobates mormonus* (Roewer) (Solpugida, Eremobatidae), *Bull Br Arachnol Soc*, 9 (9): 293–298.
Punzo, F., 1993, Diet and feeding behaviour of the solpugid *Eremobates pedipalpisetulosus* (Solpugida: Eremobatidae), *PSYSA*, 100: 151–161.
Wharton, R.A., 1987, Biology of the diurnal *Metasolpuga picta* (Kraepelin) (Solifugae, Solpugidae) compared with that of nocturnal species, *J Arachnol*, 14: 363–383.

WEBSITES
Solifugids
http://www.solpugid.com/Alexander%2Gromov.htm

Index

Page numbers in *italic* refer to illustration captions; those in **bold** refer to main subjects of boxed text.

Glossary

ADDUCTOR MUSCLES – closing muscles

ALLOMETRIC GROWTH – growth of a particular structure at a different rate than the rest of the body

ANTENNIFORM – shaped like antennae

APOLYSIS – separation of the old cuticle from the epidermis below (part of the moulting process)

APOPHYSES – outgrowths of the cuticle

APOTELE – terminal segment of the tarsus, consisting of claws and a pad (pulvilli)

AUTOTOMY – deliberate shedding of a body part

BALLOONING – using a silk thread to disperse on wind currents

BOOK LUNG – respiratory organ formed from a stack of cuticular lamellae, which resemble the pages of a book end-on

CAMOUFLAGE – concealing colouration (crypsis)

CARAPACE – tergites of the prosoma fused to form a tough dorsal shield

CHELATE – claw-like

CHELICERAE – jaws with two or three joints

CHEMORECEPTIVE HAIR – hair that detects a chemical stimulus

CLADE – group of organisms descended from a common ancestor

CLADISTICS – method of inferring evolutionary relationships based on features inherited from a shared common ancestor

CLADOGRAMS – branching tree diagrams produced using cladistic techniques

CONSPECIFICS – others of the same species

COXAE – first segment of the leg that joins it to the body

COXAL GLAND – large excretory sac located opposite the coxae of the first legs

CURSORIAL - adapted for running

CUTICLE – part of the exoskeleton which overlays the hypodermis

DENTICLES – teeth-like processes on the cuticle

DUCT – tube

ENDEMIC – native to a particular area

ENDOCUTICLE – unsclerotized layer of the cuticle

ENDOSTERNITE – lightly sclerotized plate located between the alimentary canal and nerve ganglia

EPICUTICLE – extremely thin outer layer of the cuticle

EPIGYNE – sclerotized external genital structure with complex infoldings

EXOCUTICLE – sclerotized layer of the cuticle

EXOSKELETON (INTEGUMENT) – hard outer structure that consists of the cuticle and hypodermis

EXTANT – still living

EXTRA-ORAL DIGESTION – predigestion of food outside the body by digestive fluids regurgitated from the mid-gut

GANGLION – collection of nerve cells

GONOPODS – claspers

GONOPORE – genital opening

HAEMOCYANIN – arachnid respiratory pigment

HAEMOLYMPH – arachnid blood

HYPODERMIS – cellular layer beneath the cuticle

INNERVATED – attached to nerves

INSTAR – stage of an arachnid's life in between moults

INTROMITTENT ORGAN – penis

LABIUM – lower lip

LACUNAE – empty spaces in an arachnid's circulatory system

LAMELLAE – thin plates

LUMEN – inner open space or cavity of a tubular organ

MALPIGHIAN TUBULES – slender, tube-shaped muscular glands that concentrate and store waste products

MATRIPHAGY – where young organisms feed on their own mother

MIMICRY – where a creature has evolved to look like another as a form of defense

MOULTING – periodic shedding of the cuticle

MYRMECOPHILE – an organism that lives in association with ants

NEPHROCYTES – specialized cells that absorb the products of metabolism from the haemolymph

NYMPH – immature form

OCELLUS (PLURAL: OCELLI) – 'simple eye' with a single lens

OCULARIUM – mound-like structure on the carapace that carries the eyes

OOCYTES – eggs produced by the ovary before fertilization

OPERCULUM – flap that covers the genital opening

OPISTHOSOMA – posterior section of the body behind the prosoma (also known as the abdomen)

OSTIA – valves in an arachnid heart

PALPS – second pair of appendages on the prosoma

PARTHENOGENETIC – able to reproduce asexually

PEDICEL – narrowed first segment of the opisthosoma, which forms a waist joining the prosoma to the opisthosoma

PHEROMONES – chemical substances excreted externally designed to trigger a response

PHORESY – dispersal by attaching to another organism

PHYLOGENY – the evolutionary relationship between organisms

POST-EMBRYONIC DEVELOPMENT – development of an arachnid after it has hatched from its egg

PRE-ORAL CAVITY – cavity in front of the mouth

PROSOMA – first major body section (also known as the cephalothorax)

PULVILLI – adhesive pads

RAPTORIAL – adapted for catching and holding prey

ROSTRUM – upper lip

SCAPE – an elongated structure extending outwards from an epigyne

SCLEROTIZATION – hardening of the cuticle

SEMINAL RECEPTACLES – little pouches just inside the gonopore, which store sperm until the female is ready to fertilize her eggs

SETAE – bristles or hair-like structures

SEXUAL DIMORPHISM – differences in size or shape between males and females of a particular species

SINUS – cavity

SLIT SENSILLAE – slit-shaped structures embedded in the exoskeleton, which detect mechanical stress

SOMITES – segments

SPERMATOCYTES – male reproductive cells that will become spermatozoa

SPERMATOPHORE – a package containing spermatozoa, which the male passes to the female during copulation

SPIRACLES – small openings on the surface, which lead to the closed respiratory system of tracheae

STERNITE – ventral plate of a body segment

STERNUM – prosomal sternites that have developed into a shield or plate

SUPERCOOLING – the synthesis of antifreeze chemicals called cryoprotectants which protect against internal freezing down to -7°C

SUBESOPHAGEAL – below the oesophagus

SUPRAESOPHAGEAL – above the oesophagus

TACTILE HAIR – large, articulated, innervated hair that detects movement

TERGITE – dorsal plate of a body segment

THERMOREGULATION – the regulation of body temperature within acceptable limits

TORPOR – state of inactivity

TRACHEAE – narrow tubes that run from spiracles on the surface and branch throughout the body, allowing oxygen to permeate the tissues

TRICHOBOTHRIUM – extremely sensitive sensory hair that is innervated and suspended in a membrane in the exoskeleton

TROGLOPHILE – organism that can complete its lifecycle in a cave, but is not confined to this habitat

TROGLOBITE – organism that is specifically adapted to life in caves and is not found anywhere else

TUBERCLES – outgrowths of the cuticle

VASA DEFERENTIA – ducts that convey sperm from the testes to the exterior

VIVIPAROUS – gives birth to live young

WARNING COLOURATION – conspicuous markings indicating that an animal is distasteful or venomous

Picture credits

CHAPTER 1 p.4 © NHMPL; p.5 © The Natural History Museum. Redrawn from the original in Foelix, Rainer F, *Biology of Spiders.* Oxford University Press, 1996; p.6 © NHMPL; p.8 © The Natural History Museum. Redrawn from the original in Pinto-da-Rocha, R, Machado, G, and Giribet, G, (Editors), *Harvestmen, The Biology of Opiliones.* Harvard University Press, 2007; p.10 © Chu & Foelix; p.11 © The Natural History Museum. Redrawn from the original in Foelix, Rainer F, *Biology of Spiders.* Oxford University Press, 1996; p.12 clockwise from top left: © The Natural History Museum. Redrawn from the original in Foelix, Rainer F, *Biology of Spiders.* Oxford University Press, 1996; © The Natural History Museum. Redrawn from the original in Hillyard, P D and Sankey J H P, *Harvestmen.* Published for The Linnean Society of London and The Estuarine and Brackish-water Sciences Association by E.J. Brill, 1989; © The Natural History Museum. Redrawn from the original in Polis, Gary A. (Ed), *The Biology of Scorpions.* Original © 1990 by the Board of Trustees of the Leland Stanford Jr. University; © The Natural History Museum. Redrawn from the original in Weygoldt, Peter, *Whip Spiders (Chelicerata: Amblypygi) Their Biology, Morphology and Systematics,* Apollo Books, 2000; © The Natural History Museum. Redrawn from the original in Foelix, Rainer F, *Biology of Spiders.* Oxford University Press, 1996; p.13 © The Natural History Museum.Redrawn from the original in Matthiesen, *Reproductive system and embryos of Brazilian scorpions.* Anais da Academia Brasileira de Ciências, 1970; p.14 top © The Natural History Museum. Redrawn from the original in Foelix, Rainer F, *Biology of Spiders.* Oxford University Press, 1996; p.14 bottom © The Natural History Museum. Redrawn from the original in Weygoldt, Peter, *Whip Spiders (Chelicerata: Amblypygi) Their Biology, Morphology and Systematics,* Apollo Books, 2000; p.15 © The Natural History Museum. Redrawn from the original in Preston-Mafham, Rod & Ken, *Spiders of the World.* Blandford Press Ltd, 1984; p.20 © Tom McHugh/Science Photo Library; p.22 left © The Natural History Museum. Redrawn from the original in Weygoldt, Peter, Courtship behaviour and sperm transfer in the Giant Whip Scorpion, Mastigoproctus Giganteus (Lucas) (Uropygi, Thelyphonidae), *Behaviour,* 1970, 36: 1-8; bottom left, right © The Natural History Museum. Redrawn from the original, in Weygoldt, Peter, Mating and Spermatophore Morphology in Whip Spiders, *Zoologischer Anzeiger,* 1997/8, 236: 259-276. © Elsevier.

CHAPTER 2 All images in this chapter © George Beccaloni except p.30 © David Maitland/NHPA/Photoshot; p.32 right © Marshal Hedlin; p.33 © Jan Beccaloni; p.40 top © The Natural History Museum. Redrawn from the original in Roberts, Michael J, *Collins Field Guide, Spiders of Britain & Northern Europe.* Harper Collins, 1996; p.50 © The Natural History Museum. Redrawn from the original in Foelix, Rainer F, *Biology of Spiders.* Oxford University Press, 1996; p.56 © Jan Beccaloni; p.83 © The Natural History Museum. Redrawn from the original in Foelix, Rainer F, *Biology of Spiders.* Oxford University Press, 1996; p.85 top © Jan Beccaloni.

CHAPTER 3 p.90 © Piotr Naskrecki/Minden Pictures/FLPA; p.91 © Piotr Naskrecki/NHPA/Photoshot; p.92 left, right © The Natural History Museum. Redrawn from the original in Weygoldt, Peter, *Whip Spiders (Chelicerata: Amblypygi) Their Biology, Morphology and Systematics.* Apollo Books, 2000; p.93, 94 top © Piotr Naskrecki/NHPA/Photoshot; p.94 bottom © George Beccaloni; p.97 Robert Pickett/Papiliophotos.com; p.99, 101 © George Beccaloni; p.103 © Piotr Naskrecki/Minden Pictures/FLPA; pp.107, 108 © George Beccaloni.

CHAPTER 4 p.110 © Carsten Kamenz; p.112 © The Natural History Museum. Redrawn from the original in Schultz, Jeffrey W, Muscular anatomy of the giant whipscorpion Mastigoproctus giganteus (Lucas) (Arachnida: Uropygi) and it's evolutionary significance. *Zoological Journal of the Linnean Society,* 1993, 108: 335-365, p.113 © George Beccaloni; pp.114-116 © Carsten Kamenz; p.117 © George Beccaloni; p.118 © Nature's Images/Science Photo Library; p.121 © Piotr Naskrecki/NHPA/Photoshot; pp.123, 124 © George Beccaloni; p.127 © Carsten Kamenz.

CHAPTER 5 p.128 © Hans Hendrickx; p.129 © Mark Harvey, Western Australian Museum; p.130 left © The Natural History Museum. Redrawn from the original in Rowland, J. M., *Classification, phylogeny and zoogeography of the American arachnids of the order Schizomida.* PhD dissertation, Texas Tech University, Lubbock, Texas, 1976.; p.130 right © The Natural History Museum. Redrawn from the original in Briggs, Thomas S and Hom, Kevin, A New Schizomid Whip-Scorpion from California with notes on others. *The Pan-Pacific Entomologist,* Pacific Coast Entomological Society, Vol.42, No.4; pp.131, 134 © Piotr Naskrecki/NHPA/Photoshot.

CHAPTER 6 p.138 © Vilda/Rollin Verlinde; p.140 © The Natural History Museum. Redrawn from the original in van der Hammen, L, Comparative Studies in Chelicerata II. Epimerata (Palpigradi and Actinotrichida). Rijksmuseum van Natuurlijke Historie, Leiden; p.141 © G. Csizsmárová and L. Kováč.

CHAPTER 7 p.146 © Marshal Hedin; p.147 © Gonzalo Giribet; p.128 top © The Natural History Museum; middle, bottom © Giovanni Talarico; p.149 © Piotr Naskrecki/NHPA/Photoshot; p.150 © Giovanni Talarico; p.151 © Piotr Naskrecki/NHPA/Photoshot; p.152 © Giovanni Talarico; p.153 top © The Natural History Museum. Redrawn from the original by Giovanni Talarico; p.155 © Giovanni Talarico ; p.156 © Piotr Naskrecki/NHPA/Photoshot.

CHAPTER 8 p.158 © Erbe, Pooley: USDA, ARS, EMU; p.160 © NHMPL; p.161 left © David Evans Walter; right © Crown copyright courtesy of CSL/Science Photo Library; p.162 left © Michael McAloon; top right © Ronald Schmaeschke; bottom right © © NHMPL; p.163 top © Michael McAloon; middle © Andrew Syred/ Science Photo Library; bottom © NHMPL; p.164 © NHMPL; p.165 © The Natural History Museum. Redrawn from the original in Baker, Anne S, *Mites and ticks of domestic animals an identification guide and information source.* The Natural History Museum, 1999; p.166 © Andrew Syred/ Science Photo Library; p.172 © NHMPL; p.173 © Stephen Dalton/ NHPA; p.175 © Nigel Cattlin/FLPA; p.177 © M I WALKER/NHPA/ Photoshot; p.178 © NHMPL; p.180 © Nigel Cattlin/FLPA; p.181 © NHMPL; p.182 © Daniel Heuclin/NHPA/Photoshot; p.184 © Michael McAloon; p.185 © Clouds Hill Imaging Ltd/ Science Photo Library; p.186 © Eye of Science/ Science Photo Library; p.187 © Anthony Gould; p.189 © Steve Gschmeissner/Science Photo Library; p.190 © Luis Fernández Garcia; p.192 © A.N.T. Photo Library/NHPA/Photoshot; p.194 © Stephen Dalton/ NHPA/Photoshot; p.195 Nigel Cattlin/FLPA; p.196 © Nigel Cattlin/FLPA; p.198 © USDA; p.203 NHMPL; p.204 © Ken Griffiths/ NHPA/Photoshot; p.206 © Marcelo de Campos Pereira.

CHAPTER 9 p.208 © Glauco Machado; p.210 © Ingo Arndt/ FLPA; p.211 top © Marshal Hedin; middle © Gonzalo Giribet; bottom © Abel Pérez González; p.212 © Marshal Hedin; p.213 © Abel Pérez González; p.214 © The Natural History Museum. Redrawn from the original in Hillyard, P D and Sankey J H P, *Harvestmen.* Published for The Linnean Society of London and The Estuarine and Brackish-water Sciences Association by E.J. Brill, 1989; p.216 top © George Beccaloni; bottom © Abel Pérez González; p.219 © The Natural History Museum. Redrawn from the original in Firstman, B, The relationship of the chelicerate arterial system to the evolution of the endosternite. *J Arachnol,* 1973, Vol 1 No 1, pp1-54; p.220 © Annieta de Jong; p.222 © George Beccaloni; p.223 top © Nigel Cattlin/FLPA; bottom © Pavel Krásensk; p.224, 225 © Marshal Hedin; p.226 © George Beccaloni; p.228 © Bruno A. Buzatto; p.229 © Marshal Hedin; p.230 © Thomas Marent/Minden Pictures/FLPA; p.233 © Bruno A. Buzatto; p.234 © Glauco Machado.

CHAPTER 10 p.238 © Daniel Heuclin/NHPA/Photoshot; p.240 top, bottom © George Beccaloni; p.241 top © Mark Fairhurst/UPPA/Photoshot; bottom © John Visser; p.242 top © Larry West/FLPA; bottom © John Cancalosi/UPPA/Photoshot; p.243 © Mark Fairhurst/UPPA/Photoshot; p.244 © The Natural History Museum. Redrawn from the original in Keegan, Hugh L, *Scorpions of Medical Importance.* © 1980 by The University Press of Mississippi; p.245 © Wim van Egmond; p.246 © George Beccaloni; p.248 © Stephen Dalton/NHPA/Photoshot; p.250 © Daniel Heuclin/NHPA/Photoshot; p.251 © The Natural History Museum. Redrawn from the original in Vachon, M., Etudes sue les scorpions, Algiers. Publications de l'institut Pasteur d'Algerie, 1949; p.253 © Stephen Dalton/NHPA/Photoshot; p.254 top, bottom © George Beccaloni; p.256 © NHPA/Photo Researchers; p.258 © Anthony Bannister/NHPA/Photoshot; p.261 © Frans Lanting/FLPA; p.265 © Daniel Heuclin/NHPA/ Photoshot; p.267 © George Beccaloni.

CHAPTER 11 pp.270-272 © Hans Hendrickx; p.272 bottom © The Natural History Museum. Redrawn from the original in Legg, G. & Jones, R.E., Pseudoscorpiones (Arthropoda; Arachnida). *Synopsis of the British Fauna* (New Series) 40:1–159. The Linnean Society of London, 1988; pp.273-279 © Hans Hendrickx; p.279 bottom © George Beccaloni; pp.280-282 © Hans Hendrickx; p.288 © Gonzalo Giribet.

CHAPTER 12 p.290 © James Carmichael Jr./NHPA/Photoshot; p.291 © Anja Klann; p.292 top © Bill Love/NHPA/ Photoshot; bottom © Anja Klann; p.293 top © The Natural History Museum. Redrawn from the original in Punzo, Fred, *The Biology of Camel-Spiders (Arachnida, Solifugae),* p.12, fig.2-1. Springer and Kluwer Academic Publishers, 1998. With kind permission from Springer Science and Business Media; bottom © Anja Klann; pp.294, 295 top © Anja Klann; bottom © James Carmichael Jr./NHPA/Photoshot; pp.296-300 top © Anja Klann; bottom © George Beccaloni; p.301 © Daniel Heuclin/NHPA/ Photoshot; p.303 © George Beccaloni; p.308 top © Craig Tessyman; bottom © Mark Moffett/Minden Pictures/FLPA.

NHMPL, Natural History Museum Picture Library

Every effort has been made to contact and accurately credit all copyright holders. If we have been unsuccessful, we apologise and welcome correction for future editions and reprints.

Acknowledgements

Thanks to the following experts for peer-reviewing the chapters: Jason Dunlop, Mark Harvey, Joachim Haupt, Mark Judson, Adriano Kury, Michael McAloon, Lorenzo Prendini, Tony Russell-Smith, Paul Selden, Rowley Snazell and Carlos Viquez.

Thanks to George Beccaloni, Peter Smithers, the Publishing team at the Natural History Museum, and Ray Gabriel. Also to Dirk Ahrens for translating a key German paper for me.